Student's Solutions Manual

to accompany

Elementary and Intermediate Algebra

Fifth Edition

Stefan Baratto
Clackamas Community College

Barry Bergman
Clackamas Community College

Don Hutchison
Clackamas Community College

Prepared by
Avrio Knowledge Group, Inc.

Mc Graw Hill Education

STUDENT'S SOLUTIONS MANUAL TO ACCOMPANY
ELEMENTARY AND INTERMEDIATE ALGEBRA, FIFTH EDITION

Published by McGraw-Hill Education, 2 Penn Plaza, New York, NY 10121.

This book is printed on acid-free paper.

1 2 3 4 5 6 7 8 9 0 QVS/QVS 1 0 9 8 7 6 5 4 3

ISBN: 978-0-07-757440-6
MHID: 0-07-757440-0

www.mhhe.com

Contents

Preface

This *Student's Solutions Manual* contains comprehensive, worked-out solutions to all odd-numbered exercises in the book, including the Section Exercises, Summary Exercises, Chapter Tests, and Cumulative Reviews. All steps in the solutions follow the solving style of the Examples in the book.

Use this resource to help check your progress on homework assignments, identify any errors in your solving methodology, and to help yourself prepare for exams.

Chapter 0
Prealgebra Review

Exercises 0.1

1. $\frac{3}{7}\times\frac{2}{2}=\frac{6}{14}$

$\frac{3}{7}\times\frac{3}{3}=\frac{9}{21}$

$\frac{3}{7}\times\frac{4}{4}=\frac{12}{28}$

3. $\frac{4}{9}\times\frac{2}{2}=\frac{8}{18}$

$\frac{4}{9}\times\frac{4}{4}=\frac{16}{36}$

$\frac{4}{9}\times\frac{10}{10}=\frac{40}{90}$

5. $\frac{5}{6}\times\frac{2}{2}=\frac{10}{12}$

$\frac{5}{6}\times\frac{3}{3}=\frac{15}{18}$

$\frac{5}{6}\times\frac{10}{10}=\frac{50}{60}$

7. $\frac{10}{17}\times\frac{2}{2}=\frac{20}{34}$

$\frac{10}{17}\times\frac{3}{3}=\frac{30}{51}$

$\frac{10}{17}\times\frac{10}{10}=\frac{100}{170}$

9. $\frac{9}{16}\times\frac{2}{2}=\frac{18}{32}$

$\frac{9}{16}\times\frac{3}{3}=\frac{27}{48}$

$\frac{9}{16}\times\frac{10}{10}=\frac{90}{160}$

11. $\frac{7}{9}\times\frac{2}{2}=\frac{14}{18}$

$\frac{7}{9}\times\frac{5}{5}=\frac{35}{45}$

$\frac{7}{9}\times\frac{20}{20}=\frac{140}{180}$

< Objective 1 >

13. $\frac{10}{15}=\frac{2\times5}{3\times5}=\frac{2}{3}$

15. $\frac{10}{14}=\frac{2\times5}{2\times7}=\frac{5}{7}$

17. $\frac{12}{18}=\frac{2\times2\times3}{2\times3\times3}=\frac{2}{3}$

19. $\frac{35}{40}=\frac{5\times7}{2\times2\times2\times5}=\frac{7}{2\times2\times2}=\frac{7}{8}$

21. $\frac{11}{44}=\frac{11}{2\times2\times11}=\frac{1}{2\times2}=\frac{1}{4}$

23. $\frac{11}{33}=\frac{11}{3\times11}=\frac{1}{3}$

25. $\frac{24}{27}=\frac{2\times2\times2\times3}{3\times3\times3}=\frac{2\times2\times2}{3\times3}=\frac{8}{9}$

27. $\frac{32}{40}=\frac{2\times2\times2\times2\times2}{2\times2\times2\times5}=\frac{2\times2}{5}=\frac{4}{5}$

29. $\frac{75}{105}=\frac{3\times5\times5}{3\times5\times7}=\frac{5}{7}$

31. $\frac{24}{30}=\frac{2\times2\times2\times3}{2\times3\times5}=\frac{2\times2}{5}=\frac{4}{5}$

33. $\frac{105}{135}=\frac{3\times5\times7}{3\times3\times3\times5}=\frac{7}{3\times3}=\frac{7}{9}$

< Objective 2 >

35. $\frac{3}{7}\times\frac{4}{5}=\frac{3\times2\times2}{7\times5}=\frac{12}{35}$

37. $\frac{3}{4}\times\frac{7}{5}=\frac{3\times7}{2\times2\times5}=\frac{21}{20}$

39. $\frac{3}{5}\times\frac{5}{7}=\frac{3\times5}{5\times7}=\frac{3}{7}$

41. $\frac{6}{13}\times\frac{4}{9}=\frac{2\times3\times2\times2}{13\times3\times3}=\frac{2\times2\times2}{13\times3}=\frac{8}{39}$

43. $\frac{3}{11}\times\frac{7}{9}=\frac{3\times7}{11\times3\times3}=\frac{7}{11\times3}=\frac{7}{33}$

45. $\frac{4}{21}\times\frac{7}{12}=\frac{2\times2\times7}{3\times7\times2\times2\times3}=\frac{1}{3\times3}=\frac{1}{9}$

47. $\frac{1}{7}\div\frac{3}{5}=\frac{1}{7}\times\frac{5}{3}=\frac{5}{7\times3}=\frac{5}{21}$

49. $\frac{2}{5}\div\frac{3}{4}=\frac{2}{5}\times\frac{4}{3}=\frac{2\times2\times2}{5\times3}=\frac{8}{15}$

51. $\frac{8}{9}\div\frac{4}{3}=\frac{8}{9}\times\frac{3}{4}=\frac{2\times2\times2\times3}{3\times3\times2\times2}=\frac{2}{3}$

53. $\frac{7}{10}\div\frac{5}{9}=\frac{7}{10}\times\frac{9}{5}=\frac{7\times3\times3}{2\times5\times5}=\frac{63}{50}$

55. $\frac{8}{15}\div\frac{2}{5}=\frac{8}{15}\times\frac{5}{2}=\frac{2\times2\times2\times5}{3\times5\times2}=\frac{2\times2}{3}=\frac{4}{3}$

57. $\frac{8}{21}\div\frac{24}{35}=\frac{8}{21}\times\frac{35}{24}=\frac{2\times2\times2\times5\times7}{3\times7\times2\times2\times2\times3}$

$=\frac{5}{3\times3}=\frac{5}{9}$

59. $30=2\times3\times5$

$50=2\times5\times5$

$\text{LCD}=2\times3\times5\times5=150$

61. $48=2\times2\times2\times2\times3$

$80=2\times2\times2\times2\times5$

$\text{LCD}=2\times2\times2\times2\times3\times5=240$

63. $3=1\times3$

$4=2\times2$

$5=1\times5$

$\text{LCD}=2\times2\times3\times5=60$

65. $8=2\times2\times2$

$10=2\times5$

$15=3\times5$

$\text{LCD}=2\times2\times2\times3\times5=120$

67. $5=1\times5$

$10=2\times5$

$25=5\times5$

$\text{LCD}=2\times5\times5=50$

< Objective 3 >

69. The LCD of 5 and 4 is 20.

$\frac{2}{5}\times\frac{4}{4}=\frac{8}{20}$

$\frac{1}{4}\times\frac{5}{5}=\frac{5}{20}$

$\frac{2}{5}+\frac{1}{4}=\frac{8}{20}+\frac{5}{20}=\frac{13}{20}$

71. The LCD of 5 and 15 is 15.

$\frac{2}{5}\times\frac{3}{3}=\frac{6}{15}$

$\frac{2}{5}+\frac{7}{15}=\frac{6}{15}+\frac{7}{15}=\frac{13}{15}$

73. The LCD of 8 and 12 is 24.

$\frac{3}{8}\times\frac{3}{3}=\frac{9}{24}$

$\frac{5}{12}\times\frac{2}{2}=\frac{10}{24}$

$\frac{3}{8}+\frac{5}{12}=\frac{9}{24}+\frac{10}{24}=\frac{19}{24}$

75. The LCD of 30 and 18 is 90.

$$\frac{7}{30}\times\frac{3}{3}=\frac{21}{90}$$
$$\frac{5}{18}\times\frac{5}{5}=\frac{25}{90}$$
$$\frac{7}{30}+\frac{5}{18}=\frac{21}{90}+\frac{25}{90}=\frac{46}{90}=\frac{23}{45}$$

77. The LCD of 15 and 18 is 90.

$$\frac{7}{15}\times\frac{6}{6}=\frac{42}{90}$$
$$\frac{13}{18}\times\frac{5}{5}=\frac{65}{90}$$
$$\frac{7}{15}+\frac{13}{18}=\frac{42}{90}+\frac{65}{90}=\frac{107}{90}$$

79. The LCD of 5, 10, and 15 is 30.

$$\frac{1}{5}\times\frac{6}{6}=\frac{6}{30}$$
$$\frac{1}{10}\times\frac{3}{3}=\frac{3}{30}$$
$$\frac{1}{15}\times\frac{2}{2}=\frac{2}{30}$$
$$\frac{1}{5}+\frac{1}{10}+\frac{1}{15}=\frac{6}{30}+\frac{3}{30}+\frac{2}{30}=\frac{11}{30}$$

81. $\frac{8}{9}-\frac{3}{9}=\frac{5}{9}$

83. $\frac{6}{7}-\frac{2}{7}=\frac{4}{7}$

85. The LCD of 8 and 3 is 24.

$$\frac{7}{8}\times\frac{3}{3}=\frac{21}{24}$$
$$\frac{2}{3}\times\frac{8}{8}=\frac{16}{24}$$
$$\frac{7}{8}-\frac{2}{3}=\frac{21}{24}-\frac{16}{24}=\frac{5}{24}$$

87. The LCD of 9 and 5 is 45.

$$\frac{2}{9}\times\frac{2}{2}=\frac{4}{18}$$
$$\frac{11}{18}-\frac{2}{9}=\frac{11}{18}-\frac{4}{18}=\frac{7}{18}$$

89. The LCD of 3 and 11 is 33.

$$\frac{2}{3}\times\frac{11}{11}=\frac{22}{33}$$
$$\frac{7}{11}\times\frac{3}{3}=\frac{21}{33}$$
$$\frac{2}{3}-\frac{7}{11}=\frac{22}{33}-\frac{21}{33}=\frac{1}{33}$$

91. The LCD of 42 and 36 is 252.

$$\frac{5}{42}\times\frac{6}{6}=\frac{30}{252}$$
$$\frac{1}{36}\times\frac{7}{7}=\frac{7}{252}$$
$$\frac{5}{42}-\frac{1}{36}=\frac{30}{252}-\frac{7}{252}=\frac{23}{252}$$

93. $\frac{1}{3}+\frac{1}{3}+\frac{1}{2}=\frac{2}{6}+\frac{2}{6}+\frac{3}{6}=\frac{7}{6}=1\frac{1}{6}$

$\frac{7}{6}$ cups or $1\frac{1}{6}$ cups of flour is used in the pancake recipe.

95. $\frac{3}{4}+\frac{1}{2}+\frac{2}{3}=\frac{9}{12}+\frac{6}{12}+\frac{8}{12}=\frac{23}{12}=1\frac{11}{12}$

Carol walked $\frac{23}{12}$ mi or $1\frac{11}{12}$ mi.

97. $\frac{1}{4}-\frac{3}{16}=\frac{4}{16}-\frac{3}{16}=\frac{1}{16}$

$\frac{1}{16}$ lb was lost in cooking.

99. False

101. sometimes

Exercises 0.2

< Objectives 1, 3, and 4 >

1. True

3. True

5. True

7. False

9. False

11. False

13. True

15. True

17. False

19. True

21. True

23. False

25. The absolute value of –10 is 10.

27. $|-20| = 20$

29. The absolute value of –7 is 7.

31. The opposite of 30 is –30.

33. $-(-6) = 6$

35. $|50| = 50$

37. The absolute value of the opposite of 3 is 3.

39. The opposite of the absolute value of –7 is –7.

41. –3, 2, and 0 are integers.

43. 0 and 2 are whole numbers.

45. –2, 0, and 1 are integers.

47. 0 and 1 are whole numbers.

< Objective 2 >

49.

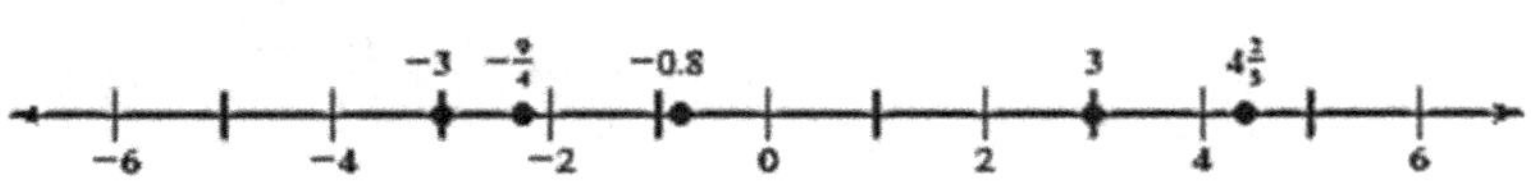

51.

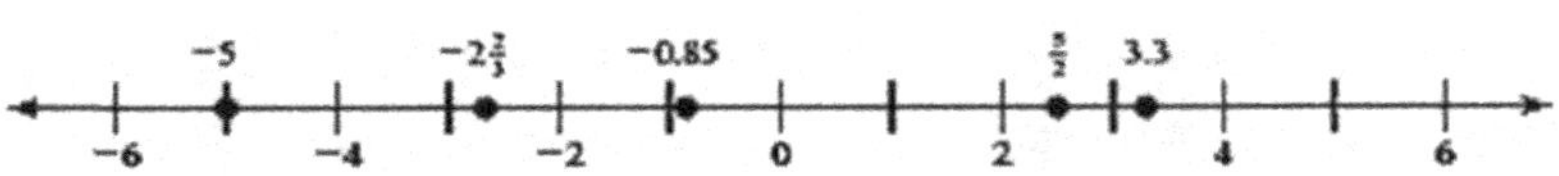

53. $-5 > -9$

55. $-20 < -10$

57. $|3| = 3$

59. $-4 < |-4|$

61. –$50

63. −10°F

65. +25,000 people

67. never

69. sometimes

71. –4 in.

73. #4, #3, #2, #1, #5

75. Above and Beyond.

77. **(a)** $-(-3) = 3$

(b) $-(-(-3)) = -3$

(c) $-(-(-(-3))) = 3$

(d) Above and Beyond

(e) $-(-(-(-(-(-(-7)))))) = -7$

Exercises 0.3

< Objectives 1–3 >

1. $-6+(-5)=-11$

3. $11+(-7)=4$

5. $4+(-6)=-2$

7. $7+9=16$

9. $(-11)+5=-6$

11. $-8+(-7)=-15$

13. $-12+4=-8$

15. $-9+10=1$

17. $-4+4=0$

19. $7+(-13)=-6$

21. $-8+5=-3$

23. $6+(-6)=0$

25. $\frac{45}{16}-\frac{9}{16}=\frac{36}{16}=\frac{9}{4}$

27. $\frac{29}{8}+\left(-\frac{17}{8}\right)=\frac{12}{8}=\frac{3}{2}$

29. $-\frac{73}{16}+\frac{119}{16}=\frac{46}{16}=\frac{23}{8}$

31. $4+(-7)+(-5)=-8$

33. $-2+(-6)+(-4)=-12$

35. $-3+(-7)+5+(-2)=-7$

37. $11-13=11+(-13)=-2$

39. $9-3=9+(-3)=6$

41. $-8-3=-8+(-3)=-11$

43. $-12-8=-12+(-8)=-20$

45. $-2-(-3)=-2+3=1$

47. $-5-(-5)=-5+5=0$

49. $28-(-22)=28+22=50$

51. $-15-(-25)=-15+25=10$

53. $-25-(-15)=-25+15=-10$

55. $-(-20)-(-15)=20+15=35$

57. $48-(-15)=48+15=63$

59. $\frac{10}{2}-\left(-\frac{7}{2}\right)=\frac{10}{2}+\frac{7}{2}=\frac{17}{2}$

61. $-\frac{7}{8}-\left(-\frac{19}{8}\right)=-\frac{7}{8}+\frac{19}{8}=\frac{12}{8}=\frac{3}{2}$

63. $-7-(-5)-6=-7+5+(-6)=-8$

65. $-10-8-(-7)=-10+(-8)+7=-11$

67. $18-(-9)=27$
The drop in temperature was 27°F.

69. $34+12-27-6+15=28$
Micki got off the elevator on the 28th floor.

71. $600-(-325)=925$
They are 925 ft apart.

73. $99.8+2.5-3.7=98.6$
Jose's temperature was 98.6°F.

75. $95-71=24$
The difference between the highest and the lowest of his scores was 24 points.

77. always

79. never

81. $-11.392-13.491=-24.883$

83. $-7.259-4.235=-11.494$

85. $-18.271-(-12.569)=-5.702$

87. $-17.346-(-28.293)=10.947$

89. Above and Beyond

91. Above and Beyond

93. Above and Beyond

95. **(a)** $|b|+a$ is positive.
(b) $b+(-a)$ is negative.
(c) $(-b)+a$ is positive.
(d) $-b+|-a|$ is positive.

Exercises 0.4

< Objectives 1 and 2 >

1. $7\bullet 8=56$

3. $(4)(-3)=-12$

5. $(-8)(9)=-72$

7. $(-5)\left(\frac{1}{3}\right)=-\frac{5}{3}$

9. $(-10)(0)=0$

11. $(-8)(-8)=64$

13. $(-4)\left(\frac{5}{7}\right)=-\frac{20}{7}$

15. $(-9)(-12)=108$

17. $(-20)(1)=-20$

19. $(-1.3)(6)=-7.8$

21. $-10(-15)=150$

23. $\left(-\frac{7}{10}\right)\left(-\frac{5}{14}\right)=\frac{7\bullet 5}{2\bullet 5\bullet 2\bullet 7}=\frac{1}{4}$

25. $\left(\frac{3}{5}\right)\left(-\frac{10}{27}\right)=-\frac{3\bullet 2\bullet 5}{5\bullet 3\bullet 9}=-\frac{2}{9}$

27. $\left(-\frac{5}{8}\right)\left(-\frac{4}{15}\right)=\frac{5\bullet 4}{2\bullet 4\bullet 3\bullet 5}=\frac{1}{6}$

29. $(-5)(3)(-8)=120$

31. $(-5)(-9)(-3)=-135$

33. $(2)(-5)(-3)(-5)=-150$

35. $(-4)(-3)(-6)(-2)=144$

37. $5(-6+9)=5\bullet(-6)+5\bullet 9=-30+45=15$

39. $-8(-9+15)=(-8)(-9)+(-8)\bullet 15$
$=72+(-120)=-48$

41. $-4(-5-3)=(-4)(-5)-(-4)\bullet 3$
$=20-(-12)=20+12=32$

43. $-4(-6-3)=(-4)(-6)-(-4)\bullet 3$
$=24-(-12)=24+12=36$

< Objective 3 >

45. $15 \div (-3) = -5$

47. $\frac{54}{9} = 6$

49. $\frac{-50}{5} = -10$

51. $\frac{-24}{-3} = 8$

53. $\frac{90}{-6} = -15$

55. $18 \div (-1) = -18$

57. $\frac{0}{-9} = 0$

61. $-7 \div 0$ is undefined.

63. $\frac{-150}{6} = -25$

65. $-45 \div (-9) = 5$

67. $-\frac{8}{11} \div \frac{18}{55} = -\frac{8}{11} \cdot \frac{55}{18} = -\frac{20}{9}$

69. $\frac{7}{10} \div \left(-\frac{14}{25}\right) = \frac{7}{10}\left(-\frac{25}{14}\right) = -\frac{5}{4}$

71. $\frac{-75}{15} = -5$

73. $42 \div 3 = 14$
She has been dieting for 14 weeks.

75. $8.91 \div 4.5 = 1.98$
The cost is $1.98 per pound.

77. $(70 - 58) \div 5 = 12 \div 5 = 2.4$
The temperature changed by −2.4°F each hour.

79. $\frac{5-15}{2+3} = \frac{-10}{5} = -2$

81. $\frac{-6+18}{-2-4} = \frac{12}{-6} = -2$

83. $\frac{(5)(-12)}{(-3)(5)} = \frac{-60}{-15} = 4$

85. False

87. never

89. With a graphing calculator:
[(−)] 5.634 [÷] 2.398 [ENTER]
The result is −2.349 (rounded to the nearest thousandth).

91. With a graphing calculator:
[(−)] 18.137 [÷] [(] [(−)] 5.236 [)] [ENTER]
The result is 3.464 (rounded to the nearest thousandth).

93. With a graphing calculator:
32.245 [÷] [(] [(−)] 48.298 [)] [ENTER]
The result is −0.668 (rounded to the nearest thousandth).

95. Above and Beyond

97. Above and Beyond

Exercises 0.5

< Objective 1 >

1. $3 \cdot 3 \cdot 3 \cdot 3 \cdot 3 = 3^5$

3. $7 \cdot 7 \cdot 7 \cdot 7 \cdot 7 = 7^5$

5. $8 \cdot 8 \cdot 8 \cdot 8 \cdot 8 \cdot 8 = 8^6$

7. $(-2)(-2)(-2) = (-2)^3$

< Objective 2 >

9. $3^2 = 3 \cdot 3 = 9$

11. $2^4 = 2 \cdot 2 \cdot 2 \cdot 2 = 16$

13. $(-8)^3 = (-8)(-8)(-8) = -512$

15. $-8^3 = -(8 \cdot 8 \cdot 8) = -512$

17. $-5^2 = -(5 \cdot 5) = -25$

19. $(-4)^2 = (-4)(-4) = 16$

21. $-(-2)^5 = -(-2)(-2)(-2)(-2)(-2) = 32$

23. $10^3 = 10 \cdot 10 \cdot 10 = 1{,}000$

25. $10^6 = 1{,}000{,}000$

< Objective 3 >

27. $2 \times 4^3 = 2 \times 64 = 128$

29. $3 \times 4^2 = 3 \times 16 = 48$

31. $5 + 2^2 = 5 + 4 = 9$

33. $3^4 \times 2^4 = 81 \times 16 = 1{,}296$

35. $4 + 3 \cdot 5 = 4 + 15 = 19$

37. $(7 + 2) \cdot 6 = 9 \cdot 6 = 54$

39. $-12 - 8 \div 4 = -12 - 2 = -14$

41. $(12 - 8) \div 4 = 4 \div 4 = 1$

43. $8 \cdot 7 + 2 \cdot 2 = 56 + 4 = 60$

45. $(7 \cdot 5) + 3 \cdot 2 = 35 + 6 = 41$

47. $3 \cdot 5^2 = 3 \cdot 25 = 75$

49. $(3 \cdot 5)^2 = 15^2 = 225$

51. $4 \cdot 3^2 - 2 = 4 \cdot 9 - 2 = 36 - 2 = 34$

53. $7(2^3 - 5) = 7(8 - 5) = 7(3) = 21$

55. $3 \cdot 2^4 - 26 \cdot 2 = 3 \cdot 16 - 26 \cdot 2 = 48 - 52 = -4$

57. $(2 \cdot 4)^2 - 8 \cdot 3 = 8^2 - 8 \cdot 3 = 64 - 24 = 40$

59. $5(3 + 4)^2 = 5(7)^2 = 5(49) = 245$

61. $(5 \cdot 3 + 4)^2 = (15 + 4)^2 = (19)^2 = 361$

63. $5[3(2 + 5) - 5] = 5[3(7) - 5] = 5[21 - 5]$
$= 5[16] = 80$

65. $-2\left[(3 - 5)^2 - (-4 + 2)^3 \cdot (8 \div 4 \cdot 2)\right]$
$= -2\left[(-2)^2 - (-2)^3 \cdot (2 \cdot 2)\right]$
$= -2[4 - (-8) \cdot 4] = -2[4 - (-32)]$
$= -2[36] = -72$

67. $4(2 + 3)^2 - 125 = 4(5)^2 - 125 = 4(25) - 125$
$= 100 - 125 = -25$

69. $(4 \cdot 2 + 3)^2 - 25 = (8 + 3)^2 - 25 = 11^2 - 25$
$= 121 - 25 = 96$

71. $\left[-20 - 4^2 + (-4)^2 + 2\right] \div 9$
$= [-20 - 16 + 16 + 2] \div 9 = [-18] \div 9 = -2$

73. $4 \cdot 8 \div 2 - 5^2 = 4 \cdot 8 \div 2 - 25 = 32 \div 2 - 25$
$= 16 - 25 = -9$

75. $15 + 5 - 3 \cdot 2 + (-2)^3 = 15 + 5 - 6 + (-8) = 6$

77. The phrase "doubled five times" in exponential form can be expressed as 2^5.

79. $8-2\frac{3}{4}-2\frac{1}{2}=\frac{32}{4}-\frac{11}{4}-\frac{10}{4}=\frac{11}{4}=2\frac{3}{4}$

They still need to drive $2\frac{3}{4}$ hr.

81. True

83. always

85. With a graphing calculator:
1.2 [^] 3 [÷] 2.0736 [×] 2.4 [+] 1.6935
[−] 2.4896 [ENTER]
The result is 1.2 (rounded to the nearest tenth).

87. With a graphing calculator:
1.23 [×] 3.169 [−] 2.05194 [+] [(] 5.128
[×] 3.15 [−] 10.1742 [)] [ENTER]
The result is 7.8 (rounded to the nearest tenth).

89. $14\frac{1}{4}-12\frac{3}{8}=\frac{114}{8}-\frac{99}{8}=\frac{15}{8}=1\frac{7}{8}$

The interest rate increased by $1\frac{7}{8}\%$.

91. $\frac{1{,}789-676}{21}=\frac{1{,}113}{21}=53$
On average, 53 gal/day of gas was used.

93. $36\div(4+2)-4=2$

Summary Exercises

1. $\frac{5}{7}\times\frac{2}{2}=\frac{10}{14}$
$\frac{5}{7}\times\frac{4}{4}=\frac{20}{28}$
$\frac{5}{7}\times\frac{10}{10}=\frac{50}{70}$

3. $\frac{4}{9}\times\frac{2}{2}=\frac{8}{18}$
$\frac{4}{9}\times\frac{4}{4}=\frac{16}{36}$
$\frac{4}{9}\times\frac{6}{6}=\frac{24}{54}$

5. $\frac{7}{15}\times\frac{5}{21}=\frac{7\times5}{3\times5\times3\times7}=\frac{1}{3\times3}=\frac{1}{9}$

7. $\frac{5}{17}\div\frac{15}{34}=\frac{5}{17}\times\frac{34}{15}=\frac{2}{3}$

9. $\frac{7}{8}+\frac{15}{24}=\frac{21}{24}+\frac{15}{24}=\frac{36}{24}=\frac{3}{2}$

11. $\frac{11}{18}-\frac{2}{9}=\frac{11}{18}-\frac{4}{18}=\frac{7}{18}$

13. $\frac{16}{3}\times\frac{16}{4}=\frac{64}{3}$

The kitchen floor is $\frac{64}{3}$ sq yd, or 22 sq yd (to the next whole square yard).
$22\times9=198$
It costs $198 to cover the kitchen floor.

15. $18\div\frac{3}{8}=\frac{18}{1}\times\frac{8}{3}=48$
48 lots can be formed.

17. The absolute value of 12 is 12.

19. $-|-3|=-3$

21. $-|-4|=-4$

23. The absolute value of −16 is 16.

25. $-3<-1$

27. $-|-7|<-(-2)$

29. $15+(-7)=8$

31. $-23-(-12)=-23+12=-11$

33. $-\frac{9}{13}-\frac{4}{39}=-\frac{27}{39}+\left(-\frac{4}{39}\right)=-\frac{31}{39}$

35. $7+(-4)+8+(-7)=4$

37. $-35+30=-5$

39. $3-(-2)=3+2=5$

41. $\frac{23}{4}-\left(-\frac{3}{4}\right)=\frac{23}{4}+\frac{3}{4}=\frac{26}{4}=\frac{13}{2}$

43. $8-12-(-5)=8+(-12)+5=1$

45. $7-(-4)-7-4=7+4+(-7)+(-4)=0$

47. $625-69.74-29.95-57.65=\$467.66$
She had left \$467.66 of her original deposit.

49. $(-18)(-2)=36$

51. $(-5)(3)=-15$

53. $(-4)^2=(-4)(-4)=16$

55. $(-6)(-5)(4)(-3)=-360$

57. $-4(8-7)=-4(8)-(-4)(7)=-32-(-28)$
$=-32+28=-4$

59. $-8(5-2)=-8(5)-(-8)(2)=-40-(-16)$
$=-40+16=-24$

61. $(-48)\div 12=-4$

63. $-2\div 0$ is undefined.

65. $-\frac{7}{9}\div\left(-\frac{2}{3}\right)=-\frac{7}{9}\bullet\left(-\frac{3}{2}\right)=\frac{7}{6}$

67. $8\div(-4)=-2$

69. $12\times 3{,}500=42{,}000$
The agency lost \$42,000 in a year.

71. $3^3=3\bullet 3\bullet 3$

73. $2^6=2\bullet 2\bullet 2\bullet 2\bullet 2\bullet 2$

75. $18-12\div 2=18-6=12$

77. $6\bullet 2^3=6\bullet 8=48$

79. $5\bullet 3^2-4=5\bullet 9-4=45-4=41$

81. $5(4-2)^2=5(2)^2=5\bullet(4)=20$

83. $(5\bullet 4-2)^2=(20-2)^2=(18)^2=324$

85. $3\bullet 5-2^2=3\bullet 5-4=15-4=11$

87. $14(5)+4(-2)=70-8=62$
The exam grade of the student is 62 points.

Chapter Test 0

1. $\frac{27}{99}=\frac{3\bullet 3\bullet 3}{3\bullet 3\bullet 11}=\frac{3}{11}$

3. $\frac{4}{15}+\frac{7}{10}=\frac{8}{30}+\frac{21}{30}=\frac{29}{30}$

5. $13+(-11)+(-5)=-3$

7. $28\div(-4)=-7$

9. $(-7)(5)=-35$

11. $-9-8-(-5)=-9+(-8)+5=-12$

13. $23-4\bullet 12\div 3+|-4|=23-48\div 3+4$
$=23-16+4=11$

15. $\frac{3}{8}\times\frac{6}{11}=\frac{3\times 3\times 2}{2\times 2\times 2\times 11}=\frac{3\times 3}{2\times 2\times 11}=\frac{9}{44}$

17. $-7<-5$

19. $\frac{14-2.8}{0.35}=32$

32 lots can be formed.

Chapter 1
From Arithmetic to Algebra

Exercises 1.1

< Objective 1 >

1. The sum of c and d is written as $c+d$.

3. w plus z is written as $w+z$.

5. x increased by 5 is written as $x+5$.

7. 10 more than y is written as $y+10$.

9. a minus b is written as $a-b$.

11. 7 decreased by b is written as $7-b$.

13. 6 less than r is written as $r-6$.

15. w times z is written as wz.

17. The product of 5 and t is written as $5t$.

19. The product of 8, m, and n is written as $8mn$.

21. The product of 8 and the quantity m plus n is written as $8(m+n)$.

23. Twice the sum of x and y is written as $2(x+y)$.

25. The sum of twice x and y is written as $2x+y$.

27. Twice the difference of x and y is written as $2(x-y)$.

29. The quantity a plus b times the quantity a minus b is written as $(a+b)(a-b)$.

31. The product of m and 3 less than m is written as $m(m-3)$.

33. 5 divided by x is written as $\frac{5}{x}$.

35. The sum of a and b, divided by 7 is written as $\frac{a+b}{7}$.

37. The difference of p and q, divided by 4 is written as $\frac{p-q}{4}$.

39. The sum of a and 3, divided by the difference of a and 3 is written as $\frac{a+3}{a-3}$.

41. 5 more than a number is written as $x+5$.

43. 7 less than a number is written as $x-7$.

45. 9 times a number is written as $9x$.

47. 6 more than 3 times a number is written as $3x+6$.

49. Twice the sum of a number and 5 is written as $2(x+5)$.

51. The product of 2 more than a number and 2 less than that same number is written as $(x+2)(x-2)$.

53. The quotient of a number and 7 is written as $\frac{x}{7}$.

55. The sum of a number and 5, divided by 8 is written as $\frac{x+5}{8}$.

57. 6 more than a number divided by 6 less than that same number is written as $\frac{x+6}{x-6}$.

59. Four times the length of a side is written as $4s$.

61. π times the radius r squared times the height h is written as $\pi r^2 h$.

63. One-half the product of the height h and the sum of two unequal sides b_1 and b_2 is written as $\frac{1}{2}h(b_1+b_2)$.

< Objective 2 >

65. $2(x+5)$ is an expression. Its meaning is clear.

67. $4+\div m$ is not an expression. The two operations in a row have no meaning.

69. $2b=6$ is not an expression. The equals sign is not an operation sign.

71. $2a(3b+5)$ is an expression. Its meaning is clear.

< Objective 3 >

73. The other number expressed in terms of x, is $35-x$.

75. The expression for the interest earned is Prt.

77. The expression for the number of student tickets sold is $400-x$.

79. True

81. never

83. The formula that describes the number of carriage bolts sold last month is $H-284$.

85. The average number of customers in the system is $\frac{a}{h-a}$.

87. Above and Beyond

Exercises 1.2

< Objective 1 >

1. $3c-2b=3(-4)-2(5)=-12-10=-22$

3. $7c+6b=7(-4)+6(5)=-28+30=2$

5. $-b^2+b=-(5)^2+(5)=-25+5=-20$

7. $3a^2=3(-2)^2=3(4)=12$

9. $c^2-2d=(-4)^2-2(6)=16-12=4$

11. $2a^2+3b^2=2(-2)^2+3(5)^2=2(4)+3(25)$
$=8+75=83$

13. $2(c-d)=2[(-4)-6]=2(-10)=-20$

15. $4(2a-d)=4[2(-2)-6]=4(-4-6)$
$=4(-10)=-40$

17. $a(b+3c)=-2[5+3(-4)]=-2(5-12)$
$=-2(-7)=14$

19. $\frac{6d}{c}=\frac{6(6)}{-4}=\frac{36}{-4}=-9$

21. $\frac{3d+2c}{b}=\frac{3(6)+2(-4)}{5}=\frac{18+(-8)}{5}=\frac{10}{5}$
$=2$

23. $\dfrac{2b-3a}{c+2d}=\dfrac{2(5)-3(-2)}{-4+2(6)}=\dfrac{10+6}{-4+12}=\dfrac{16}{8}=2$

25. $b^2-d^2=5^2-6^2=25-36=-11$

27. $(b-d)^2=(5-6)^2=(-1)^2=1$

29. $(d-b)(d+b)=(6-5)(6+5)=(1)(11)$
$=11$

31. $c^3-a^3=(-4)^3-(-2)^3=-64-(-8)=-56$

33. $(c-a)^3=[(-4)-(-2)]^3=(-2)^3=-8$

35. $(d-b)(d^2+db+b^2)$
$=(6-5)[6^2+(6)(5)+5^2]$
$=(1)(36+30+25)=(1)(91)=91$

37. $b^2+a^2=(5)^2+(-2)^2=25+4=29$

39. $(b+a)^2=[5+(-2)]^2=(3)^2=9$

41. $a^2+2ad+d^2=(-2)^2+2(-2)(6)+(6)^2$
$=4-24+36=16$

43. $x^2-2y^2+z^2=(-2)^2-2(-3)^2+(4)^2$
$=4-2(9)+16=4-18+16$
$=2$

45. $2xy-(x^2-2yz)$
$=2(-2)(-3)-[(-2)^2-2(-3)(4)]$
$=2(6)-[4-2(-12)]=12-(4+24)$
$=12-28=-16$

47. $2y(z^2-2xy)+yz^2$
$=2(-3)[(4)^2-2(-2)(-3)]+(-3)(4)^2$
$=-6(16-12)+(-3)(16)=-24-48=-72$

< Objective 3 >

49. $R_T=\dfrac{R_1R_2}{R_1+R_2}=\dfrac{(9)(15)}{(9)+(15)}=\dfrac{135}{24}=5.625$
The total resistance is 5.625 Ω.

51. $P=2L+2W=2(10)+2(5)=20+10=30$
The perimeter of the rectangle is 30 in.

53. $P=\dfrac{I}{r\bullet t}=\dfrac{(150)}{(0.04)(2)}=1{,}875$
The principal was \$1,875.

55. $F=\dfrac{9}{5}C+32=\dfrac{9}{5}(-10)+32=14$
The Fahrenheit temperature is 14°F.

57. $0.08t+5.25=0.08(173)+5.25=19.09$
The monthly bill is \$19.09.

59. $x-7=2y+5$
$(22)-7\stackrel{?}{=}2(5)+5$
$15=15$
True

61. $2(x+y)=2x+y$
$2[(-4)+(-2)]\stackrel{?}{=}2(-4)+(-2)$
$2(-6)\stackrel{?}{=}-8+(-2)$
$-12\neq-10$
False

63. False

65. never

< Objective 2 >

67. With a graphing calculator:
[(−)] [2.34] [+] [(] [(−)] [3.14] [×] [4.12] [)]
[ENTER]
Display: −15.3.

69. With a graphing calculator:
[(−)] 3.14 [∧] 2 [−] [(] 2 [×] [(−)] 2.34 [∧] 2 [)] [ENTER]
Display: −1.1.

71. With a graphing calculator:
[(] [(−)] 2.34 [×] [(−)] 3.14 [)] [÷] [(] 4.12 [−] [(−)] 2.34 [)] [ENTER]
Display: 1.1.

73. With a graphing calculator:
[(] 2 [×] [(−)] 2.34 [+] [(−)] 3.14 [)] [÷] [(] 2 [×] [(−)] 2.34 [+] 4.12 [)] [ENTER]
Display: $13.96 \approx 14$.

75. $x^2 - 4x^3 + 3x = (3)^2 - 4(3)^3 + 3(3) = -90$
With a graphing calculator:
[(] [(−)] 3.14 [∧] 2 [×] 4.12 [∧] 2 [)] [÷] [(] [(−)] 2.34 [×] [(−)] 3.14 [)] [(] [ENTER]
Display: 22.8.

77. With a graphing calculator:
27 [∧] 2 [−] [(] 4 [×] 27 [∧] 3 [)] [+] [(] 3 [×] 27 [)] [ENTER]
Display: −77,922.

79. With a graphing calculator:
232 [+] [(−)] 487 [×] 58 [∧] 2 [ENTER]
Display: −1,638,036.

81. With a graphing calculator:
[(] 58 [−] 487 [)] [∧] 2 [−] 232 [∧] 2 [ENTER]
Display: 130,217.

83. With a graphing calculator:
[(] [(−)] 487 [∧] 2 [−] 58 [∧] 2 [)] [÷] [(] 58 [∧] 2 [−] 232 [∧] 2 [)] [ENTER]
Display: −4.6.

85. $-2t^2 + 13t + 1 - 2(1)^2 + 13(1) + 1 = 11 + 1$
$= 12$
(When $t = 1$ hr)
The concentration of antihistamine 1 hour after being administered is 12 mcg/mL.

87. $\dfrac{rT}{5{,}252} = \dfrac{(1{,}180)(3)}{5{,}252} = \dfrac{3{,}540}{5{,}252} = 0.674$
(rounded to the nearest thousandth)

89. Above and Beyond

91. **(a)**

Value of *x*	0	1	2	3	4	5
Value of expression	0	80	96	72	32	0

(b) The expression appears to be largest at $x = 2$.
(c)

Value of *x*	1.5	1.6	1.7	1.8	1.9	2.0	2.1	2.2	2.3	2.4	2.5
Value of expression	94.5	95.744	96.492	96.768	96.596	96	95.004	93.632	91.908	89.856	87.5

(d) The expression appears to be largest at $x = 1.8$.
(e) The expression appears to be largest at $x = 1.81$.

93. Above and Beyond

95. Above and Beyond

Exercises 1.3

< Objective 1 >

1. $5a+2$ has two terms: $5a$ and 2.

3. $5x$ has one term: $5x$.

5. $3x^2+3x-7$ or $3x^2+3x+(-7)$ has three terms: $3x^2$, $3x$, and -7.

7. In the group of terms $5ab$, $3b$, $3a$, $4ab$, the like terms are $5ab$ and $4ab$.

9. In the group of terms $4xy^2$, $2x^2y$, $5x^2$, $-3x^2y$, $5y$, $6x^2y$, the like terms are $2x^2y$, $-3x^2y$, and $6x^2y$.

< Objective 2 >

11. $6p+9p=15p$

13. $7b^3+10b^3=(7+10)b^3=17b^3$

15. $21xyz+7xyz=(21+7)xyz=28xyz$

17. $9z^2-3z^2=9z^2+(-3z^2)=6z^2$

19. $5a^3-5a^3=5a^3+(-5a^3)=0$

21. $16p^2q-17p^2q=16p^2q+(-17p^2q)=-p^2q$

23. $6p^2q-21p^2q=6p^2q+(-21p^2q)=-15p^2q$

25. $10x^2-7x^2+3x^2=10x^2+(-7x^2)+3x^2$
$=6x^2$

27. $-6c+3d+5c=-c+3d$

29. $4x+4y-7x-5y=(4-7)x+(4-5)y$
$=-3x-y$

31. $2a-7b-3+2a-3b+2$
$=(2+2)a+([(-7)-3])b+(-3)+2$
$=4a-10b-1$

33. $-(2a+3b)=-2a-3b$

35. $5a-(2b-3c)=5a-2b+3c$

37. $3x-(4y+5x)=3x-4y-5x=-2x-4y$

39. $5p-(-3p+2q)=5p+3p-2q=8p-2q$

< Objective 3 >

41. $(6a-5)+(3a+9)=6a-5+3a+9$
$=9a+4$

43. $(-7p^2+9p)+(4p^2-5p)$
$=-7p^2+9p+4p^2-5p=-3p^2+4p$

45. $(3x^2-2x)+(-5x^2+2x)$
$=3x^2-2x-5x^2+2x=-2x^2$

47. $(2x^2+5x-3)+(3x^2-7x+4)$
$=2x^2+5x-3+3x^2-7x+4=5x^2-2x+1$

49. $(2b^2+8)+(5b+8)=2b^2+8+5b+8$
$=2b^2+5b+16$

51. $(8y^3-5y^2)+(5y^2-2y)$
$=8y^3-5y^2+5y^2-2y=8y^3-2y$

53. $(3x^2-7x^3)+(-5x^2+4x^3)$
$=3x^2-7x^3-5x^2+4x^3=-3x^3-2x^2$

55. $(4x^2-2+7x)+(5-8x-6x^2)$
$=4x^2-2+7x+5-8x-6x^2=-2x^2-x+3$

< Objective 4 >

57. $(3x-5)-(x+2)=3x-5-x-2=2x-7$

59. $(4m^2-5m)-(3m^2-2m)$
$=4m^2-5m-3m^2+2m=m^2-3m$

61. $(4y^2+5y)-(6y^2+5y)$
$=4y^2+5y-6y^2-5y=-2y^2$

63. $(3x^2-5x-2)-(x^2-4x-3)$
$=3x^2-5x-2-x^2+4x+3=2x^2-x+1$

65. $(8a^2-9a)-(3a+7)=8a^2-9a-3a-7$
$=8a^2-12a-7$

67. $(-9p^2+4p)-(2p-5p^2)$
$=-9p^2+4p-2p+5p^2=-4p^2+2p$

69. $(3x^2-8x+7)-(x^2-5-8x)$
$=3x^2-8x+7-x^2+5+8x=2x^2+12$

71. $[(4b-2)+(5b+3)]-(3b+2)$
$=9b+1-3b-2=6b-1$

73. $[(2x^2-3x+5)+(3x^2+5x-7)]$
$-(5x^2+7x-6)$
$=5x^2+2x-2-5x^2-7x+6=-5x+4$

75. $[(4x^2-5)+(2x-7)]-(2x^2-3x)$
$=4x^2+2x-12-2x^2+3x=2x^2+5x-12$

77. $(2y^2-8y)-[(3y^2-3y)+(5y^2+3y)]$
$=2y^2-8y-(8y^2)=-6y^2-8y$

79. $[(9x^2-3x+5)-(3x^2+2x-1)]$
$-(x^2-2x-3)$
$=(9x^2-3x+5-3x^2-2x+1)-x^2+2x+3$
$=6x^2-5x+6-x^2+2x+3=5x^2-3x+9$

81. $P=2L+2W=2(8x+9)+2(6x-7)$
$=16x+18+12x-14$
$=(16+12)x+18-14=28x+4$
The expression for the perimeter of the rectangle is $(28x+4)$.

83. The profit is given by the revenue minus the cost.
$90x-x^2-(150+25x)$
$=90x-x^2-150-25x=-x^2+65x-150$
Thus, the profit is $-x^2+65x-150$.

85. $(3y^2+3y-2)-(y^2-8)$
$=3y^2+3y-2-y^2+8=2y^2+3y+6$
The remaining piece of beam measures $(2y^2+3y+6)$ m.

87. $P=x+(3x+3)+(2x^2-5x+1)$
$=x+3x+3+2x^2-5x+1=2x^2-x+4$
The perimeter of the triangle measures $(2x^2-x+4)$ ft.

89. $P=6y+2+10+3y+10+5+(5y+2)+8y$
$=22y+29$
The perimeter measures $(22y+29)$ cm.

91. False

93. always

95. $7x^2-5y^3=7(7.1695)^2-5(3.128)^3$
With a scientific calculator:
7 [×] 7.1695 [y^x] 2 [−] 5 [×] 3.128 [y^x] 3 [=]
The result is 206.8 (rounded to nearest tenth).

97. $4x^2y + 2xy^2 - 5x^3y$

$= 4(1.29)^2(2.56) + 2(1.29)(2.56)^2$

$-5(1.29)^3(2.56)$

With a scientific calculator:

4 [×] 1.29 [y^x] 2 [×] 2.56 [+] 2 [×] 1.29 [×] 2.56 [y^x] 2 [−] 5 [×] 1.29 [y^x] 3 [×] 2.56 [=]

The result is 6.5 (rounded to nearest tenth).

99. $54p + 32p = 86p$

The total load that the two beams can support is 86*p*.

101. $\dfrac{703w}{h^2} = \dfrac{703(190)}{(69)^2} = \dfrac{133{,}570}{4{,}761} = 28.1$

The 69-in, 190-lb man has a BMI of 28.1.

103. Above and Beyond

105. Above and Beyond

107. Above and Beyond

109. Above and Beyond

Exercises 1.4

< Objective 1 >

1. $x + 7 = 12 \qquad (5)$

$5 + 7 \stackrel{?}{=} 12$

$12 = 12$

Because $12 = 12$ is a true statement, 5 is a solution of the equation.

3. $x - 15 = 6 \qquad (-21)$

$-21 - 15 \stackrel{?}{=} 6$

$-36 \neq 6$

Because the two sides do not name the same number, the statement is not true, and −21 is not a solution.

5. $5 - x = 2 \qquad (4)$

$5 - 4 \stackrel{?}{=} 2$

$1 \neq 2$

Because the two sides do not name the same number, the statement is not true, and 4 is not a solution.

7. $8 - x = 5 \qquad (-3)$

$8 - (-3) \stackrel{?}{=} 5$

$11 \neq 5$

Because the two sides do not name the same number, the statement is not true, and −3 is not a solution.

9. $3x + 4 = 13 \qquad (8)$

$3(8) + 4 \stackrel{?}{=} 13$

$24 + 4 \stackrel{?}{=} 13$

$28 \neq 13$

Because the two sides do not name the same number, the statement is not true, and 8 is not a solution.

11. $4x - 5 = 7 \qquad (2)$

$4(2) - 5 \stackrel{?}{=} 7$

$3 \neq 7$

Because the two sides do not name the same number, the statement is not true, and 2 is not a solution.

13. $5 - 2x = 10 \qquad \left(-\dfrac{5}{2}\right)$

$5 - 2\left(-\dfrac{5}{2}\right) \stackrel{?}{=} 10$

$5 + 5 \stackrel{?}{=} 10$

$10 = 10$

Because $10 = 10$ is a true statement, $-\dfrac{5}{2}$ is the solution of the equation.

15. $6-\frac{3}{4}x=4+\frac{1}{4}x$ (2)

$$6-\frac{3}{4}(2)\stackrel{?}{=}4+\frac{1}{4}(2)$$
$$\frac{9}{2}=\frac{9}{2}$$

Because $\frac{9}{2}=\frac{9}{2}$ is a true statement, 2 is the solution of the equation.

17. $x+3+2x=5+x+8$ (5)

$$5+3+2(5)\stackrel{?}{=}5+5+8$$
$$8+10\stackrel{?}{=}18$$
$$18=18$$

Because $18=18$ is a true statement, 5 is a solution of the equation.

19. $\frac{3}{4}x=18$ (20)

$$\frac{3}{4}(20)\stackrel{?}{=}18$$
$$15\neq 18$$

Because the two sides do not name the same number, the statement is not true, and 20 is not a solution.

21. $\frac{3}{5}x+5=11$ (10)

$$\frac{3}{5}(10)+5\stackrel{?}{=}11$$
$$6+5\stackrel{?}{=}11$$
$$11=11$$

Because $11=11$ is a true statement, 10 is a solution of the equation.

23. $2x+1=9$ is a linear equation.

25. $7x+2x+8-3$ is an expression.

27. $3x+5=9$ is a linear equation.

< Objective 2 >

29.
$$\begin{array}{rcl} x+7 &=& 9 \\ -7 & & -7 \\ \hline x &=& 2 \end{array}$$
The solution is 2.

Check: $(2)+7\stackrel{?}{=}9$
$9=9$

31.
$$\begin{array}{rcl} x-8 &=& 3 \\ +8 & & +8 \\ \hline x &=& 11 \end{array}$$
The solution is 11.

Check: $(11)-8\stackrel{?}{=}3$
$3=3$

33.
$$\begin{array}{rcl} x-8 &=& -10 \\ +8 & & +8 \\ \hline x &=& -2 \end{array}$$
The solution is −2.

Check: $-2-8\stackrel{?}{=}-10$
$-10=-10$

35.
$$\begin{array}{rcl} 11 &=& x+5 \\ -5 & & -5 \\ \hline 6 &=& x \end{array}$$
The solution is 6.

Check: $11\stackrel{?}{=}(6)+5$
$11=11$

37.
$$\begin{array}{rcl} 4x &=& 3x+4 \\ -3x & & -3x \\ \hline x &=& 4 \end{array}$$
The solution is 4.

Check: $4(4)\stackrel{?}{=}3(4)+4$
$16\stackrel{?}{=}12+4$
$16=16$

39.
$$\begin{array}{rcl} 11x &=& 10x-10 \\ -10x & & -10x \\ \hline x &=& -10 \end{array}$$

Check: $11(-10)\stackrel{?}{=}10(-10)-10$
$-110=-110$

The solution is −10.

41. $4x-10=5(x-2)$

$$4x-10=5x-10$$
$$-x=0$$
$$x=0$$

Check: $4\cdot 0-10\stackrel{?}{=}5(0-2)$
$-10=-10$

The solution is −10.

43. $\frac{9}{5}x+\frac{5}{6}=\frac{4}{5}x+\frac{1}{6}$

$\frac{5}{5}x=-\frac{4}{6}$

$x=-\frac{2}{3}$

Check: $\frac{9}{5}\left(-\frac{2}{3}\right)+\frac{5}{6}\overset{?}{=}\frac{4}{5}\left(-\frac{2}{3}\right)+\frac{1}{6}$

$-\frac{18}{15}+\frac{5}{6}\overset{?}{=}-\frac{8}{15}+\frac{1}{6}$

$-\frac{11}{30}=-\frac{11}{30}$

The solution is $-\frac{2}{3}$.

45.
$$\begin{array}{rcr} 5x-7= & & 4x-2 \\ \underline{-4x} & & \underline{-4x} \\ x-7= & & -3 \\ \underline{+7} & & \underline{+7} \\ x & = & 4 \end{array}$$

Check: $5(4)-7\overset{?}{=}4(4)-3$

$13=13$

The solution is 4.

47. $\frac{5}{3}x=9+\frac{2}{3}x$

$-\frac{2}{3}x \quad -\frac{2}{3}x$

$\frac{3}{3}x=9$

$x=9$

Check: $\frac{5}{3}(9)\overset{?}{=}9+\frac{2}{3}(9)$

$15=15$

The solution is 9.

49. $3+6x+2=3x+11+2x$

$6x+5=5x+11$

$x+5=11$

$x=6$

Check: $3+6(6)+2\overset{?}{=}3(6)+11+2(6)$

$41=41$

The solution is 6.

51. $4x+7+3x=5x+13+x$

$7x+7=6x+13$

$x+7=13$

$x=6$

Check: $4(6)+7+3(6)\overset{?}{=}5(6)+13+(6)$

$49=49$

The solution is 6.

53. $4(3x+4)=11x-2$

$12x+16=11x-2$

$x+16=-2$

$x=-18$

Check: $4[3(-18)+4]\overset{?}{=}11(-18)-2$

$4(-50)\overset{?}{=}-198-2$

$-200=-200$

The solution is -18.

55. $3(7x+2)=5(4x+1)+17$

$21x+6=20x+5+17$

$x+6=22$

$x=16$

Check: $3(7\times16+2)\overset{?}{=}5(4\times16+1)+17$

$3(114)\overset{?}{=}5(65)+17$

$342=342$

The solution is 16.

57. $\frac{5}{4}x-1=\frac{1}{4}x+7$

$x-1=7$

$x=8$

Check: $\frac{5}{4}(8)-1\overset{?}{=}\frac{1}{4}(8)+7$

$10-1\overset{?}{=}2+7$

$9=9$

The solution is 8.

59. $\frac{9}{2}x-\frac{3}{4}=\frac{7}{2}x+\frac{5}{4}$

$x-\frac{3}{4}=\frac{5}{4}$

$x=\frac{8}{4}=2$

Check: $\frac{9(2)}{2}-\frac{3}{4}\overset{?}{=}\frac{7(2)}{2}+\frac{5}{4}$

$9-\frac{3}{4}\overset{?}{=}7+\frac{5}{4}$

$\frac{36}{4}-\frac{3}{4}\overset{?}{=}\frac{28}{4}+\frac{5}{4}$

$\frac{33}{4}=\frac{33}{4}$

The solution is 2.

61. $0.56x=9-0.44x$

$x=9$

Check: $0.56(9)\overset{?}{=}9-0.44(9)$

$5.04\overset{?}{=}9-3.96$

$5.04=5.04$

The solution is 9.

63. $0.12x+0.53x-8=-0.92x+0.57x+4$

$0.65x-8=-0.35x+4$

$x-8=4$

$x=12$

Check:

$0.12(12)+0.53(12)-8\overset{?}{=}-0.92(12)+0.57(12)+4$

$1.44+6.36-8\overset{?}{=}-11.04+6.84+4$

$-0.2=-0.2$

The solution is 12.

65. $x+3=7$

67. $3x-7=2x$

69. $2(x+5)=x+18$

< Objective 3 >

71. Let x be the number.

$x+7=33$

$x=26$

The number is 26.

73. Let x be the number.

$x-15=7$

$x=22$

The number is 22.

75. Let x be the number of votes received by the losing candidate. Then $x+1,840$ is the total number of votes cast.

$x+1,840=3,260$

$x=1,420$

The losing candidate received 1,420 votes.

77. Let x be the cost of the dryer. Then $x+360$ is the cost of the washer and dryer combined.

$x+360=650$

$x=290$

The dryer costs $290.

79. **(c)**

$$\begin{array}{rcl} 8x+5 & = & 9x-4 \\ +4 & & +4 \\ \hline 8x+9 & = & 9x \end{array}$$

81. **(a)**

$$\begin{array}{rcl} 12x-6 & = & 8x+14 \\ -8x & & -8x \\ \hline 4x-6 & = & 14 \end{array}$$

83. True

85. True

87. sometimes

89. Let x be the number of hex bolts sold. Then $x-284$ is the number of carriage bolts sold.

$x-284=2{,}680$

$x=2{,}964$

K Jones Manufacturing sold 2,964 hex bolts.

91. Let d be the diameter of the cylinder. Then the required stroke length is $2d+2$.

93. Above and Beyond

95. Above and Beyond

97. Above and Beyond

99. Above and Beyond

Exercises 1.5

< Objective 1 >

1. $5x=20$

$\frac{5x}{5}=\frac{20}{5}$

$x=4$

The solution is 4.

Check: $5(4)\stackrel{?}{=}20$

$20=20$

3. $7x=42$

$\frac{7x}{7}=\frac{42}{7}$

$x=6$

The solution is 6.

Check: $7(6)\stackrel{?}{=}42$

$42=42$

5. $63=9x$

$\frac{63}{9}=\frac{9x}{9}$

$7=x$

The solution is 7.

Check: $63\stackrel{?}{=}9(7)$

$63=63$

7. $4x=-16$

$\frac{4x}{4}=\frac{-16}{4}$

$x=-4$

The solution is –4.

Check: $4(-4)\stackrel{?}{=}-16$

$-16=-16$

9. $-9x=72$

$\frac{-9x}{-9}=\frac{72}{-9}$

$x=-8$

The solution is –8.

Check: $-9(-8)\stackrel{?}{=}72$

$72=72$

11. $6x=-54$

$\frac{6x}{6}=\frac{-54}{6}$

$x=-9$

The solution is –9.

Check: $6(-9)\stackrel{?}{=}-54$

$-54=-54$

13. $-4x=-12$

$\frac{-4x}{-4}=\frac{-12}{-4}$

$x=3$

The solution is 3.

Check: $-4(3)\stackrel{?}{=}-12$

$-12=-12$

15. $-21=24x$

$\frac{-21}{24}=\frac{24x}{24}$

$-\frac{7}{8}=x$

The solution is $-\frac{7}{8}$.

Check: $-21\stackrel{?}{=}24\left(-\frac{7}{8}\right)$

$-21=-21$

17. $-6x=-54$

$\frac{-6x}{-6}=\frac{-54}{-6}$

$x=9$

The solution is 9.

Check: $-6(9)\stackrel{?}{=}-54$

$-54=-54$

19. $\frac{x}{4}=2$

$4\left(\frac{x}{4}\right)=4\cdot 2$

$x=8$

The solution is 8.

Check: $\frac{(8)}{4}\overset{?}{=}2$

$2=2$

21. $\frac{x}{5}=3$

$(5)\frac{x}{5}=3(5)$

$x=15$

The solution is 15.

Check: $\frac{15}{5}\overset{?}{=}3$

$3=3$

23. $6=\frac{x}{7}$

$(7)6=\frac{x}{7}(7)$

$42=x$

The solution is 42.

Check: $6\overset{?}{=}\frac{42}{7}$

$6=6$

25. $\frac{x}{5}=-4$

$(5)\frac{x}{5}=-4(5)$

$x=-20$

The solution is –20.

Check: $-\frac{20}{5}\overset{?}{=}-4$

$-4=-4$

27. $-\frac{x}{3}=8$

$-3\left(-\frac{x}{3}\right)=8(-3)$

$x=-24$

The solution is –24.

Check: $\frac{(-24)}{3}\overset{?}{=}8$

$8=8$

29. $\frac{2}{3}x=6$

$3\left(\frac{2}{3}x\right)=6(3)$

$\frac{2x}{2}=\frac{18}{2}$

$x=9$

The solution is 9.

Check: $\frac{2}{3}(9)\overset{?}{=}6$

$\frac{2}{\cancel{3}}\left(\overset{3}{\cancel{9}}\right)\overset{?}{=}6$

$6=6$

31. $\frac{3}{4}x=-16$

$\frac{4}{3}\left(\frac{3}{4}x\right)=\frac{4}{3}(-16)$

$x=-\frac{64}{3}$

The solution is $-\frac{64}{3}$.

Check: $\frac{3}{4}\left(-\frac{64}{3}\right)\overset{?}{=}-16$

$-16=-16$

33. $-\frac{2}{5}x=10$

$-5\left(-\frac{2}{5}x\right)=10(-5)$

$\frac{2x}{2}=\frac{-50}{2}$

$x=-25$

Check: $-\frac{2}{5}(-25)\overset{?}{=}10$

$-\frac{2}{\cancel{5}}\left(-\overset{5}{\cancel{25}}\right)\overset{?}{=}10$

$10=10$

The solution is –25.

35. $5x+4x=36$

$9x=36$

$\frac{9x}{9}=\frac{36}{9}$

$x=4$

The solution is 4.

Check: $5(4)+4(4)\overset{?}{=}36$

$20+16\overset{?}{=}36$

$36=36$

37. $\frac{7}{9}x-5=\frac{3}{9}x+11$

$\frac{7}{9}x=\frac{3}{9}x+16$

$\frac{4}{9}x=16$

$x=36$

Check: $\frac{7}{9}(36)-5\overset{?}{=}\frac{3}{9}(36)+11$

$28-5\overset{?}{=}12+11$

$23=23$

The solution is 36.

39. $4(x+5)+7x=-3(x-2)$

$$4x+20+7x=-3x+6$$
$$11x+20=-3x+6$$
$$11x=-3x-14$$
$$14x=-14$$
$$x=-1$$

Check: $4[(-1)+5]+7(-1)\stackrel{?}{=}-3[(-1)-2]$

$$4(4)-7\stackrel{?}{=}-3(-3)$$
$$9=9$$

The solution is –1.

41. $3.2x=12.8$

$$\frac{3.2x}{3.2}=\frac{12.8}{3.2}$$
$$x=4$$

The solution is 4.

Check: $3.2(4)\stackrel{?}{=}12.8$

$$12.8=12.8$$

43. $-4.5x=13.5$

$$\frac{-4.5x}{-4.5}=\frac{13.5}{-4.5}$$
$$x=-3$$

The solution is –3.

Check: $-4.5(-3)\stackrel{?}{=}13.5$

$$13.5=13.5$$

45. $1.3x+2.8x=12.3$

$$4.1x=12.3$$
$$\frac{4.1x}{4.1}=\frac{12.3}{4.1}$$
$$x=3$$

Check: $1.3(3)+2.8(3)\stackrel{?}{=}12.3$

$$3.9+8.4\stackrel{?}{=}12.3$$
$$12.3=12.3$$

The solution is 3.

47. $9.3x+(-6.2x)=12.4$

$$3.1x=12.4$$
$$\frac{3.1x}{3.1}=\frac{12.4}{3.1}$$
$$x=4$$

Check: $9.3(4)+(-6.2(4))\stackrel{?}{=}12.4$

$$37.2-24.8\stackrel{?}{=}12.4$$
$$12.4=12.4$$

The solution is 4.

49. Let x be the number.
$6x=72$

51. Let x be the number.
$\frac{x}{7}=6$

53. Let x be the number.
$\frac{1}{3}x=8$

55. Let x be the number.
$\frac{3}{4}x=18$

57. Let x be the number.
$\frac{2x}{5}=12$

< Objective 2 >

59. Let x be the total theater audience.

$$\frac{3}{4}x=87$$
$$x=116$$

There were originally 116 patrons.

61. $110=3.14d$

$$d\approx 35$$

The diameter of the tree is approximately 35 in.

63. $$200=\frac{8}{5}x$$
$$\frac{5}{8}200=\frac{5}{8}\bullet\frac{8}{5}x$$
$$125=x$$
The equivalent distance is 125 mi.

65. False

67. sometimes

69. $$\text{hp}=\frac{d^2n}{2.5}$$
$$194.4=\frac{(9)^2n}{2.5}$$
$$486=81n$$
$$6=n$$
The engine has 6 cylinders.

71. $p=\frac{t}{d}$

73. Above and Beyond

Exercises 1.6

< Objective 1 >

1. $$3x+1=13$$
$$3x+1-1=13-1$$
$$3x=12$$
$$\frac{3x}{3}=\frac{12}{3}$$
$$x=4$$
The solution is 4.

Check: $3(4)+1\overset{?}{=}13$
$13=13$

3. $$3x-2=7$$
$$3x-2+2=7+2$$
$$3x=9$$
$$\frac{3x}{3}=\frac{9}{3}$$
$$x=3$$
The solution is 3.

Check: $3(3)-2\overset{?}{=}7$
$7=7$

5. $$4-7x=18$$
$$4-7x-4=18-4$$
$$-7x=14$$
$$\frac{-7x}{-7}=\frac{14}{-7}$$
$$x=-2$$
The solution is –2.

Check: $4-7(-2)\overset{?}{=}18$
$18=18$

7. $$3-4x=-9$$
$$3-4x-3=-9-3$$
$$-4x=-12$$
$$\frac{-4x}{-4}=\frac{-12}{-4}$$
$$x=3$$
The solution is 3.

Check: $3-4(3)\overset{?}{=}-9$
$-9=-9$

9. $$\frac{x}{2}+1=5$$
$$\frac{x}{2}+1-1=5-1$$
$$\frac{x}{2}=4$$
$$2\left(\frac{x}{2}\right)=2\bullet4$$
$$x=8$$
The solution is 8.

Check: $\frac{(8)}{2}+1\overset{?}{=}5$
$5=5$

11. $$\frac{3}{4}x+8=32$$
$$\frac{3}{4}x+8-8=32-8$$
$$\frac{3}{4}x=24$$
$$\frac{4}{3}\left(\frac{3}{4}x\right)=\frac{4}{3}\bullet24$$
$$x=32$$
The solution is 32.

Check: $\frac{3}{4}(32)+8\overset{?}{=}32$
$32=32$

13.
$$\begin{aligned} 5x &= 2x+9 \\ 5x-2x &= 2x+9-2x \\ 3x &= 9 \\ x &= 3 \end{aligned}$$

Check: $5(3)\stackrel{?}{=}2(3)+9$

$15=15$

The solution is 3.

15.
$$\begin{aligned} 9x+2 &= 3x+38 \\ 9x+2-3x &= 3x+38-3x \\ 6x+2 &= 38 \\ 6x+2-2 &= 38-2 \\ 6x &= 36 \\ x &= 6 \end{aligned}$$

Check: $9(6)+2\stackrel{?}{=}3(6)+38$

$56=56$

The solution is 6.

17.
$$\begin{aligned} 4x-8 &= x-14 \\ 3x-8 &= -14 \\ 3x &= -6 \\ x &= -2 \end{aligned}$$

Check: $4(-2)-8\stackrel{?}{=}(-2)-14$

$-16=-16$

The solution is –2.

19.
$$\begin{aligned} 5(3x+4) &= 10(x+2) \\ 15x+20 &= 10x+20 \\ 5x &= 0 \\ x &= 0 \end{aligned}$$

Check: $5(3(0)+4)\stackrel{?}{=}10((0)+2)$

$5(4)\stackrel{?}{=}10(2)$

$20=20$

The solution is 0.

21.
$$\begin{aligned} 5x+4 &= 7x-8 \\ 5x &= 7x-12 \\ -2x &= -12 \\ x &= 6 \end{aligned}$$

Check: $5(6)+4\stackrel{?}{=}7(6)-8$

$34=34$

The solution is 6.

23.
$$\begin{aligned} 6x+7-4x &= 8+7x-26 \\ 2x+7 &= 7x-18 \\ 7 &= 5x-18 \\ 25 &= 5x \\ 5 &= x \end{aligned}$$

Check: $6(5)+7-4(5)\stackrel{?}{=}8+7(5)-26$

$30+7-20\stackrel{?}{=}8+35-26$

$17=17$

The solution is 5.

25.
$$\begin{aligned} 6x-3+5x+11 &= 8x-12 \\ 11x+8 &= 8x-12 \\ 3x+8 &= -12 \\ 3x &= -20 \\ x &= -\frac{20}{3} \end{aligned}$$

Check:

$$6\left(-\frac{20}{3}\right)-3+5\left(-\frac{20}{3}\right)+11\stackrel{?}{=}8\left(-\frac{20}{3}\right)-12$$

$$-40-3-\frac{100}{3}+11\stackrel{?}{=}-\frac{160}{3}-12$$

$$-32-\frac{100}{3}\stackrel{?}{=}-\frac{160}{3}-12$$

$$-\frac{96}{3}-\frac{100}{3}\stackrel{?}{=}-\frac{160}{3}-\frac{36}{3}$$

$$-\frac{196}{3}=-\frac{196}{3}$$

The solution is $-\frac{20}{3}$.

27. $$5(8-x)=3x$$
$$40-5x=3x$$
$$40=8x$$
$$5=x$$

Check: $$5(8-(5))\stackrel{?}{=}3(5)$$
$$15=15$$

The solution is 5.

29. $$7(2x-1)-5x=x+25$$
$$14x-7-5x=x+25$$
$$9x-7=x+25$$
$$8x-7=25$$
$$8x=32$$
$$x=4$$

Check: $$7(2(4)-1)-5(4)\stackrel{?}{=}(4)+25$$
$$7(7)-20\stackrel{?}{=}29$$
$$29=29$$

The solution is 4.

31. $$2(2x-1)=3(x+1)$$
$$4x-2=3x+3$$
$$x-2=3$$
$$x=5$$

Check: $$2(2(5)-1)\stackrel{?}{=}3((5)+1)$$
$$18=18$$

The solution is 5.

33. $$8x-3(2x-4)=17$$
$$8x-6x+12=17$$
$$2x+12=17$$
$$2x=5$$
$$x=\frac{5}{2}$$

Check: $$8\left(\frac{5}{2}\right)-3\left(2\left(\frac{5}{2}\right)-4\right)\stackrel{?}{=}17$$
$$20-3(5-4)\stackrel{?}{=}17$$
$$17=17$$

The solution is $\frac{5}{2}$.

35. $$7(3x+4)=8(2x+5)+13$$
$$21x+28=16x+40+13=16x+53$$
$$5x+28=53$$
$$5x=25$$
$$x=5$$

Check: $$7(3(5)+4)\stackrel{?}{=}8(2(5)+5)+13$$
$$7(19)\stackrel{?}{=}8(15)+13$$
$$133=133$$

The solution is 5.

37. $$9-4(3x+1)=3(6-3x)-9$$
$$9-12x-4=18-9x-9$$
$$5-12x=9-9x$$
$$5=9+3x$$
$$-4=3x$$
$$-\frac{4}{3}=x$$

Check:
$$9-4\left(3\left(-\frac{4}{3}\right)+1\right)\stackrel{?}{=}3\left(6-3\left(-\frac{4}{3}\right)\right)-9$$
$$9-4(-4+1)\stackrel{?}{=}3(6+4)-9$$
$$9+12\stackrel{?}{=}30-9$$
$$21=21$$

The solution is $-\frac{4}{3}$.

39. $$5.3x-7=2.3x+5$$
$$5.3x-7-2.3x=2.3x+5-2.3x$$
$$3x-7=5$$
$$3x=12$$
$$x=4$$

Check: $$5.3(4)-7\stackrel{?}{=}2.3(4)+5$$
$$21.2-7\stackrel{?}{=}9.2+5$$
$$14.2=14.2$$

The solution is 4.

41.
$$\frac{2x}{3}-\frac{5}{3}=3$$
$$3\left(\frac{2x}{3}-\frac{5}{3}\right)=3\cdot 3$$
$$3\left(\frac{2x}{3}\right)-3\left(\frac{5}{3}\right)=9$$
$$2x-5=9$$
$$x=7$$

43.
$$\frac{x}{6}+\frac{x}{5}=11$$
$$30\left(\frac{x}{6}+\frac{x}{5}\right)=30\cdot 11$$
$$30\left(\frac{x}{6}\right)+30\left(\frac{x}{5}\right)=330$$
$$5x+6x=330$$
$$x=30$$

45.
$$\frac{2x}{5}-\frac{x}{3}=\frac{7}{15}$$
$$15\left(\frac{2x}{5}-\frac{x}{3}\right)=15\cdot\frac{7}{15}$$
$$15\left(\frac{2x}{5}\right)-15\left(\frac{x}{3}\right)=7$$
$$6x-5x=7$$
$$x=7$$

47.
$$\frac{x}{5}-\frac{x-7}{3}=\frac{1}{3}$$
$$15\left(\frac{x}{5}-\frac{x-7}{3}\right)=15\cdot\frac{1}{3}$$
$$3x-5(x-7)=5$$
$$3x-5x+35=5$$
$$-2x=-30$$
$$x=15$$

49.
$$\frac{5x-3}{4}-2=\frac{x}{3}$$
$$12\left(\frac{5x-3}{4}-2\right)=12\cdot\frac{x}{3}$$
$$3(5x-3)-12\cdot 2=4x$$
$$15x-9-24=4x$$
$$11x=33$$
$$x=3$$

51.
$$\frac{3x-2}{3}-\frac{2x-5}{5}=\frac{7}{15}$$
$$15\left(\frac{3x-2}{3}-\frac{2x-5}{5}\right)=15\cdot\frac{7}{15}$$
$$15\left(\frac{3x-2}{3}\right)-15\left(\frac{2x-5}{5}\right)=7$$
$$5(3x-2)-3(2x-5)=7$$
$$15x-10-6x+15=7$$
$$9x+5=7$$
$$9x=2$$
$$x=\frac{2}{9}$$

< Objective 2 >

53. $3(x-1)=2x+3$
$$3x-3=2x+3$$
$$x=6$$
Conditional equation; the solution is 6.

55. $3(x-1)=3x+3$
$$3x-3=3x+3$$
$$-3=3$$
Contradiction; the solution is $\{\ \}$.

57. $3(x-1)=3x-3$
$$3x-3=3x-3$$
Identity; the solution is $\mathbb{R}$.

59. $3x-(x-3)=2(x+1)+2$

$$3x-x+3=2x+2+2$$
$$2x+3=2x+4$$
$$3=4$$

Contradiction; the solution is $\{\ \}$.

61. $\frac{x}{2}-\frac{x}{3}=\frac{x}{6}$

$$6\left(\frac{x}{2}-\frac{x}{3}\right)=6\bullet\frac{x}{6}$$
$$3x-2x=x$$
$$x=x$$

Identity; the solution is $\mathbb{R}$.

63. $2x+3=7$

65. $4x-7=41$

67. $\frac{2}{3}x+5=21$

69. $3x=x+12$

< Objective 3 >

71. Let x be the number.

$$2x+16=24$$
$$2x=8$$
$$x=4$$

The number is 4.

73. Let x be the number.

$$5x-12=78$$
$$5x=90$$
$$x=18$$

The number is 18.

75. Let x be an integer. Then $x+1$ is the next integer.

$$x+(x+1)=71$$
$$2x+1=71$$
$$2x=70$$
$$x=35$$
$$x+1=36$$

The numbers are 35 and 36.

77. Let x be an integer. Then $x+1$ is the next integer and $x+2$ is the following integer.

$$x+(x+1)+(x+2)=90$$
$$3x+3=90$$
$$3x=87$$
$$x=29$$
$$x+1=30$$
$$x+2=31$$

The numbers are 29, 30, and 31.

79. Let x be an even integer. Then $x+2$ is the next consecutive even integer.

$$x+(x+2)=66$$
$$2x+2=66$$
$$2x=64$$
$$x=32$$
$$x+2=34$$

The numbers are 32 and 34.

81. Let x be an odd integer. Then $x+2$ is the next consecutive odd integer.

$$x+(x+2)=52$$
$$2x+2=52$$
$$2x=50$$
$$x=25$$
$$x+2=27$$

The numbers are 25 and 27.

83. Let x be an integer. Then $x+1$ is the next consecutive integer.
$$4x=3(x+1)+9=3x+3+9$$
$$x=12$$
$$x+1=13$$
The numbers are 12 and 13.

85. Let x be the number of votes the losing candidate received. Then the winner received $x+160$ votes.
$$x+(x+160)=3,260$$
$$2x+160=3,260$$
$$2x=3,100$$
$$x=1,550$$
$$x+160=1,710$$
The winning candidate received 1,710 votes; the losing candidate received 1,550 votes.

87. Let x be the cost of the dryer. Then $x+90$ is the cost of the washer.
$$x+(x+90)=950$$
$$2x+90=950$$
$$2x=860$$
$$x=430$$
$$x+90=520$$
The washer costs \$520 and the dryer costs \$430.

89. Let x be Dan's age. Then $2x$ is Diane's age.
$$2x+x=27$$
$$3x=27$$
$$x=9$$
$$2x=18$$
Diane is 18 years old and Dan is 9 years old.

91. Let x be the amount in the bank earning $2\frac{1}{2}\%$ interest. Then $21,000-x$ is the amount in the bank earning $3\frac{1}{4}\%$ interest.
$$0.025x+0.0325(21,000-x)=615$$
$$0.025x+682.5-0.0325x=615$$
$$-0.0075x=-67.5$$
$$x=9,000$$
$$21,000-x=12,000$$
Johanna deposited \$12,000 at $3\frac{1}{4}\%$ interest and \$9,000 at $2\frac{1}{2}\%$ interest.

93. Let x be Sam's gross monthly pay. His payroll deductions amount to $0.24x$. His net monthly income is $x-0.24x=0.76x$. Therefore, he gives $0.05(0.76x)=0.038x$ to charity each month.
$$0.038x=190$$
$$x=5,000$$
Sam's gross monthly income is \$5,000 per month.

95. $95=\frac{9}{5}C+32$
$$63=\frac{9}{5}C$$
$$35=C$$
The corresponding temperature is 35°C.

97. False

99. sometimes

101. $230x-52=191$
$$230x=243$$
$$x=\frac{243}{230}$$
Using a calculator:

243 [÷] 230 [=]

$x=1.06$ (rounded to the nearest hundredth)

103. $360-29(2x+1)=2,464$

$$360-58x-29=2,464$$
$$-58x=2,464-331$$
$$-58x=2,133$$
$$x=-\frac{2,133}{58}$$

Using a calculator:

[+/−] 2133 [÷] 58 [=]

$x=-36.78$ (rounded to the nearest hundredth)

105. $23.12x-34.2=34.06$

$$23.12x=68.26$$
$$x=\frac{68.26}{23.12}$$

Using a calculator:

68.26 [÷] 23.12 [=]

$x=2.95$ (rounded to the nearest hundredth)

107. $3.2(0.5x-5.1)=-6.4(9.7x+15.8)$

$$1.6x-16.32=-62.08x-101.12$$
$$60.48x=-84.8$$
$$x=-\frac{84.8}{60.48}$$

Using a calculator:

[+/−] 84.8 [÷] 60.48 [=]

$x=-1.33$ (rounded to the nearest hundredth)

109. $7=\frac{t+16}{4}$

$$4\cdot 7=4\cdot\frac{t+16}{4}$$
$$28=t+16$$
$$12=t$$

The child is 12 years old.

111. Let x be the size of the folder. Then $x-.36x$ is the size of the compressed folder.

$$x-0.36x=11.2$$
$$0.64x=11.2$$
$$x=\frac{11.2}{0.64}$$
$$x=17.5$$

The folder was 17.5 MB before compressing.

113. Above and Beyond

115. A solution to an equation is a value for which the original equation is true.

117. Multiplying by 0 always equals $0=0$.

119. **(a)** $2x+3=0$

$$2x=-3$$
$$x=-\frac{3}{2}$$

(b) $4x+7=0$

$$4x=-7$$
$$x=-\frac{7}{4}$$

(c) $6x-1=0$

$$6x=1$$
$$x=\frac{1}{6}$$

(d) $5x-2=0$

$$x=\frac{2}{5}$$

(e) $-3x+8=0$

$$x=\frac{8}{3}$$

(f) $-5x-9=0$

$$x=-\frac{9}{5}$$

(g) $ax+b=0$

$$ax=-b$$
$$x=-\frac{b}{a}$$

Exercises 1.7

< Objective 1 >

1. $5 < 10$

3. $7 > -2$

5. $0 < 4$

7. $-2 > -5$

9. $x < 3$: x is less than 3.

11. $x \geq -4$: x is greater than or equal to –4.

13. $-5 \leq x$: -5 is less than or equal to x.

< Objective 2 >

15. $x > 2$

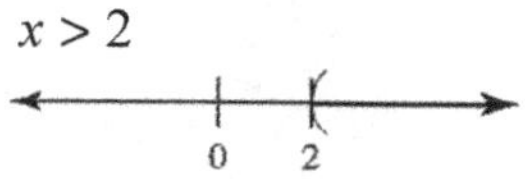

17. $x < 6$

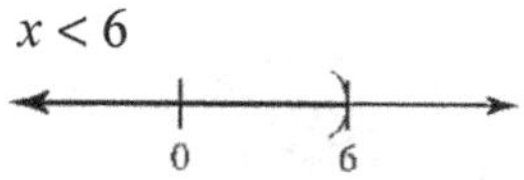

19. $x > 1$

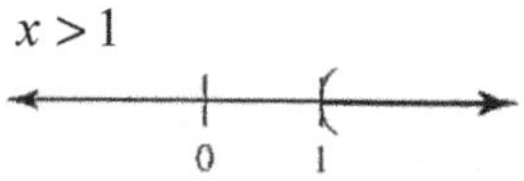

21. $x < 8$

23. $x > -5$

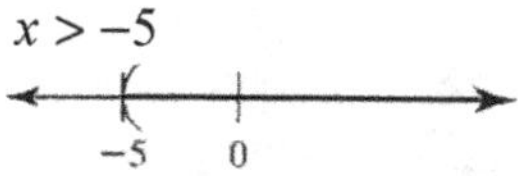

25. $x \geq 9$

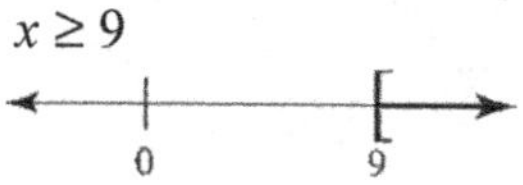

27. $x < 0$

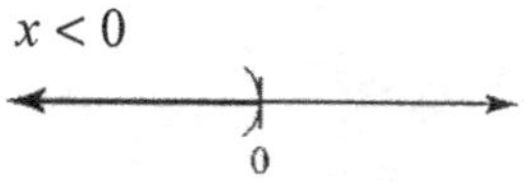

< Objective 3 >

29.
$$x - 8 \leq 3$$
$$x - 8 + 8 \leq 3 + 8$$
$$x \leq 11$$

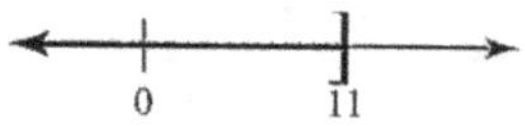

31.
$$x + 8 \geq 10$$
$$x + 8 - 8 \geq 10 - 8$$
$$x \geq 2$$

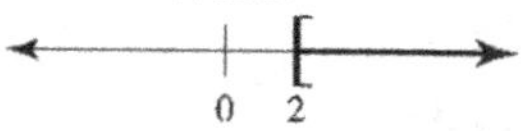

33.
$$5x < 4x + 7$$
$$5x - 4x < 4x + 7 - 4x$$
$$x < 7$$

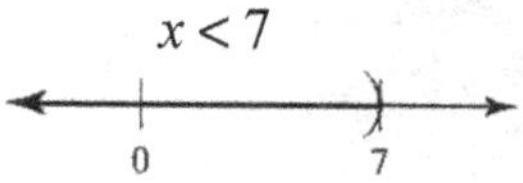

35.
$$6x - 8 \leq 5x$$
$$6x - 8 + 8 \leq 5x + 8$$
$$6x \leq 5x + 8$$
$$6x - 5x \leq 5x + 8 - 5x$$
$$x \leq 8$$

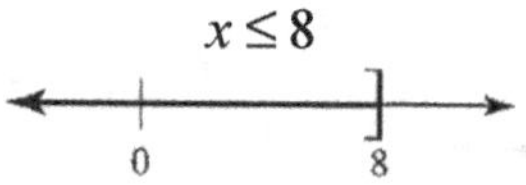

37.
$$8x + 1 \geq 7x + 9$$
$$8x + 1 - 1 \geq 7x + 9 - 1$$
$$8x \geq 7x + 8$$
$$8x - 7x \geq 7x + 8 - 7x$$
$$x \geq 8$$

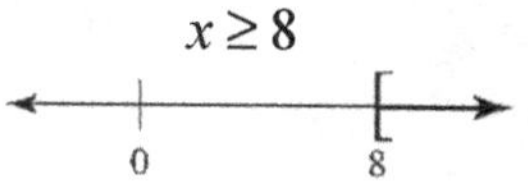

39.
$$7x + 5 < 6x - 4$$
$$7x + 5 - 5 < 6x - 4 - 5$$
$$7x < 6x - 9$$
$$7x - 6x < 6x - 9 - 6x$$
$$x < -9$$

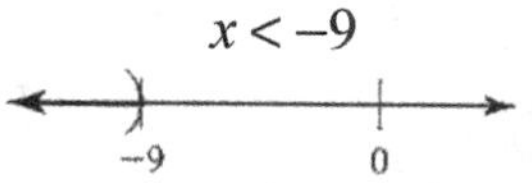

41.
$$\frac{3}{4}x-5\ge 7-\frac{1}{4}x$$
$$\frac{3}{4}x-5+5+\frac{1}{4}x\ge 7-\frac{1}{4}x+5+\frac{1}{4}x$$
$$x\ge 12$$

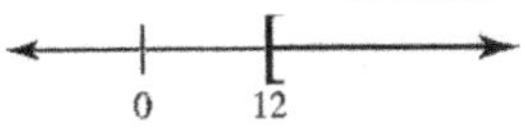

43.
$$11+0.63x>9-0.37x$$
$$11+0.63x+0.37x-11>9-0.37x+0.37x-11$$
$$x>-2$$

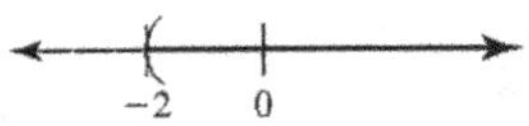

45.
$$3x\le 9$$
$$\frac{1}{3}(3x)\le\frac{1}{3}\bullet 9$$
$$x\le 3$$

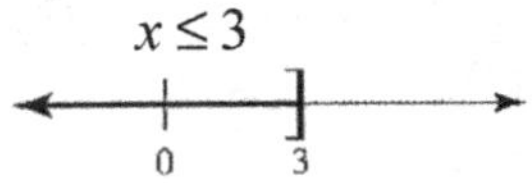

47.
$$5x>-35$$
$$\frac{1}{5}(5x)>\frac{1}{5}(-35)$$
$$x>-7$$

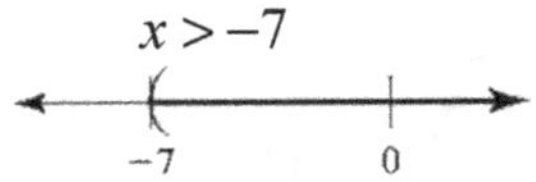

49.
$$-6x\ge 18$$
$$-\frac{1}{6}(-6x)\le-\frac{1}{6}\bullet 18$$
$$x\le -3$$

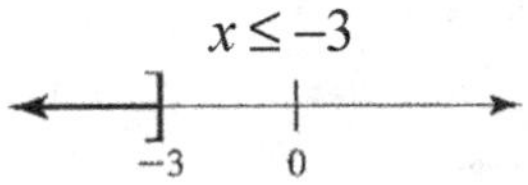

51.
$$-2x<-12$$
$$\frac{-2x}{-2}>\frac{-12}{-2}$$
$$x>6$$

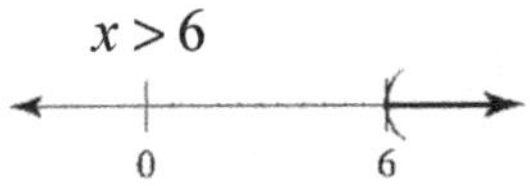

53.
$$\frac{x}{4}>5$$
$$4\bullet\frac{x}{4}>4\bullet 5$$
$$x>20$$

55.
$$-\frac{x}{2}\ge -3$$
$$-2\left(-\frac{x}{2}\right)\le -2(-3)$$
$$x\le 6$$

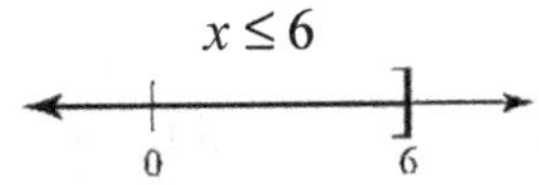

57.
$$\frac{2x}{3}<6$$
$$\frac{3}{2}\left(\frac{2x}{3}\right)<\frac{3}{2}(6)$$
$$x<9$$

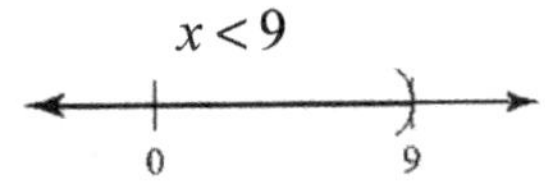

59.
$$5x>3x+8$$
$$5x-3x>3x+8-3x$$
$$2x>8$$
$$x>4$$

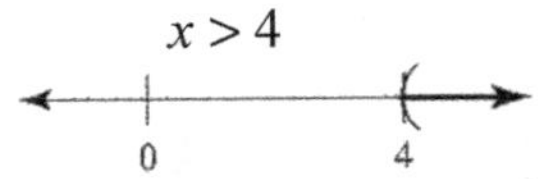

61.
$$5x-2<3x$$
$$2x-2<0$$
$$2x<2$$
$$x<1$$

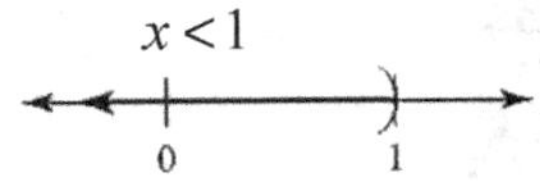

63.
$$3-2x>5$$
$$-2x>2$$
$$x<-1$$

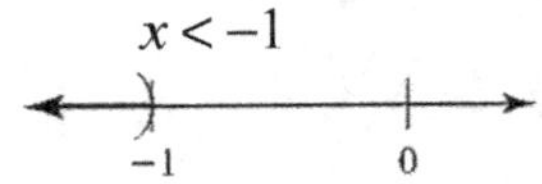

65. $2x \ge 5x + 18$

$-3x \ge 18$

$x \le -6$

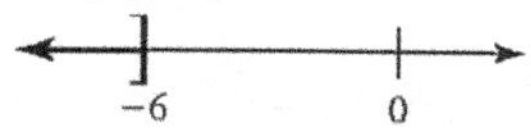

67. $\frac{1}{3}x - 5 \le \frac{5}{3}x + 11$

$\frac{1}{3}x - 16 \le \frac{5}{3}x$

$-16 \le \frac{4}{3}x$

$-12 \le x$

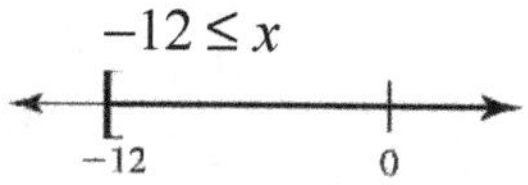

69. $0.34x + 21 \ge 19 - 1.66x$

$0.34x \ge -2 - 1.66x$

$2x \ge -2$

$x \ge -1$

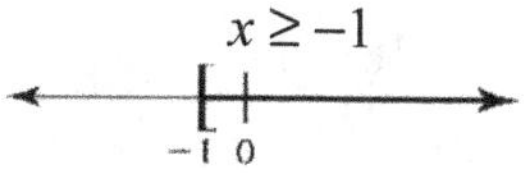

71. $7x - 5 < 3x + 2$

$7x < 3x + 7$

$4x < 7$

$x < \frac{7}{4}$

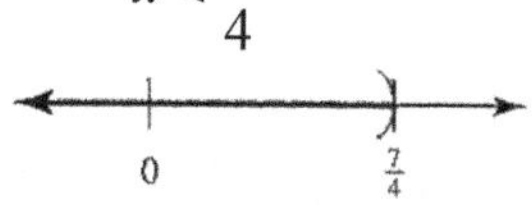

73. $5x + 7 > 8x - 17$

$5x + 24 > 8x$

$24 > 3x$

$8 > x$

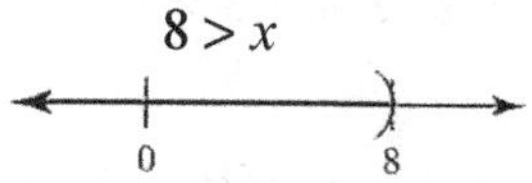

75. $3x - 2 \le 5x + 3$

$3x - 5 \le 5x$

$-5 \le 2x$

$-\frac{5}{2} \le x$

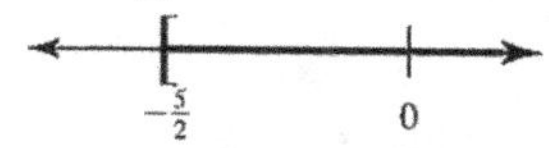

77. $x + 5 > 3$

79. $2x - 4 \le 7$

81. $4x - 15 > x$

83. Let P be the number of giant pandas left in the bamboo forests of China.
$P < 1000$

85. Let x be the grade on the fourth test.

$\frac{72 + 81 + 79 + x}{4} \ge 80$

$232 + x \ge 320$

$x \ge 88$

87. Let S represent Juanita's sales.

$0.05S > 500$

$S > 10000$

Sales > 10000

89. Let x be the number of courses.

$45 + 290x \le 1500$

$290x \le 1455$

$x \le 5.02$

The greatest number of courses for which Samantha can register is 5 courses.

91. Let L be the length of the rectangle.

$2 \cdot 40 + 2L \le 180$

$2L \le 100$

$L \le 50$

The maximum length of the rectangle can be 50 cm.

93. Let x be the cost of jeans.
$29+2x \le 125$
$2x \le 96$
$x \le 48$
Joyce can spend up to \$48 on each pair of jeans.

95. "x is nonnegative" matches **(a)** $x \ge 0$

97. "x is no more than 5" matches **(c)** $x \le 5$

99. "x is at least 5" matches **(b)** $x \ge 5$

101. True

103. True

105. always

107. always

109. Let x be the number of studs
$7.19x \le 450$
$x \le 62.58$
No more than 62 studs can be bought.

111. The solution set is $\mathbb{R}$. The original inequality will be true for any value of the variable since the inequality is equivalent to $7 > -5$.

113. Above and Beyond

115. Above and Beyond.

Summary Exercises

1. 8 more than y is written as $y+8$.

3. The product of 8 and a is written as $8a$.

5. The product of x and 7 less than x is written as $x(x-7)$.

7. The quotient when a plus 2 is divided by a minus 2 is written as $\frac{a+2}{a-2}$.

9. The quotient of 9 and a number is written as $\frac{9}{x}$.

11. $3x+w=3(-3)+2=-9+2=-7$

13. $x+y-3z=(-3)+(6)-3(-4)=-3+6+12$
$=15$

15. $5(x^2-w^2)=5\left[(-3)^2-(2)^2\right]=5(9-4)$
$=5(5)=25$

17. $\frac{2x-4z}{y-z}=\frac{2(-3)-4(-4)}{6-(-4)}=\frac{-6+16}{10}=\frac{10}{10}=1$

19. $-4x^2-2zw^2+4z$
$=-4(-3)^2-2(-4)(2)^2+4(-4)$
$=-4(9)-2(-4)(4)+4(-4)$
$=-36+32-16=-20$

21. $4a^3, -3a^2$

23. $5m^2, -4m^2, m^2$

25. $9x+7x=16x$

27. $9xy-6xy=3xy$

29. $7a+3b+12a-2b=19a+b$

31. $5x^3+17x^2-2x^3-8x^2=3x^3+9x^2$

33. $(2a^3+12a^3)-4a^3=14a^3-4a^3=10a^3$

35. There is a $(37-x)$ ft piece of board left.

37. The length is $(w+4)$ m.

39. The pieces are x and $(25-x)$ ft long.

41. $P = 2(2x) + 2(-x+4) = 4x - 2x + 8$
$= 2x + 8$
The perimeter of the rectangle is $(2x+8)$ m.

43. $5x - 3 = 7$ (2)
$5(2) - 3 \stackrel{?}{=} 7$
$7 = 7$
Because $7 = 7$ is a true statement, 2 is a solution of the equation.

45. $7x - 2 = 2x + 8$ (2)
$7(2) - 2 \stackrel{?}{=} 2(2) + 8$
$14 - 2 \stackrel{?}{=} 4 + 8$
$12 = 12$
Because $12 = 12$ is a true statement, 2 is a solution of the equation.

47. $x + 3 = 5$
$x + 3 - 3 = 5 - 3$
$x = 2$
Check: $(2) + 3 \stackrel{?}{=} 5$
$5 = 5$
The solution is 2.

49. $5x = 4x - 5$
$5x - 4x = 4x - 5 - 4x$
$x = -5$
Check: $5(-5) \stackrel{?}{=} 4(-5) - 5$
$-25 = -25$
The solution is –5.

51. $9x - 7 = 8x - 6$
$9x - 7 + 7 = 8x - 6 + 7$
$9x = 8x + 1$
$9x - 8x = 8x + 1 - 8x$
$x = 1$
Check: $9(1) - 7 \stackrel{?}{=} 8(1) - 6$
$2 = 2$
The solution is 1.

53. $4(2x+3) = 7x + 5$
$8x + 12 = 7x + 5$
$8x = 7x - 7$
$x = -7$
Check: $4[2(-7)+3] \stackrel{?}{=} 7(-7) + 5$
$4(-11) \stackrel{?}{=} -44$
$-44 = -44$
The solution is −7.

55. $5x = 35$
$\frac{5x}{5} = \frac{35}{5}$
$x = 7$
Check: $5(7) \stackrel{?}{=} 35$
$35 = 35$
The solution is 7.

57. $-9x = 36$
$\frac{-9x}{-9} = \frac{36}{-9}$
$x = -4$
Check: $-9(-4) \stackrel{?}{=} 36$
$-36 = 36$
The solution is –4.

59. $\frac{2}{3}x = 18$
$\frac{3}{2}\left(\frac{2}{3}x\right) = \frac{3}{2} \cdot 18$
$x = 27$
Check: $\frac{2}{3}(27) \stackrel{?}{=} 18$
$18 = 18$
The solution is 27.

61. $7x + 8 = 3x$
$4x = -8$
$x = -2$
Check: $7(-2) + 8 \stackrel{?}{=} 3(-2)$
$-6 = -6$
The solution is –2.

63. $4x - 7 = 2x$
$2x = 7$
$x = \frac{7}{2}$
Check: $4\left(\frac{7}{2}\right) - 7 \stackrel{?}{=} 2\left(\frac{7}{2}\right)$
$7 = 7$
The solution is $\frac{7}{2}$.

65. $\frac{x}{3}-5=1$

$$\frac{x}{3}=6$$

$$x=18$$

Check: $\frac{(18)}{3}-5\stackrel{?}{=}1$

$$1=1$$

The solution is 18.

67. $7x+4=2x+6$

$$5x=2$$

$$\frac{5x}{5}=\frac{2}{5}$$

$$x=\frac{2}{5}$$

Check: $7\left(\frac{2}{5}\right)+4\stackrel{?}{=}2\left(\frac{2}{5}\right)+6$

$$\frac{34}{5}=\frac{34}{5}$$

The solution is $\frac{2}{5}$.

69. $2x+7=4x-5$

$$12=2x$$

$$\frac{12}{2}=\frac{2x}{2}$$

$$6=x$$

Check: $2(6)+7\stackrel{?}{=}4(6)-5$

$$19=19$$

The solution is 6.

71. $\frac{10}{3}x-5=\frac{4}{3}x+7$

$$2x=12$$

$$\frac{2x}{2}=\frac{12}{2}$$

$$x=6$$

Check: $\frac{10}{3}(6)-5\stackrel{?}{=}\frac{4}{3}(6)+7$

$$15=15$$

The solution is 6.

73. $3.7x+8=1.7x+16$

$$2x=8$$

$$x=4$$

Check: $3.7(4)+8\stackrel{?}{=}1.7(4)+16$

$$22.8=22.8$$

The solution is 4.

75. $5(3x-1)-6x=3x-2$

$$15x-5-6x=3x-2$$

$$9x-5=3x-2$$

$$6x=3$$

$$x=\frac{1}{2}$$

Check: $5\left[3\left(\frac{1}{2}\right)-1\right]-6\left(\frac{1}{2}\right)\stackrel{?}{=}3\left(\frac{1}{2}\right)-2$

$$5\left(\frac{1}{2}\right)-3\stackrel{?}{=}-\frac{1}{2}$$

$$-\frac{1}{2}=-\frac{1}{2}$$

The solution is $\frac{1}{2}$.

77. $8x-5(x+3)=-10$

$$8x-5x-15=-10$$

$$3x=5$$

$$x=\frac{5}{3}$$

Check: $8\left(\frac{5}{3}\right)-5\left[\left(\frac{5}{3}\right)+3\right]\stackrel{?}{=}-10$

$$\frac{40}{3}-5\left(\frac{14}{3}\right)\stackrel{?}{=}-10$$

$$-10=-10$$

The solution is $\frac{5}{3}$.

79.
$$\frac{2x}{3}-\frac{x}{4}=5$$
$$12\left(\frac{2x}{3}-\frac{x}{4}\right)=12\bullet 5$$
$$12\left(\frac{2x}{3}\right)-12\left(\frac{x}{4}\right)=60$$
$$8x-3x=60$$
$$5x=60$$
$$x=12$$
Check: $\frac{2(12)}{3}-\frac{(12)}{4}\overset{?}{=}5$
$$8-3\overset{?}{=}5$$
$$5=5$$
The solution is 12.

81.
$$\frac{x}{2}-\frac{x+1}{3}=\frac{1}{6}$$
$$6\left(\frac{x}{2}-\frac{x+1}{3}\right)=6\bullet\frac{1}{6}$$
$$6\left(\frac{x}{2}\right)-6\left(\frac{x+1}{3}\right)=1$$
$$3x-2(x+1)=1$$
$$3x-2x-2=1$$
$$x=3$$
Check: $\frac{(3)}{2}-\frac{(3)+1}{3}\overset{?}{=}\frac{1}{6}$
$$\frac{3}{2}-\frac{4}{3}\overset{?}{=}\frac{1}{6}$$
$$\frac{1}{6}=\frac{1}{6}$$
The solution is 3.

83. Let x represent the number of hours.
$$75x+225=450$$
$$75x=225$$
$$x=3$$
The repair job took 3 hr.

85. Let x be the number.
$$4x+14=34$$
$$4x=20$$
$$x=5$$
The number is 5.

87. $x>5$

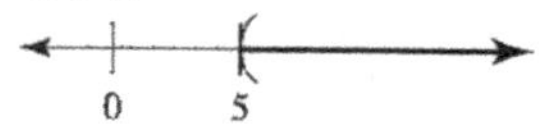

89. $x\ge 9$

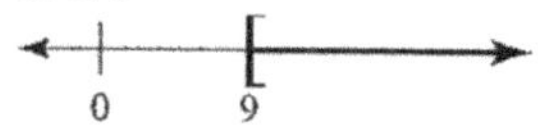

91. $x-2\le 9$
$$x\le 11$$

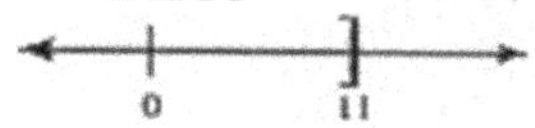

93. $4x\ge -12$
$$x\ge -3$$

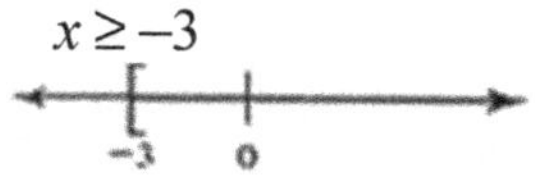

95. $2x\le 8x-3$
$$3\le 6x$$
$$\frac{1}{2}\le x$$
$$x\ge\frac{1}{2}$$

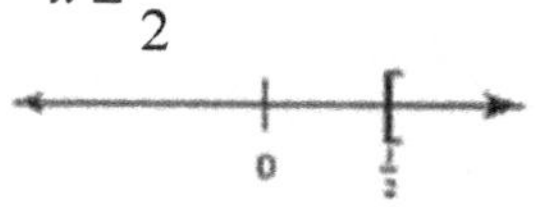

97. $5x-2\le 4x+5$
$$5x\le 4x+7$$
$$x\le 7$$

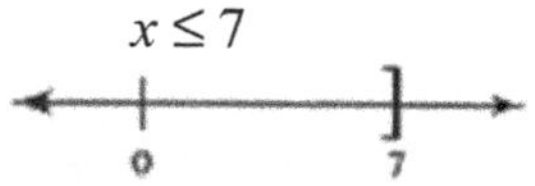

Chapter Test 1

1. The sum of x and y is written as $x+y$.

3. The product of a and b is written as ab.

5. 5 less than c is written as $c-5$.

7. 3 times the difference of m and n is written as $3(m-n)$.

9. $-4x-12=-4(-4)-12=16-12=4$

11. $4a-c=4(-2)-(-4)=-8+4=-4$

13. $8a-3b-5a+2b=3a-b$

15. $7x-3=25 \qquad (5)$

$$7(5)-3\stackrel{?}{=}25$$
$$32\neq 25$$

Because the two sides do not name the same number, the statement is not true, and 5 is not a solution.

17. $7x-12=6x$

$$7x=6x+12$$
$$x=12$$

Check: $7(12)-12\stackrel{?}{=}6(12)$

$$72=72$$

The solution is 12.

19. $5x-3(x-5)=19$

$$5x-3x+15=19$$
$$2x=4$$
$$x=2$$

Check: $5(2)-3[(2)-5]\stackrel{?}{=}19$

$$10-3(-3)\stackrel{?}{=}19$$
$$19=19$$

The solution is 2.

21. $x-5\leq 9$

$$x\leq 14$$

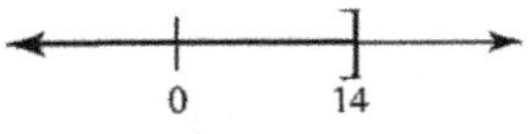

23. Let x be the number.

$$5x-7=28$$
$$5x=35$$
$$x=7$$

The number is 7.

25. Let x be the wholesale price of a gallon of milk.

$$0.25x+x=2.59$$
$$1.25x=2.59$$
$$x=2.07$$

The wholesale price of a gallon of milk is \$2.07.

Chapter 2
Functions and Graphs

Exercises 2.1

< Objective 1 >

1. $P = 4s$

$s = \frac{P}{4}$

3. $E = IR$

$R = \frac{E}{I}$

5. $V = LWH$

$H = \frac{V}{LW}$

7. $A + B + C = 180$

$B = 180 - A - C$

9. $ax + b = 0$

$ax = -b$

$x = -\frac{b}{a}$

11. $s = \frac{1}{2}gt^2$

$gt^2 = 2s$

$g = \frac{2s}{t^2}$

13. $x + 5y = 15$

$5y = 15 - x$

$y = \frac{15 - x}{5}$ or $y = -\frac{1}{5}x + 3$

15. $P = 2L + 2W$

$2L = P - 2W$

$L = \frac{P - 2W}{2}$ or $L = \frac{P}{2} - W$

17. $V = \frac{KT}{P}$

$KT = PV$

$T = \frac{PV}{K}$

19. $x = \frac{a + b}{2}$

$a + b = 2x$

$b = 2x - a$

21. $F = \frac{9}{5}C + 32$

$\frac{9}{5}C = F - 32$

$C = \frac{5}{9}(F - 32)$ or $C = \frac{5(F - 32)}{9}$

23. $S = 2\pi r^2 + 2\pi rh$

$2\pi rh = S - 2\pi r^2$

$h = \frac{S - 2\pi r^2}{2\pi r}$ or $h = \frac{S}{2\pi r} - r$

25. $2(x + 6) = 18$

27. $3(x - 5) = 21$

29. $2x + 3(x + 1) = 48$

< Objectives 2 and 3 >

31. From Exercise 5:

$$H=\frac{V}{LW}=\frac{(120)}{(8)(5)}=\frac{120}{40}=3$$

The height of the solid is 3 cm.

33. From Exercise 4:

$$r=\frac{I}{Pt}=\frac{(240)}{(2000)(4)}=\frac{240}{8000}=0.03$$

The interest rate was 3%.

35. From Exercise 21:

$$C=\frac{5}{9}(F-32)=\frac{5}{9}[(77)-32]=\frac{5}{9}(45)=25$$

The equivalent temperature is 25°C.

37. Let x be the one number. Then $x+8$ is the other number.

$$x+2(x+8)=46$$
$$x+2x+16=46$$
$$3x=30$$
$$x=10$$
$$x+8=18$$

The numbers are 10 and 18.

39. Let x be the one number. Then $x-7$ is the other number.

$$4(x-7)+2x=62$$
$$4x-28+2x=62$$
$$6x=90$$
$$x=15$$
$$x-7=8$$

The numbers are 8 and 15.

41. Let x be the one integer. Then $x+1$ is the next consecutive integer.

$$2x+3(x+1)=28$$
$$2x+3x+3=28$$
$$5x=25$$
$$x=5$$
$$x+1=6$$

The numbers are 5 and 6.

43. Let L be the length and W be the width of the rectangle. Then $L=2W+1$.

$$P=2L+2W=2(2W+1)+2W$$
$$=4W+2+2W=6W+2$$
$$74=6W+2$$
$$72=6W$$
$$12=W$$
$$L=2(12)+1=25$$

The length is 25 in. and the width is 12 in.

45. Let L be the length and W be the width of the rectangle. Then $L=3W+5$.

$$P=2L+2W=2(3W+5)+2W$$
$$74=2W+2(3W+5)=2W+6W+10$$
$$64=8W$$
$$8=W$$
$$L=3(8)+5=29$$

The length is 29 m and the width is 8 m.

47. Let s be the length of the equal-sized legs and b the length of the base. Then $b=s-3$.

$$P=2s+b=2s+(s-3)=3s-3$$
$$36=3s-3$$
$$39=3s$$
$$13=s$$
$$b=13-3=10$$

The base is 10 cm and the legs are 13 cm.

49. Let r be Patrick's speed going to the meeting.

	Rate	Time	Distance
Going	r	3	$3r$
Returning	$r-10$	4	$4(r-10)$

$$3r=4(r-10)=4r-40$$
$$40=r$$
$$r-10=30$$

Patrick's speed was 40 mi/hr while going to attend a meeting and 30 mi/hr while returning.

51. Let t be the amount of time the first car travels.

	Rate	Time	Distance
First car	50	t	$50t$
Second car	40	$t-1$	$40(t-1)$

$$50t+40(t-1)=320$$
$$50t+40t-40=320$$
$$90t=360$$
$$t=4$$

At 6 P.M. (4 hours), the two cars will be 320 miles apart.

53. Let t be the amount of time Catherine travels.

	Rate	Time	Distance
Catherine	45	t	$45t$
Max	54	$t-1$	$54(t-1)$

$$45t=54(t-1)=54t-54$$
$$54=9t$$
$$6=t$$

At 2 P.M. (6 hours), Max will catch up to Catherine.

55. From Exercise 21:

$$F=\frac{9}{5}C+32=\frac{9}{5}\cdot 35+32=95$$

The temperature is 95°F.

57. Let t be the amount of time until they meet.

	Rate	Time	Distance
Mika	45	t	$45t$
Hiroko	50	$t-1$	$50(t-1)$

$$45t+50(t-1)=425$$
$$45t+50t-50=425$$
$$95t=475$$
$$t=5$$

At 3 P.M. (5 hours), Mika and Hiroko will meet.

59. True

61. True

63. $R=\dfrac{V^2}{D}$

$$RD=V^2$$
$$D=\frac{V^2}{R}=\frac{(13.2)^2}{220}=\frac{174.24}{220}=0.792$$

The power dissipation is 0.792 W.

65. **(a)** $d=0.125w$

(b) $d=0.125w=0.125(15)=1.875$

1.875 mg yohimbine should be administered to a 15-kg fawn.

(c) $d=0.125w$

$$\frac{d}{0.125}=w$$
$$w=\frac{5.0}{0.125}=40$$

5.0-mg dosage is required by a 40-kg deer.

67. Above and Beyond

69. Above and Beyond

Exercises 2.2

< Objective 1 >

1. {Monday, Tuesday, Wednesday, Thursday, Friday, Saturday, Sunday}

3. {1, 2, 3, 6, 9, 18}

5. {2, 3, 5, 7, 11, 13, 17, 19, 23, 29}

7. {−5, −4, −3, −2, −1}

9. {0, 2, 4, 6, 8, 10, 12}

11. {3, 4, 5, 6}

13. {−3, −2}

15. {−5, −4, −3, −2, −1, 0, 1, 2}

17. {1, 3, 5, 7, ...}

19. {0, 2, 4, 6, ..., 96, 98}

21. {5, 10, 15, 20, …}

< Objectives 2 and 3 >

23. $\{x|x>10\};(10,\infty)$

25. $\{x|x\geq -5\};[-5,\infty)$

27. $\{x|2\leq x\leq 7\};[2,7]$

29. $\{x|-4\leq x\leq 4\};[-4,4]$

31. $\{-2, -1, 0, 4\}$

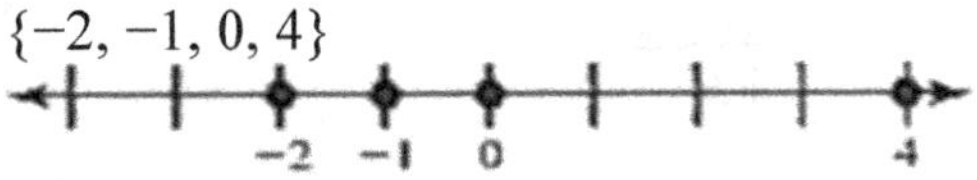

33. $\{x|x>4\}$

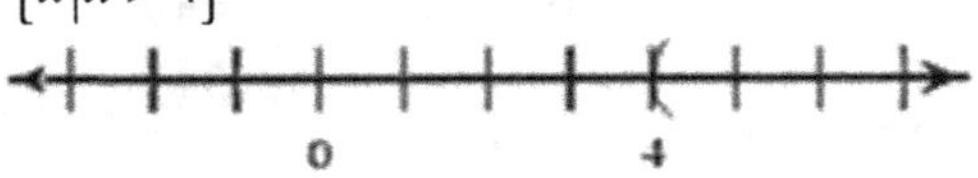

35. $\{x\,|\,x\geq -3\}$

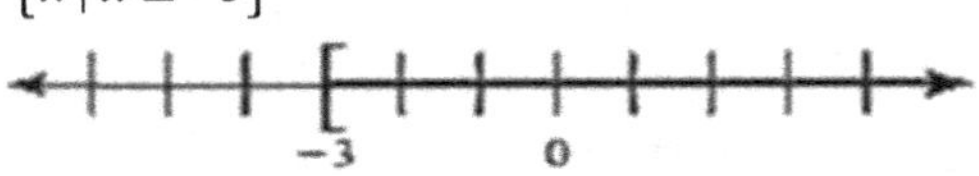

37. $\{x|2<x<7\}$

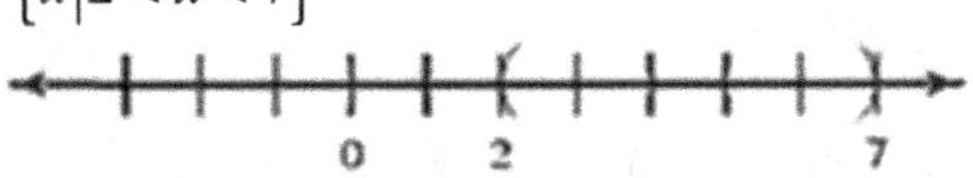

39. $\{x|-3<x\leq 5\}$

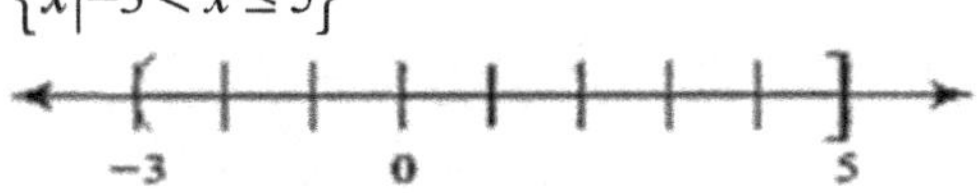

41. $\{x|-4\leq x<0\}$

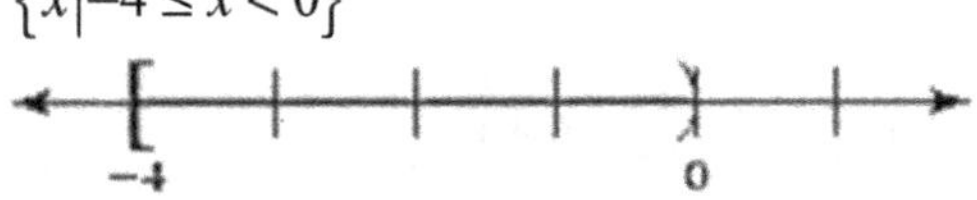

43. $\{x|-7\leq x\leq -3\}$

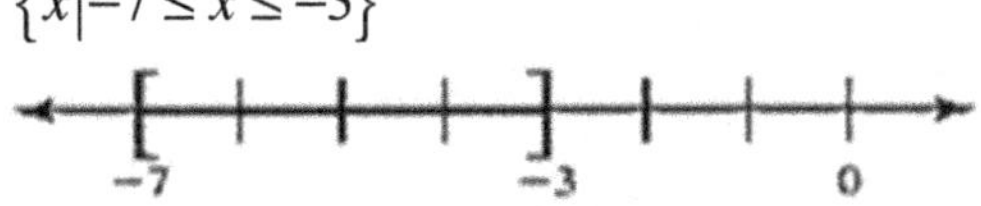

45. The set of all integers between –7 and –3, inclusive.

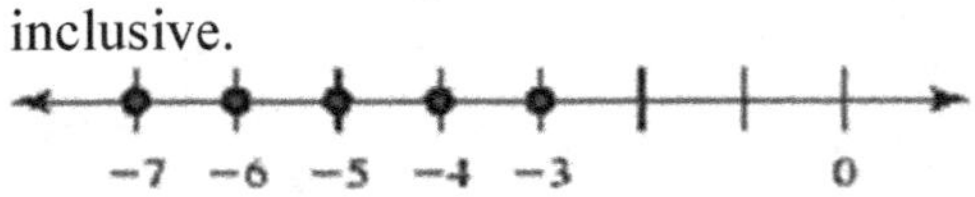

47.
$$\begin{aligned}3(x-3)&<3\\3x-9&<3\\3x-9+9&<3+9\\3x&<12\\\frac{3x}{3}&<\frac{12}{3}\\x&<4\end{aligned}$$
$\{x|x<4\};(-\infty,4)$

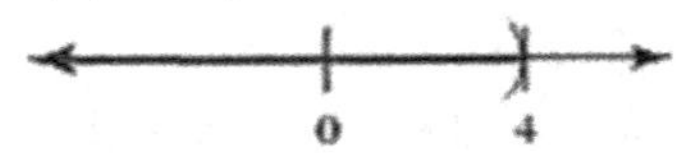

49.
$$\begin{aligned}-2(4x-5)&<16\\-8x+10&<16\\-8x+10-10&<16-10\\-8x&<6\\\frac{-8x}{-8}&<\frac{6}{-8}\\x&>-\frac{3}{4}\end{aligned}$$
$\left\{x\middle|x>-\frac{3}{4}\right\};\left(-\frac{3}{4},\infty\right)$

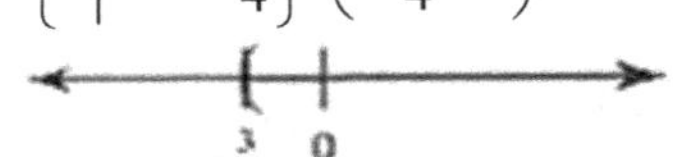

51.
$$\begin{aligned}-2(4-x)&\leq 7+2x\\-8+2x&\leq 7+2x\\-8+2x-2x&\leq 7+2x-2x\\-8&\leq 7\end{aligned}$$
This is a true statement.
$\{x|x\in\mathbb{R}\};(-\infty,\infty)$

53.
$$\begin{aligned}-2(5-x)&\leq -3(x+2)+5x\\-10+2x&\leq -3x-6+5x\\-10+2x&\leq 2x-6\\-10+2x-2x&\leq 2x-6-2x\\-10&\leq -6\end{aligned}$$
This is a true statement.
$\{x|x\in\mathbb{R}\};(-\infty,\infty)$

55. $\{x|x\leq 1\};(-\infty,1]$

57. $\{x|x>-2\};(-2,\infty)$

59. $\{x|-2\le x\le 2\};[-2,2]$

61. $\{x|-3<x<2\};(-3,2)$

63. $\{x|-2\le x<4\};[-2,4)$

65. $\{x|-2<x\le 4\};(-2,4]$

67. $\{x|-2\le x\le 4\};[-2,4]$

< Objective 4 >

69. $A\cup B=\{1,2,3,4,5,6,7,8,9\}$

71. $B\cap\varnothing=\varnothing$

73. $A\cup\varnothing=\{2,4,6,8\}$

75. $B\cap C=\{1,3,5\}$

77. $(A\cup C)\cap B=\{1,3,5\}$

79. True

81. sometimes

83. Above and Beyond

Exercises 2.3

< Objectives 1 and 2 >

1. $x+y=6$

(4, 2): $(4)+(2)\stackrel{?}{=}6$

$6=6$

(−2, 4): $(-2)+(4)\stackrel{?}{=}6$

$2\ne 6$

(0, 6): $(0)+(6)\stackrel{?}{=}6$

$6=6$

(−3, 9): $(-3)+(9)\stackrel{?}{=}6$

$6=6$

(4, 2), (0, 6), and (−3, 9) are solutions.

3. $2x-y=8$

(5, 2): $2(5)-(2)\stackrel{?}{=}8$

$8=8$

(4, 0): $2(4)-(0)\stackrel{?}{=}8$

$8=8$

(0, 8): $2(0)-(8)\stackrel{?}{=}8$

$-8\ne 8$

(6, 4): $2(6)-(4)\stackrel{?}{=}8$

$8=8$

(5, 2), (4, 0), and (6, 4) are solutions.

5. $4x+y=8$

(2, 0): $4(2)+(0)\stackrel{?}{=}8$

$8=8$

(2, 3): $4(2)+(3)\stackrel{?}{=}8$

$11\ne 8$

(0, 2): $4(0)+(2)\stackrel{?}{=}8$

$2\ne 8$

(1, 4): $4(1)+(4)\stackrel{?}{=}8$

$8=8$

(2, 0) and (1, 4) are solutions.

7. $2x-3y=6$

(0, 2): $2(0)-3(2)\overset{?}{=}6$

$-6\neq 6$

(3, 0): $2(3)-3(0)\overset{?}{=}6$

$6=6$

(6, 2): $2(6)-3(2)\overset{?}{=}6$

$6=6$

(0, −2): $2(0)-3(-2)\overset{?}{=}6$

$6=6$

(3, 0), (6, 2), and (0, −2) are solutions.

9. $3x-2y=12$

(4, 0): $3(4)-2(0)\overset{?}{=}12$

$12=12$

$\left(\frac{2}{3},-5\right)$: $3\left(\frac{2}{3}\right)-2(-5)\overset{?}{=}12$

$12=12$

(0, 6): $3(0)-2(6)\overset{?}{=}12$

$-12\neq 12$

$\left(5,\frac{3}{2}\right)$: $3(5)-2\left(\frac{3}{2}\right)\overset{?}{=}12$

$12=12$

(4, 0), $\left(\frac{2}{3},-5\right)$, and $\left(5,\frac{3}{2}\right)$ are solutions.

11. $3x+5y=15$

(0, 3): $3(0)+5(3)\overset{?}{=}15$

$15=15$

$\left(1,\frac{12}{5}\right)$: $3(1)+5\left(\frac{12}{5}\right)\overset{?}{=}15$

$15=15$

(5, 3): $3(5)+5(3)\overset{?}{=}15$

$30\neq 15$

(0, 3) and $\left(1,\frac{12}{5}\right)$ are solutions.

13. $x=3$

(3, 5), (3, 0), and (3, 7) are solutions.

15. $x+y=12$

$(4)+y=12$

$y=8;(4,8)$

$x+(5)=12$

$x=7;(7,5)$

$(0)+y=12$

$y=12;(0,12)$

$x+(0)=12$

$x=12;(12,0)$

17. $3x+y=9$

$3(3)+y=9$

$y=0;(3,0)$

$3x+(9)=9$

$3x=0$

$x=0;(0,9)$

$3x+(-3)=9$

$3x=12$

$x=4;(4,-3)$

$3(0)+y=9$

$y=9;(0,9)$

19. $5x-y=15$

$5x-(0)=15$

$x=3;(3,0)$

$5(2)-y=15$

$-y=5$

$y=-5;(2,-5)$

$5(4)-y=15$

$-y=-5$

$y=5;(4,5)$

$5x-(-5)=15$

$5x=10$

$x=2;(2,-5)$

21. $4x-2y=16$

$4x-2(0)=16$

$x=4;(4,0)$

$4x-2(-6)=16$

$4x=4$

$x=1;(1,-6)$

$4(2)-2y=16$

$-2y=8$

$y=-4;(2,-4)$

$4x-2(6)=16$

$4x=28$

$x=7;(7,6)$

23. $y=3x+9$

$(0)=3x+9$

$-9=3x$

$-3=x;(-3,0)$

$y=3\left(\frac{2}{3}\right)+9=11;\left(\frac{2}{3},11\right)$

$y=3(0)+9=9;(0,9)$

$y=3\left(-\frac{2}{3}\right)+9=7;\left(-\frac{2}{3},7\right)$

25. $y=3x-4$

$y=3(0)-4=-4;(0,-4)$

$(5)=3x-4$

$9=3x$

$x=3;(3,5)$

$(0)=3x-4$

$3x=4$

$x=\frac{4}{3};\left(\frac{4}{3},0\right)$

$y=3\left(\frac{5}{3}\right)-4=1;\left(\frac{5}{3},1\right)$

27. $x-y=10$

(0, −10), (10, 0), (5, −5), and (12, 2) are solutions.

29. $2x-y=6$

(0, −6), (3, 0), (6, 6), and (9, 12) are solutions.

31. $x+4y=8$

(8, 0), (−4, 3), and (4, 1) are solutions.

33. $5x-2y=10$

(0, −5), (4, 5), (−6, −20), and (2, 0) are solutions.

35. $\frac{1}{3}x-\frac{1}{4}y=1$

(0, 3), (1, 5), (2, 7), and (3, 9) are solutions.

37. $x=-5$

(−5, 0), (−5, 1), (−5, 2), and (−5, 3) are solutions.

39. $\frac{1}{2}x+\frac{1}{3}y=1$

$\frac{1}{2}x+\frac{1}{3}(0)=1$ $\frac{1}{2}x=1$ $x=2$	$\frac{1}{2}(0)+\frac{1}{3}y=1$ $\frac{1}{3}y=1$ $y=3$
(2, 0)	(0, 3)

41. $0.3x+0.5y=2$

$0.3x+0.5(0)=2$ $0.3x=2$ $x=\frac{20}{3}$	$0.3(0)+0.5y=2$ $0.5y=2$ $y=4$
$\left(\frac{20}{3},0\right)$	(0, 4)

43. $\frac{3}{4}x-\frac{2}{5}y=6$

$\frac{3}{4}x-\frac{2}{5}(0)=6$ $\frac{3}{4}x=6$ $x=8$	$\frac{3}{4}(0)-\frac{2}{5}y=6$ $-\frac{2}{5}y=6$ $y=-15$
(8, 0)	(0, −15)

45. $0.4x-0.7y=3$

$0.4x-0.7(0)=3$ $0.4x=3$ $x=\frac{15}{2}$	$0.4(0)-0.7y=3$ $-0.7y=3$ $y=-\frac{30}{7}$
$\left(\frac{15}{2},0\right)$	$\left(0,-\frac{30}{7}\right)$

< Objective 3 >

47. $y=0.75x+8$

$y=0.75(2)+8=9.5$

$y=0.75(5)+8=11.75$

$y=0.75(10)+8=15.5$

$y=0.75(15)+8=19.25$

$y=0.75(20)+8=23$

The hourly wages for producing 2, 5, 10, 15, and 20 units per hour are \$9.50, \$11.75, \$15.50, \$19.25, and \$23.00, respectively

49. $A=4s$

$A=4(5)=20$

$A=4(10)=40$

$A=4(12)=48$

$A=4(15)=60$

For squares whose sides are 5 cm, 10 cm, 12 cm, and 15 cm, their perimeters are 20 cm, 40 cm, 48 cm, and 60 cm, respectively.

51. $p=4x+15$

$p=4(1)+15=19$

$p=4(5)+15=35$

$p=4(10)+15=55$

$p=4(12)+15=63$

When 1, 5, 10, and 12 units are sold, the price for each unit is \$19, \$35, \$55, and \$63, respectively.

53. $y=162x+4,365$

$y=162(1)+4,365=4,527$

$y=162(2)+4,365=4,689$

$y=162(3)+4,365=4,851$

$y=162(4)+4,365=5,013$

$y=162(6)+4,365=5,337$

x	1	2	3	4	6
y	4,527	4,689	4,851	5,013	5,337

55. False

57. True

59. False

61. sometimes

63.

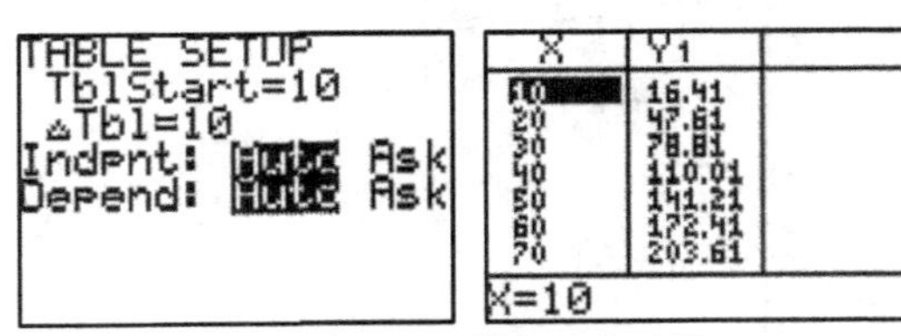

The ordered pairs are (10, 16. 41), (20, 47.61), (30, 78.81), (40, 110.01), and (50, 141.21).

65. $s = \frac{3}{4}L + 1$

$s = \frac{3}{4}(12) + 1 = 10$

$s = \frac{3}{4}(20) + 1 = 16$

$s = \frac{3}{4}(24) + 1 = 19$

10, 16, and 19 studs are required to build walls of length 12 ft, 20 ft, and 24 ft, respectively.

67. **(a)** $d = 7.5w$

$d = 7.5(30) = 225$

The dosage for a 30-kg child is 225 mg.

(b) $d = 7.5w$

$150 = 7.5w$

$\frac{150}{7.5} = w$

$20 = w$

The weight of a child that requires a 150-mg dose is 20 kg.

69. $F = kx$

$F = 72(3) = 216 \text{ lb}$

$F = 72(5) = 360 \text{ lb}$

The force exerted is 216 lb if $x = 3$ and 360 lb if $x = 15$ ft.

71. $x + y + z = 0 \qquad (2, -3, \ )$

$2 + (-3) + z = 0$

$2 - 3 + z = 0$

$-1 + z = 0$

$z = 1$

(2, −3, 1) is the ordered-triple solution.

73. $x + y + z = 0 \qquad (1, \ , 5)$

$1 + y + 5 = 0$

$y + 6 = 0$

$y = -6$

(1, −6, 5) is the ordered-triple solution.

75. $2x + y + z = 2 \qquad (-2, \ , 1)$

$2(-2) + y + 1 = 2$

$-4 + y + 1 = 2$

$y - 3 = 2$

$y = 5$

(−2, 5, 1) is the ordered-triple solution.

77. Above and Beyond

Exercises 2.4

< Objective 1 >

1. $A(4, 4)$

3. $C(2, 0)$

5. $E(-4, -5)$

7. $S(-6, -6)$

9. $U(-3, 5)$

< Objective 2 >

11. $M(5, 3)$

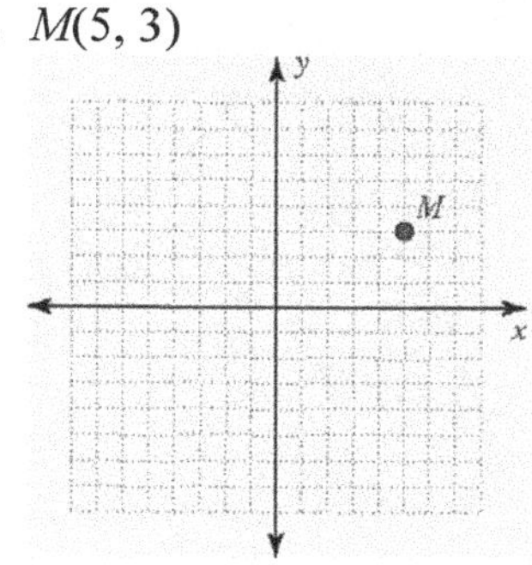

13. $P(-4, 5)$

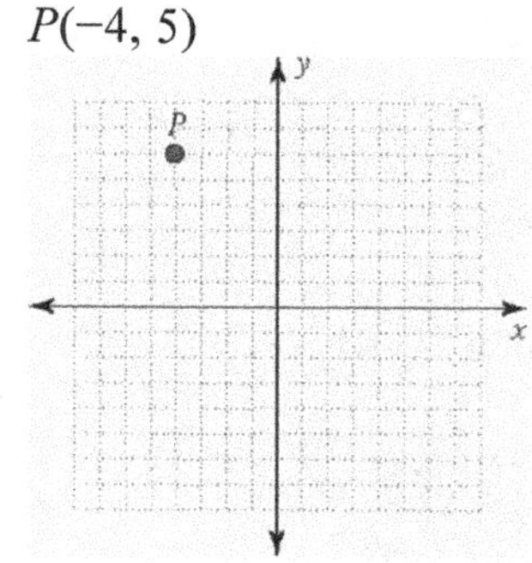

15. $R(-4, -6)$

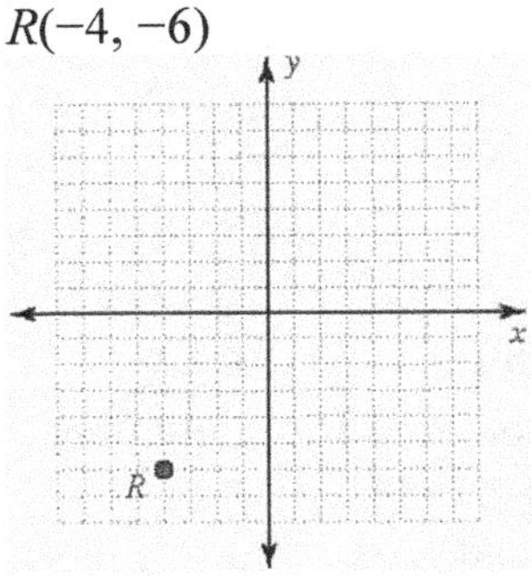

17. $F(-3, -1)$

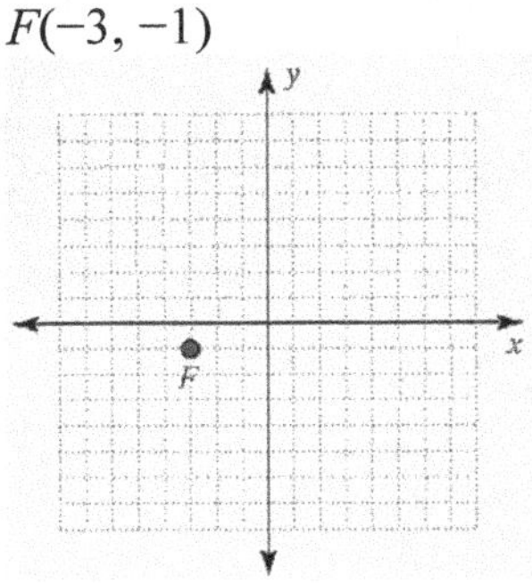

19. $H(4, -3)$

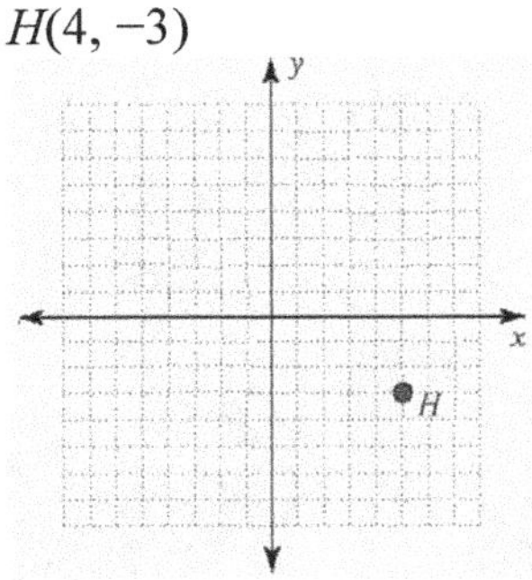

21. $J(-5, 3)$

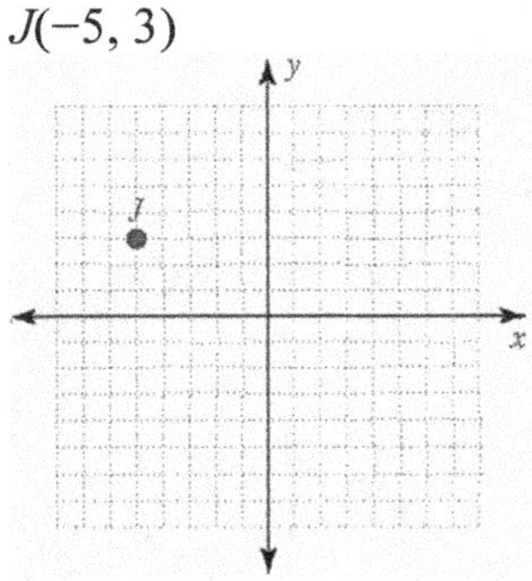

23. (4, 5) is in quadrant I.

25. (−6, −8) is in quadrant III.

27. (5, 0) is on the x-axis.

29. (−4, 7) is in quadrant II.

31. (0, −4) is on the y-axis.

33. $\left(5\frac{3}{4}, -3\right)$ is in quadrant IV.

< Objective 3 >

35. The ordered pairs are (1, 30), (2, 45), (3, 60), (4, 60), (5, 75), (6, 90), and (7, 95).

37. The ordered pairs are (7, 100), (15, 70), (20, 80), (30, 70), (40, 50), (50, 40), (60, 30), (70, 40), and (80, 25).

39. **(a)-(b)** The ordered pairs are $A(1500, 350)$ and $B(2300, 430)$.

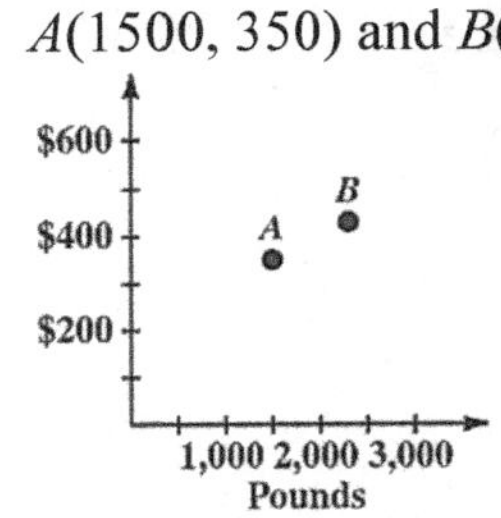

41. The ordered pairs are (1, 6000), (2, 5000), (3, 8000), and (4, 9000).

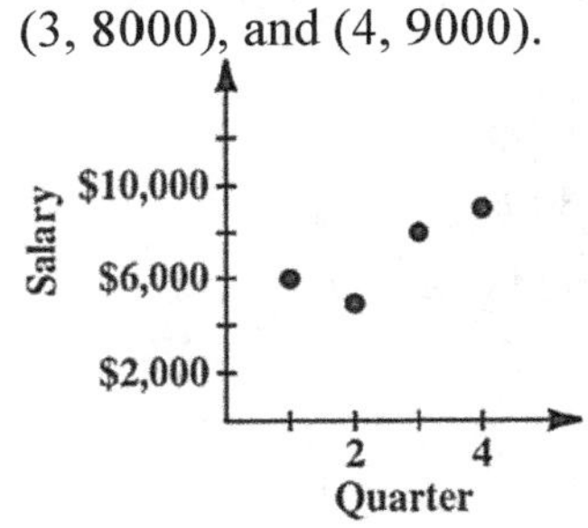

43. always

45. sometimes

47. The ordered pairs are (−10, 4.6), (10, 14.9), (30, 48.3), (50, 47.1), (70, 71.1), and (90, 99.2).

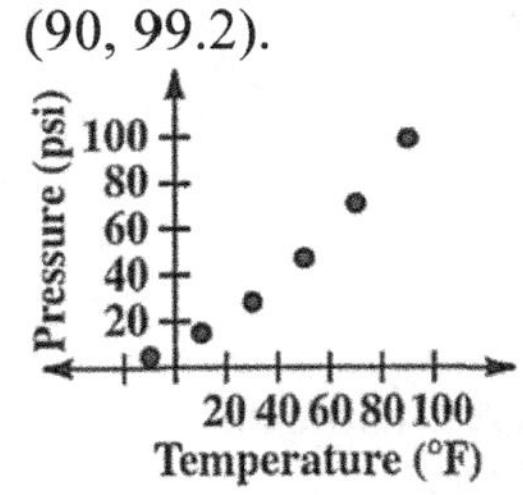

49. The ordered pairs are (0, 7.8), (0.5, 7.14), (1, 9.25), (2, 12.5), (7, 20.25), and (9, 21.25).

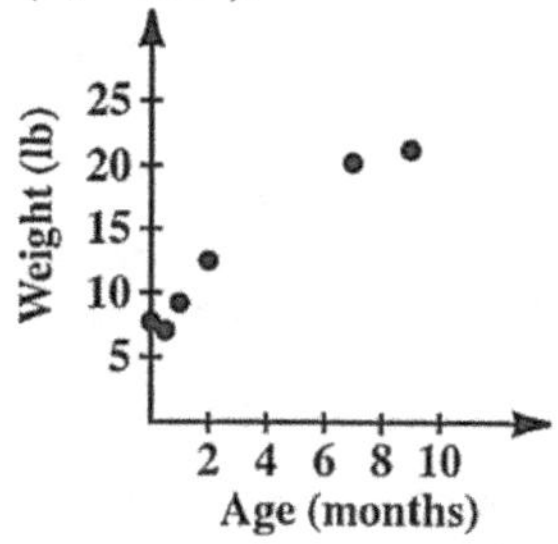

51. The points appear to lie on a line. Answers may vary; as an example, (2, −6) would also lie on the line.

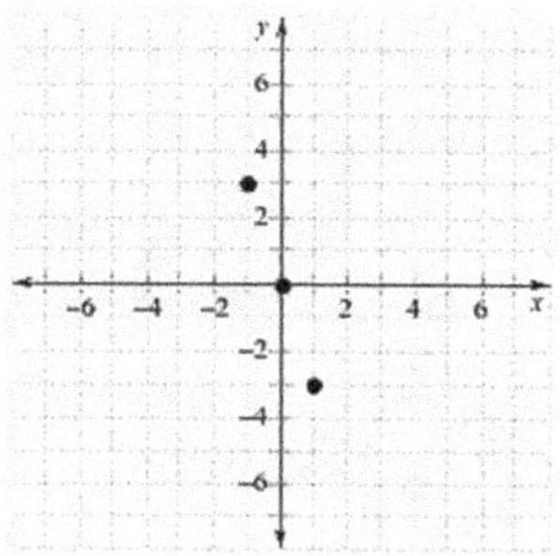

53. Above and Beyond

55. Above and Beyond

57. Above and Beyond

Exercises 2.5

< Objective 1 >

1. Domain: {Colorado, Edmonton, Calgary, Vancouver}
Range: {21, 5, 18, 17}

3. Domain: {Chamber, Testament, Rainmaker, Street Lawyer}
Range: $\left\{\pi, 2\pi, \frac{1}{2}, 6\right\}$

5. Domain: {1, 3, 5, 7, 9}
Range: {2, 4, 6, 8, 10}

7. Domain: {1}
Range: {2, 3, 4, 5, 6}

9. Domain: {−3, −2, −1, 4, 5}
Range: {3, 4, 5, 6}

11. {(1, 12,961.45), (2, 13,178.25), (3, 13,195.16), (4, 13,251.28), (5, 13,232.21)}

13. {(−2, −8), (−1, −1), (0, 0), (1, 1), (2, 8)}
Domain: {−2, −1, 0, 1, 2}
Range: {−8, −1, 0, 1, 8}

15. {(10, 90), (20, 180), (30, 270), (40, 360)}
Domain: {10, 20, 30, 40}
Range: {90, 180, 270, 360}

< Objective 2 >

17. {(1, 6), (2, 8), (3, 9)} is a function.

19. {(−1, 4), (−2, 5), (−3, 7)} is a function.

21. {(1, 3), (1, 2), (1, 1)} is not a function.

23. {(−3, 5), (6, 3), (6, 9)} is not a function.

25. The table defines a function.

27. The table does not define a function.

29. The table defines a function.

< Objective 3 >

31. $f(x)=x^2-x-2$

(a) $f(0)=0^2-0-2=-2$

(b) $f(-2)=(-2)^2-(-2)-2=4+2-2$
$=4$

(c) $f(1)=1^2-1-2=1-1-2=-2$

33. $f(x)=3x^2+x-1$

(a) $f(-2)=3(-2)^2+(-2)-1$
$=3(4)-2-1=12-2-1=9$

(b) $f(0)=3(0)^2+0-1=-1$

(c) $f(1)=3(1)^2+1-1=3+1-1=3$

35. $f(x)=x^3-2x^2+5x-2$

(a) $f(-3)=(-3)^3-2(-3)^2+5(-3)-2$
$=-27-18-15-2=-62$

(b) $f(0)=0^3-2(0)^2+5(0)-2=-2$

(c) $f(1)=1^3-2(1)^2+5(1)-2$
$=1-2+5-2=2$

37. $f(x)=-3x^3+2x^2-5x+3$

(a) $f(-2)$
$=-3(-2)^3+2(-2)^2-5(-2)+3$
$=24+8+10+3=45$

(b) $f(0)=-3(0)^3+2(0)^2-5(0)+3=3$

(c) $f(3)=-3(3)^3+2(3)^2-5(3)+3$
$=-81+18-15+3=-75$

39. $f(x)=2x^3+4x^2+5x+2$

(a) $f(-1)=2(-1)^3+4(-1)^2+5(-1)+2$
$=-1$

(b) $f(0)=2(0)^3+4(0)^2+5(0)+2=2$

(c) $f(1)=2(1)^3+4(1)^2+5(1)+2=13$

< Objective 4 >

41. $y=-3x+2$
$f(x)=-3x+2$

43. $y=4x-8$
$f(x)=4x-8$

45. $3x+2y=6$
$2y=-3x+6$
$y=-\frac{3}{2}x+3$
$f(x)=-\frac{3}{2}x+3$

47. $-2x+6y=9$
$6y=2x+9$
$y=\frac{1}{3}x+\frac{3}{2}$
$f(x)=\frac{1}{3}x+\frac{3}{2}$

49. $-5x-8y=-9$
$-8y=5x-9$
$y=-\frac{5}{8}x+\frac{9}{8}$
$f(x)=-\frac{5}{8}x+\frac{9}{8}$

51. $f(a)=5a-1$

53. $f(x+1)=5(x+1)-1=5x+5-1=5x+4$

55. $f(x+h)=5(x+h)-1=5x+5h-1$

57. $g(m)=-3m+2$

59. $g(x+2)=-3(x+2)+2=-3x-6+2$
$=-3x-4$

61. $g(x+h)=-3(x+h)+2=-3x-3h+2$

63. $k(t)=\frac{1}{2}t+3$

65. $k(x-6)=\frac{1}{2}(x-6)+3=\frac{1}{2}x-3+3=\frac{1}{2}x$

67. $k(x+h)=\frac{1}{2}(x+h)+3=\frac{1}{2}x+\frac{1}{2}h+3$

69. $f(2{,}500)=50(2{,}500)-15{,}000=110{,}000$
The profit earned from the sales of 2,500 units is $110,000.

71. **(a)** $C(10) = 36(10) + 52 = 412$

It costs \$4.12 to make a 10-minute call at 10 A.M.

(b) $C(10) = 24(10) + 33 = 273$

It costs \$2.73 to make a 10-minute call at 10 P.M.

73. **(a)** $f(55) = 0.017(55)^2 = 51.425$

A car traveling 55 mi/hr will skid 51.425 ft.

(b) $f(70) = 0.017(70)^2 = 83.3$

A car traveling 70 mi/hr will skid 83.3 ft.

75. **(a)** $f(4) = 1.5(4) + 8.3 = 14.3$

The predicted weight at the age of 4 months is 14.3 lb.

(b) $f(8) = 1.5(8) + 8.3 = 20.3$

The predicted weight at 8 months is 20.3 lb.

77. always

79. $f(x) = 3x^2 - 5x + 7$

x	f(x)
−5	107
−4	75
−3	49
−2	29
−1	15
0	7

x	$f(x)$
1	5
2	9
3	19
4	35
5	57

81. $2x^3 - 4x^2 + 5x - 9$

x	f(x)
−4	−221
−3	−114
−2	−51
−1	−20
0	−9

x	$f(x)$
1	−6
2	1
3	24
4	75

83. **(a)** Let $P(t)$ be the pitch of a 6-in. gear and let t be the number of teeth.

$$P(t) = \frac{t}{6}$$

(b) $P(30) = \frac{30}{6} = 5$

The pitch of a 6-in. gear with 30 teeth is 5.

85. **(a)** Let $C(x)$ be the cost of building a house and x be the area of the house.

$C(x) = 90x + 12{,}000$

(b) $C(1{,}800) = 90(1{,}800) + 12{,}000$
$= 174{,}000$

It costs \$174,000 to build a 1,800-ft^2 house.

Exercises 2.6

< Objective 1 >

1. {(−3, 1), (−1, 2), (−2, 3), (1, 4)} is a function.

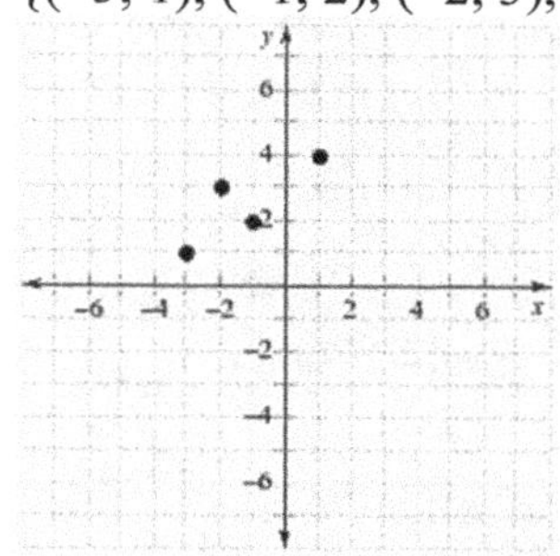

3. {(−1, 1), (2, 2), (3, 4), (5, 6)} is a function.

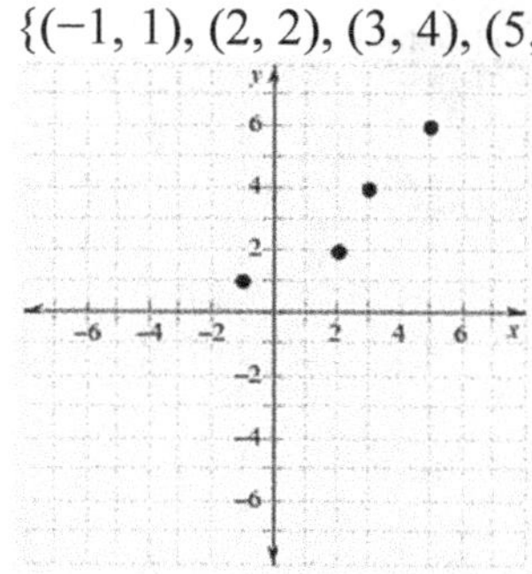

5. {(1, 2), (1, 3), (2, 1), (3, 1)} is not a function.

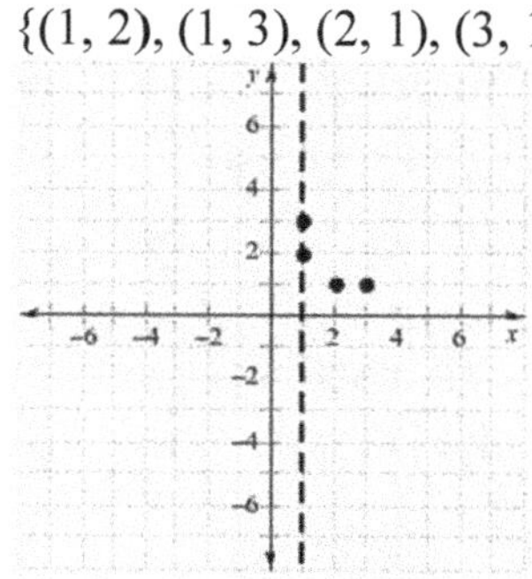

< Objective 2 >

7. This relation is a function.
$D=\{-2,-1,0,1,2\}$
$R=\{-1,0,1,2,3\}$

9. This relation is a function.
$D=\{-2,-1,0,2,3,5\}$
$R=\{-1,2,4,5\}$

11. This relation is a function.
$D=\{x|-4\le x\le 3\}=[-4,3]$
$R=\{-2\}$

13. This relation is a function.
$D=\{x|-3\le x\le 4\}=[-3,4]$
$R=\{y|-2\le y\le 5\}=[-2,5]$

15. This relation is not a function.
$D=\{x|-3\le x\le 3\}=[-3,3]$
$R=\{y|-3\le y\le 4\}=[-3,4]$

17. This relation is not a function.
$D=\{-3\}$
$R=\mathbb{R}$

19. This relation is a function.
$D=\mathbb{R}$
$R=\mathbb{R}$

21. This relation is a function.
$D=\mathbb{R}$
$R=\{y|y\ge -5\}=[-5,\infty)$

23. This relation is not a function.
$D=\{x|-6\le x\le 6\}=[-6,6]$
$R=\{y|-6\le y\le 6\}=[-6,6]$

25. This relation is a function.
$D=\mathbb{R}$
$R=\{y|y\ge 0\}=[0,\infty)$

27. This relation is a function.
$D=\mathbb{R}$
$R=\{y|y\ge 3\}=[3,\infty)$

29. This relation is not a function.
$D=\mathbb{R}$
$R=\{-4,3\}$

< Objective 3 >

31. $f(5)=-3$

33. $h(3)=5$

35. $x=-3$

37. $x=0,1$

39. $k(0)=2$

41. $g(1)=3$

43. $x=-3,6$

< Objective 4 >

45. **(a)** $f(2)=4$
(b) $f(-1)=-3$
(c) $x=4$
(d) $x=1$

47. **(a)** $f(3)=-2$
(b) $f(4)=2$
(c) $x=-2,3.75$
(d) $x=4.5$

49. **(a)** $f(3)=-5$
(b) $f(0)=4$
(c) $x=-2,2$
(d) $x=-2.5,2.5$

51. never

53. always

Summary Exercises

1. $V = LWH$

$$W = \frac{V}{LH}$$

3. $ax + by = c$

$$y = \frac{c - ax}{b}$$

5. $A = P + Prt$

$$t = \frac{A - P}{Pr}$$

7. Let x be the number.

$$3x + 7 = 25$$
$$3x = 18$$
$$x = 6$$

9. Let x be one integer. Then $x + 1$ be the next consecutive integer.

$$x + x + 1 = 85$$
$$2x = 84$$
$$x = 42$$
$$x + 1 = 43$$

The numbers are 42 and 43.

11. Let x be the average speed while going. Then, $x - 6$ is the average speed while returning. Since distance is same while going and returning

$$4x = (x - 6)\left(4\frac{1}{2}\right) = (x - 6)\left(\frac{9}{2}\right)$$
$$8x = 9x - 54$$
$$x = 54$$
$$x - 6 = 48$$
$$4x = 216$$

The average speed while going was 54 mi/hr and the average speed while returning was 48 mi/hr. The distance traveled each way was 216 mi.

13. Let x be the number of shoe pairs. For the company to break even

$$47x - 21x = 3900$$
$$26x = 3900$$
$$x = 150$$

150 pair of shoes must be produced and sold for company to break even.

15. $\{1, 3\}$

17. $\{-1, 0, 1, 2, 3\}$

19. $\{1, 3\}$

21. $\{x | x \geq 1\}$

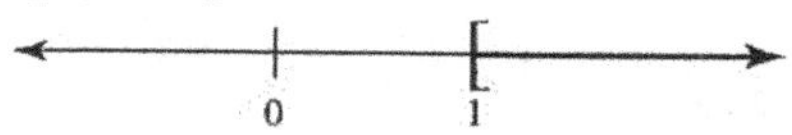

23. $\{x | -2 \leq x < 3\}$

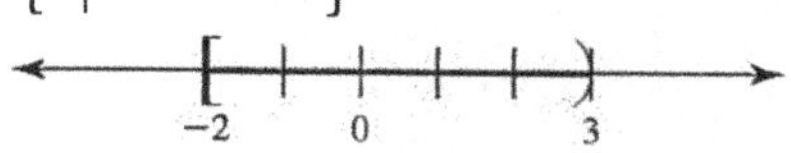

25. $\{x | x > 9\}; (9, \infty)$

27. $\{x | x \leq -5\}; (-\infty, -5]$

29. $\{x | x \leq 3\} = (-\infty, 3]$

31. $\{x | -3 \leq x < 2\} = [-3, 2)$

33. $\{x | -2 < x < 4\} = (-2, 4)$

35. $A \cup B = \{1, 2, 5, 7, 9, 11, 15\}$

37. $B \cup \varnothing = \{2, 5, 9, 11, 15\} = B$

39. $x - y = 6$

(6, 0): $(6)-(0)\stackrel{?}{=}6$

$6 = 6$

(3, 3): $(3)-(3)\stackrel{?}{=}6$

$0 \neq 6$

(3, −3): $(3)-(-3)\stackrel{?}{=}6$

$6 = 6$

(0, −6): $(0)-(-6)\stackrel{?}{=}6$

$6 = 6$

(6, 0), (3, −3), and (0, −6) are solutions.

41. $A(4, 1)$

43. $E(-1, -5)$

45. $P(4, 0)$

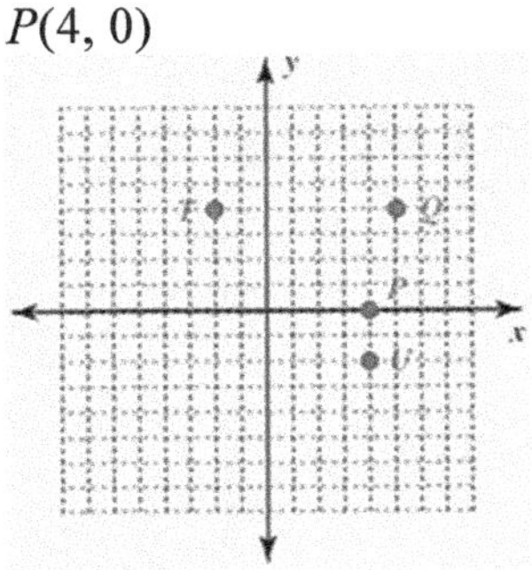

47. $T(-2, 4)$

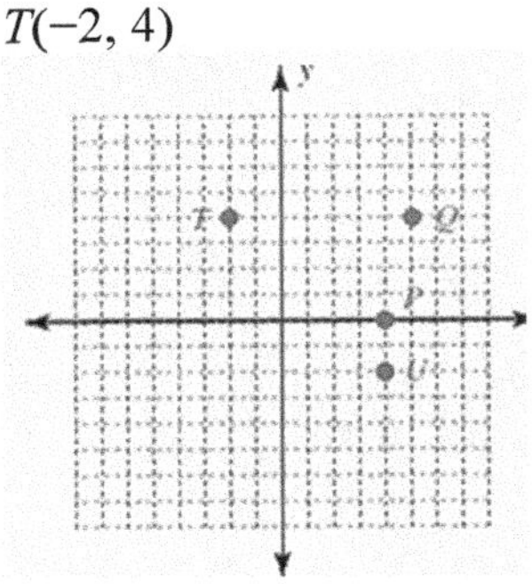

49. (−1, 4)

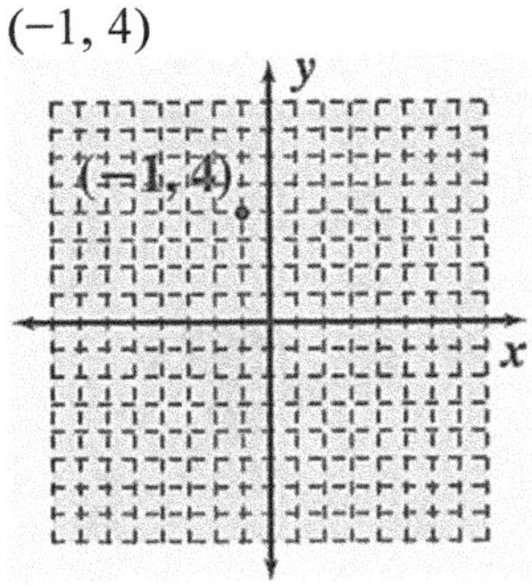

51. (6, 3)

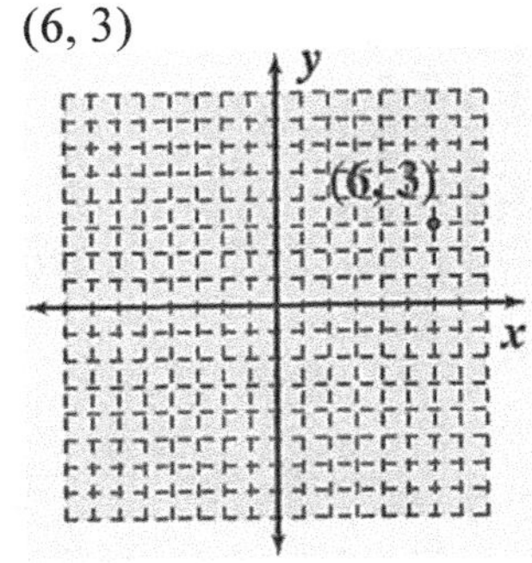

53. (3, 6) is in quadrant I.

55. (−1, −6) is in quadrant III.

57. (−5, 0) is on the x-axis.

59. $D=\{2001, 2004, 2007, 2010\}$
R={Halle Berry, Hilary Swank, Marion Cotillard, Natalie Portman}

61. D = {Brushed Nickel, Jeweled Bronze, Iron Amber, Walnut Silver}
$R=\{\$129.99, \$199.99, \$349.99, \$399.99\}$

63. $D=\{1,3,4,7,8\}$
$R=\{1,2,3,5,6\}$

65. $D=\{1\}$
$R=\{3,5,7,9,10\}$

67. This relation is a function.

69. This relation is not a function.

71. This relation is a function.

73. This relation is not a function.

75. $f(x)=x^2-3x+5$

(a) $f(0)=0^2-3(0)+5=5$

(b) $f(-1)=(-1)^2-3(-1)+5=1+3+5$
$=9$

(c) $f(1)=1^2-3(1)+5=1-3+5=3$

77. $f(x)=x^3-x^2-2x+5$

(a) $f(-1)=(-1)^3-(-1)^2-2(-1)+5$
$=-1-1+2+5=5$

(b) $f(0)=(0)^3-(0)^2-2(0)+5=5$

(c) $f(2)=2^3-2^2-2(2)+5$
$=8-4-4+5=5$

79. $f(x)=3x^2-5x+1$

(a) $f(-1)=3(-1)^2-5(-1)+1=3+5+1$
$=9$

(b) $f(0)=3(0)^2-5(0)+1=1$

(c) $f(2)=3(2)^2-5(2)+1=12-10+1$
$=3$

81. $f(x)=-2x+5$

83. $2x+3y=6$

$$3y=-2x+6$$
$$y=-\frac{2}{3}x+2$$
$$f(x)=-\frac{2}{3}x+2$$

85. $-3x+4y=12$

$$4y=3x+12$$
$$y=\frac{3}{4}x+3$$
$$f(x)=\frac{3}{4}x+3$$

87. $f(t)=-\frac{3}{4}t+2$

89. $f(x+h)=-\frac{3}{4}(x+h)+2=-\frac{3}{4}x-\frac{3}{4}h+2$

91. $f(a)=3a-2$

93. $f(x+h)=3(x+h)-2=3x+3h-2$

95. {(−3, −3), (−2, −2), (2, 2), (3, 3)} is a function.

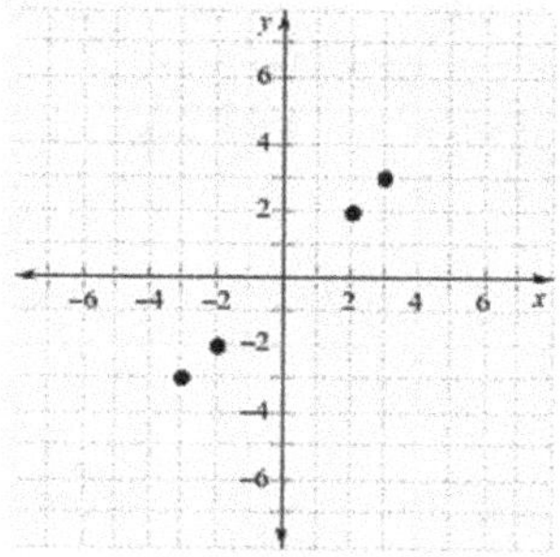

97. {(−2, 1), (−2, 3), (0, 1), (1, 2)} is not a function.

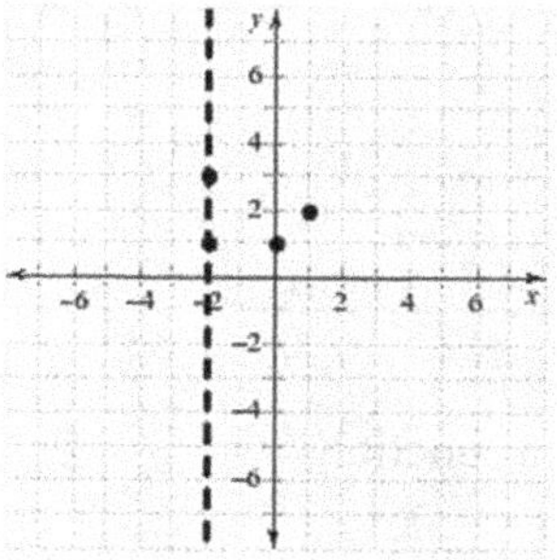

99. This relation is a function.
$D=\mathbb{R}$
$R=\{y|y\geq -4\}=[-4,\infty)$

101. This relation is not a function.
$D=\{x|-6\leq x\leq 6\}=[-6,6]$
$R=\{y|-6\leq y\leq 6\}=[-6,6]$

103. **(a)** $f(-1)=0$
(b) $f(5)=-3$
(c) $x=-5$
(d) $x=5$
(e) $x=-3.3,-2,3.4$
(f) $x=-4,1$

105. **(a)** $f(-5)=-2$
(b) $f(12)=12$
(c) $x=-5,4$
(d) $x=-2$
(e) $x=8$
(f) $x=12$

Chapter Test 2

1. 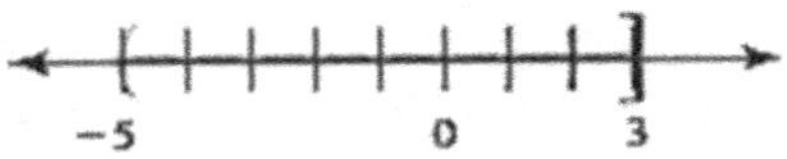

3. $A\ (3, 6)$

5. $C\ (0, -7)$

7.

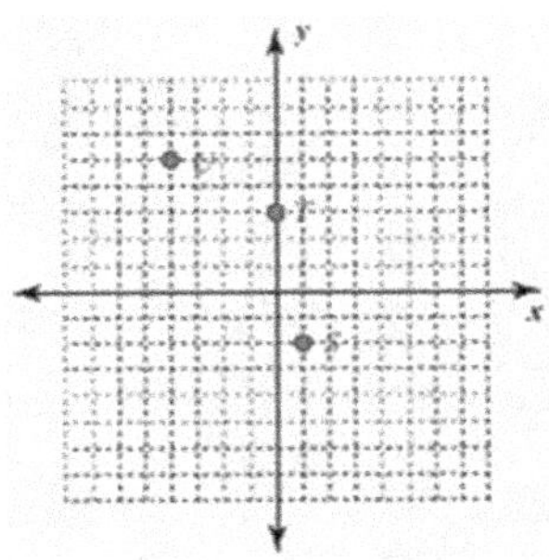

9. This relation is a function.

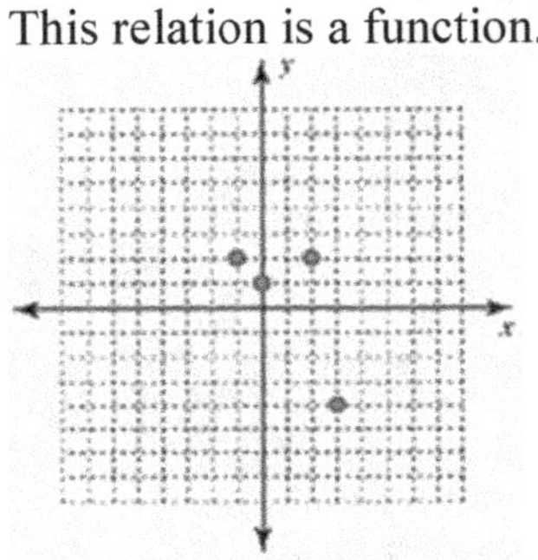

11.

$$
\begin{aligned}
4x+3y&=12\\
4(3)+3y&=12\\
12+3y-12&=12-12\\
3y&=0\\
y&=0;(3,0)\\
4x+3(4)&=12\\
4x+12-12&=12-12\\
4x&=0\\
x&=0;(0,4)\\
4x+3(3)&=12\\
4x+9-9&=12-9\\
4x&=3\\
x&=\frac{3}{4};\left(\frac{3}{4},3\right)
\end{aligned}
$$

13. $f(x)=-3x^2-2x+3$

(a) $f(-1)=-3(-1)^2-2(-1)+3=2$

(b) $f(-2)=-3(-2)^2-2(-2)+3=-5$

15. (a) $D=\{-3,1,2,3,4\}$

$R=\{-2,0,1,5,6\}$

(b) $D=${United States, Germany, Russia, China}

$R=\{50,63,65,101\}$

17. This relation is a function

19. (a) $A\cup B=\{1,2,3,5,7\}$

(b) $A\cap B=\{5\}$

21. $V=\frac{1}{3}Bh$

$Bh=3V$

$B=\frac{3V}{h}$

23. Let x be Juwan's age. Then $2x$ is Jan's age and $2x+5$ is Rick's age.

$$\begin{aligned} x+(2x)+(2x+5)&=35 \\ 5x+5&=35 \\ 5x&=30 \\ x&=6 \\ 2x&=12 \\ 2x+5&=17 \end{aligned}$$

Juwan is 6 years old, Jan is 12 years old, and Rick is 17 years old.

25. Let x be the number of notepads for break even.

$$\begin{aligned} 0.95x-0.20x&=450 \\ 0.75x&=450 \\ x&=600 \end{aligned}$$

600 notepads need to be sold for break even.

Cumulative Review: Chapters 0–2

1. $\frac{56}{88}=\frac{2\bullet 2\bullet 2\bullet 7}{2\bullet 2\bullet 2\bullet 11}=\frac{7}{11}$

3. $2\bullet 3^2-8\bullet 2=2\bullet 9-8\bullet 2=18-16=2$

5. $|12-5|=|7|=7$

7. $(-7)+(-9)=-16$

9. $(-7)(-9)=63$

11. $\frac{0}{-13}=0$

13.
$$\begin{aligned} 5-4^2\div(-8)\bullet 2&=5-16\div(-8)\bullet 2 \\ &=5-(-2)\bullet 2=5-(-4) \\ &=5+4=9 \end{aligned}$$

15. The LCD of 4 and 6 is 12.

$$\frac{3}{4}=\frac{3}{4}\bullet\frac{3}{3}=\frac{9}{12}$$
$$\frac{5}{6}=\frac{5}{6}\bullet\frac{2}{2}=\frac{10}{12}$$
$$\frac{3}{4}+\frac{5}{6}=\frac{9}{12}+\frac{10}{12}=\frac{19}{12}$$

17. $3x-y=3(-2)-3=-6-3=-9$

19. $\frac{5z-4x}{2y+z}=\frac{5(5)-4(-2)}{2(3)+5}=\frac{25+8}{11}=\frac{33}{11}=3$

21.
$$\begin{aligned} 7x-3y+2(4x-3y)&=7x-3y+8x-6y \\ &=15x-9y \end{aligned}$$

23.
$$\begin{aligned} 12x-3&=10x+5 \\ 12x-3+3&=10x+5+3 \\ 12x&=10x+8 \\ 12x-10x&=10x+8-10x \\ 2x&=8 \\ \frac{2x}{2}&=\frac{8}{2} \\ x&=4 \end{aligned}$$

25.
$$\begin{aligned} 4(x-1)-2(x-5)&=14 \\ 4x-4-2x+10&=14 \\ 2x+6&=14 \\ 2x&=8 \\ x&=4 \end{aligned}$$

27.
$$\begin{aligned} -5\le 2x+1&<7 \\ -6\le 2x&<6 \\ -3\le x&<3 \end{aligned}$$
$\{x|-3\le x<3\}$

29.
$$\begin{aligned} A&=\frac{1}{2}bh \\ bh&=2A \\ h&=\frac{2A}{b} \end{aligned}$$

31. $P = 2L + 2W$

$2W = P - 2L$

$W = \frac{P - 2L}{2} = \frac{P}{2} - L$

33. $f(0) = -2$

35. Let x be the number.

$4x - 7 = 45$

$4x = 52$

$x = 13$

The number is 13.

37. Let x be an odd integer. Then $x + 2$ is the next odd integer.

$3x = (x + 2) + 12 = x + 14$

$2x = 14$

$x = 7$

The integer is 7.

39. Let x be the width of the rectangle. Then $3x + 2$ is the length of the rectangle.

$P = 2L + 2W$

$44 = 2(3x + 2) + 2x = 6x + 4 + 2x$

$40 = 8x$

$5 = x$

$3x + 2 = 17$

The rectangle is 5 cm by 17 cm.

Chapter 3
Graphing Linear Functions

Exercises 3.1

< Objectives 1–3 >

1. $x + y = 6$

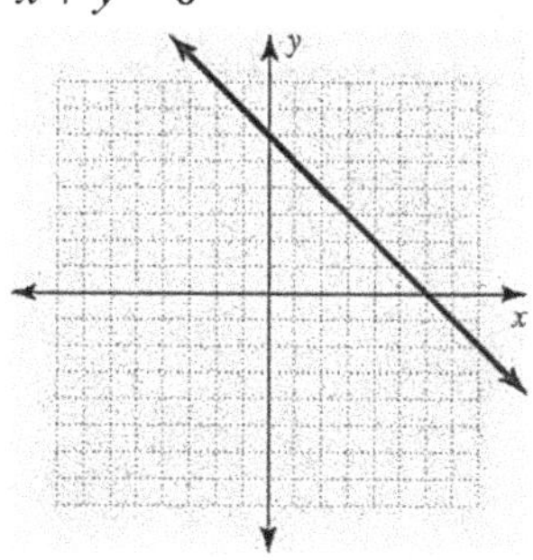

3. $x - y = -3$

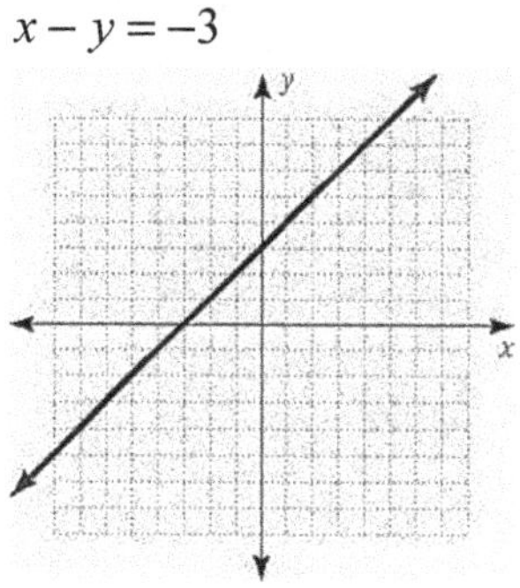

5. $3x + y = 6$

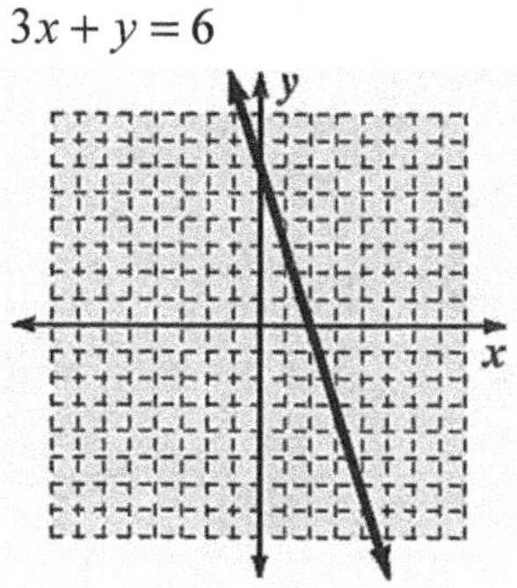

7. $3x + y = 0$

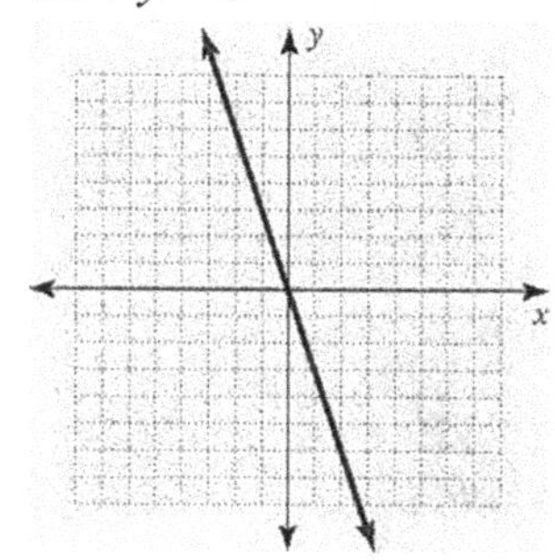

9. $x + 4y = 8$

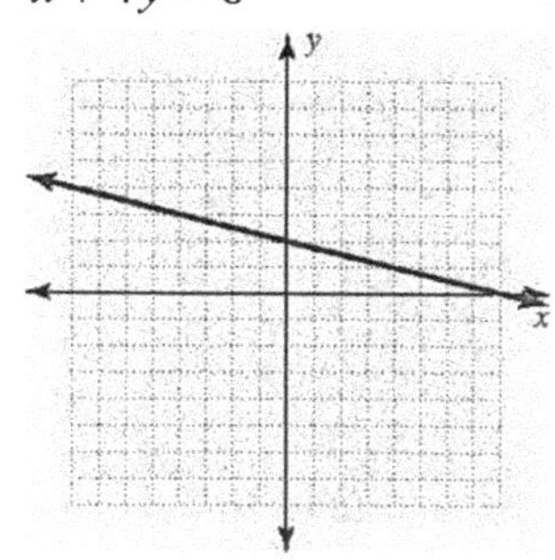

11. $y = 3x$

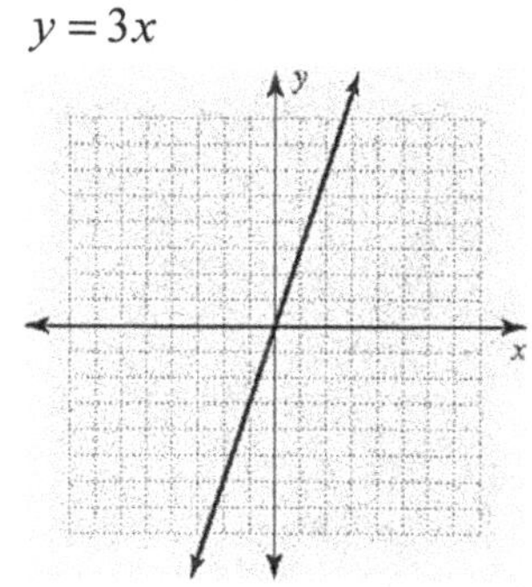

13. $y = 2x - 1$

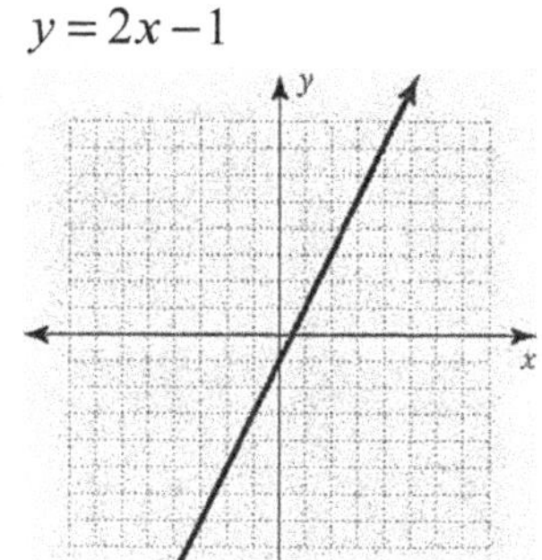

15. $y=-3x+1$

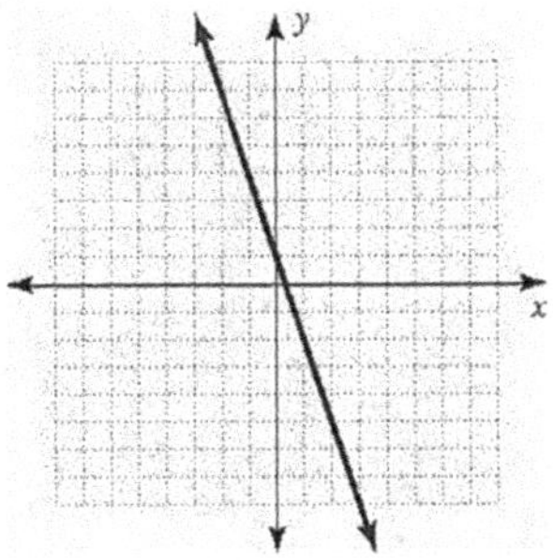

17. $y=\frac{1}{5}x$

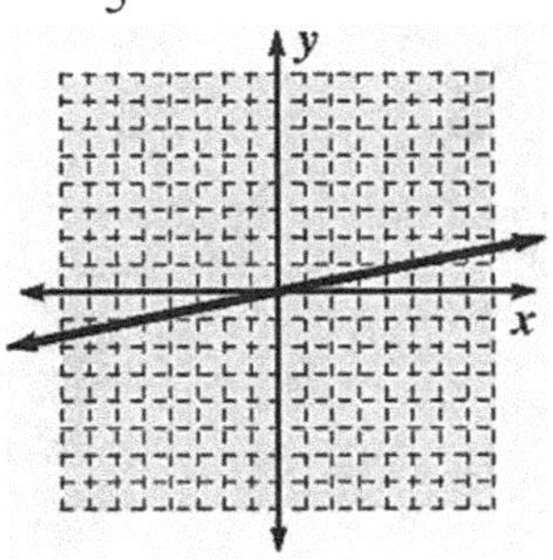

19. $y=\frac{2}{3}x-3$

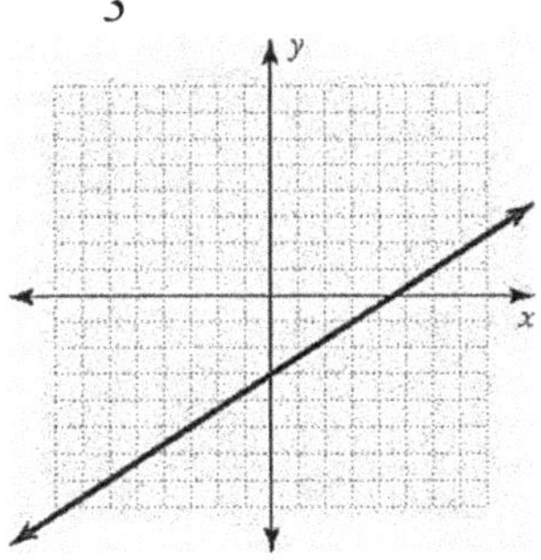

21. $x=-3$

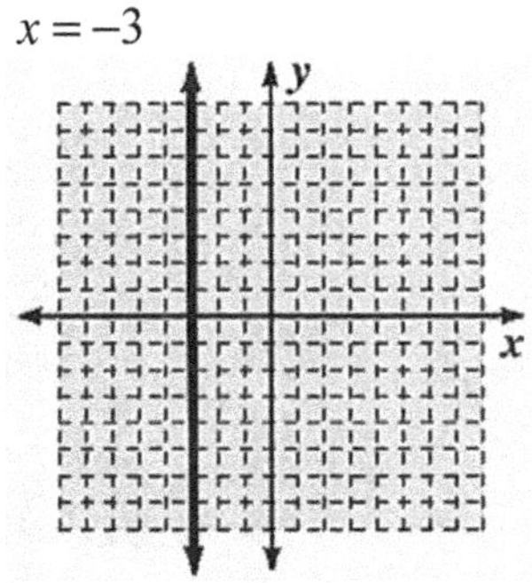

23. $y=1$

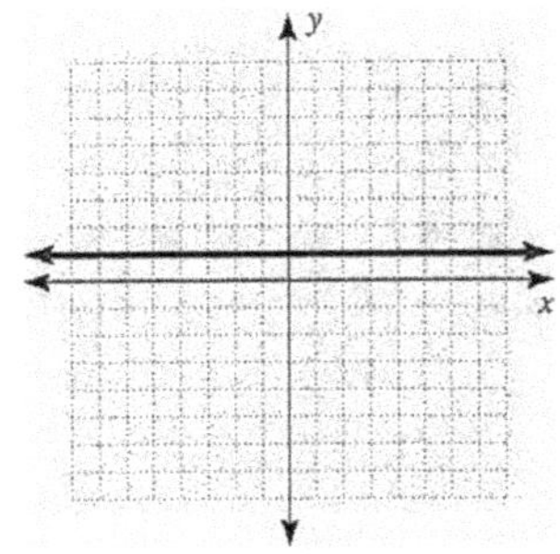

25. $x-2y=4$

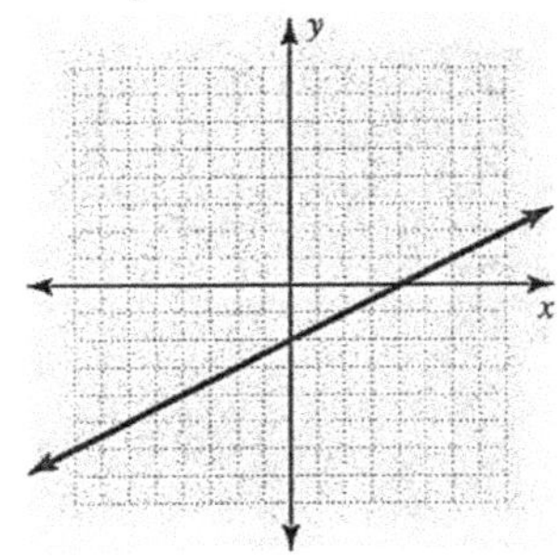

27. $5x+2y=10$

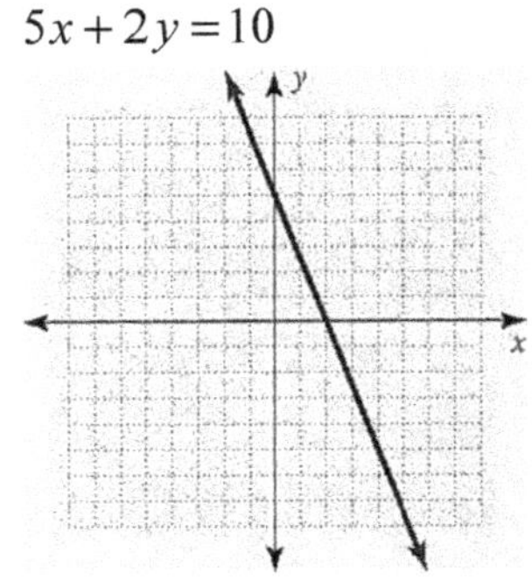

29. $3x+5y=15$

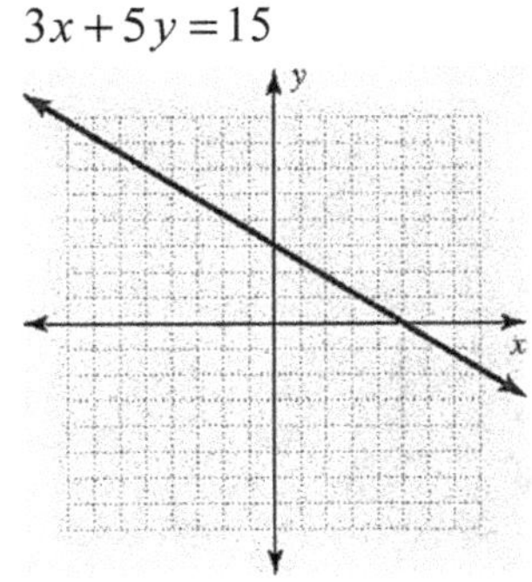

< Objective 4 >

31. $x+3y=6$

$$3y=-x+6$$

$$y=-\frac{1}{3}x+2$$

$$f(x)=-\frac{1}{3}x+2$$

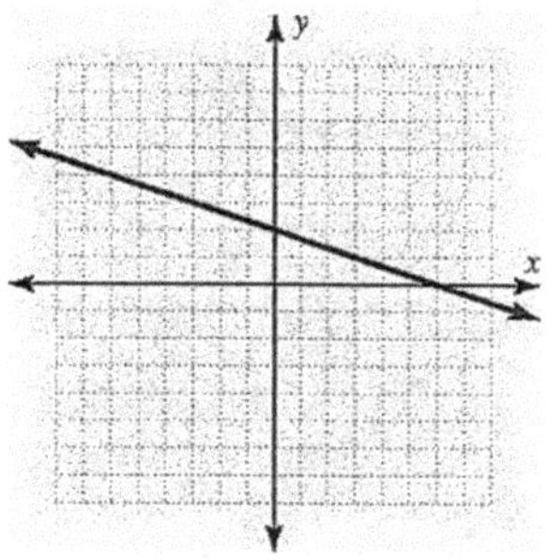

33. $3x+4y=12$

$$4y=-3x+12$$

$$y=-\frac{3}{4}x+3$$

$$f(x)=-\frac{3}{4}x+3$$

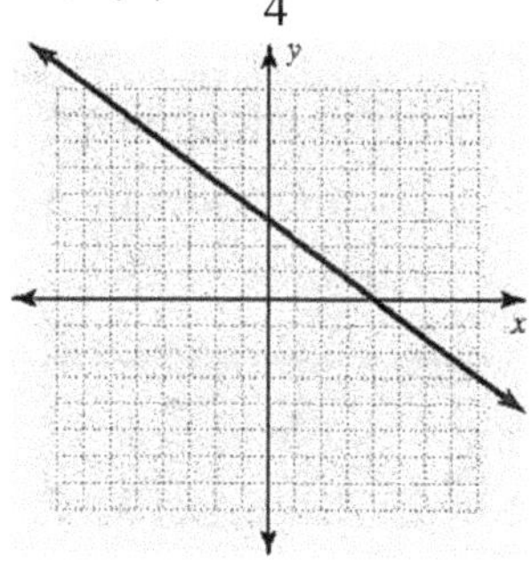

35. $5x-4y=20$

$$-4y=-5x+20$$

$$y=\frac{5}{4}x-5$$

$$f(x)=\frac{5}{4}x-5$$

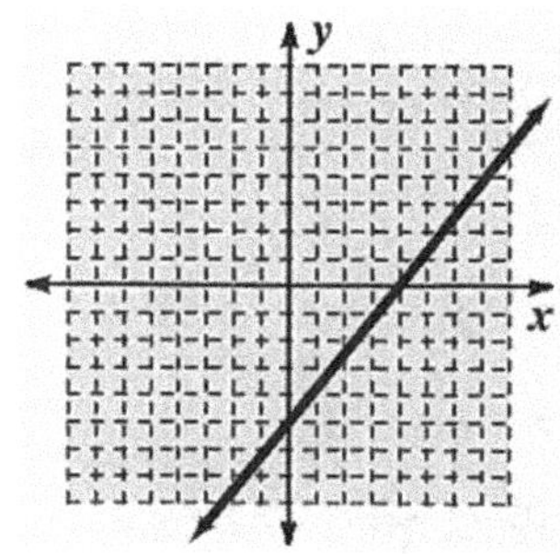

37. $y=2x$

39. $y=x+3$

41. $y=3x-3$

43. $x-4y=12$

45. $x+y=4$

$x-y=2$

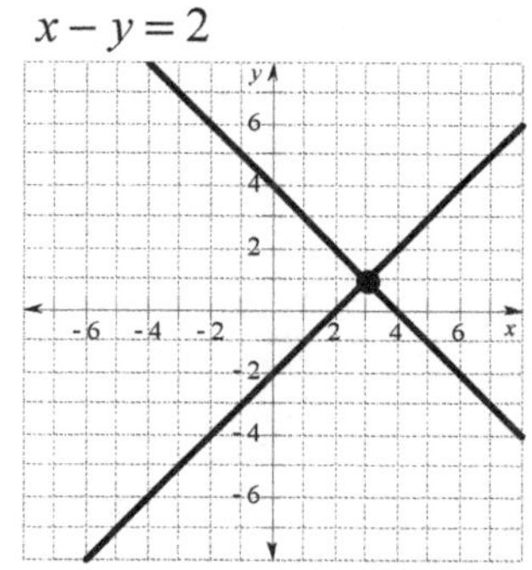

The lines intersect at (3, 1).

47. $f(x)=0.10x+200$

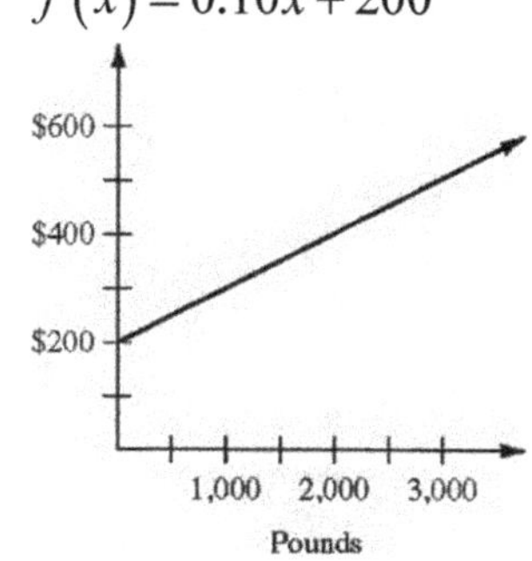

49. **(a)** $f(x)=15x-100$

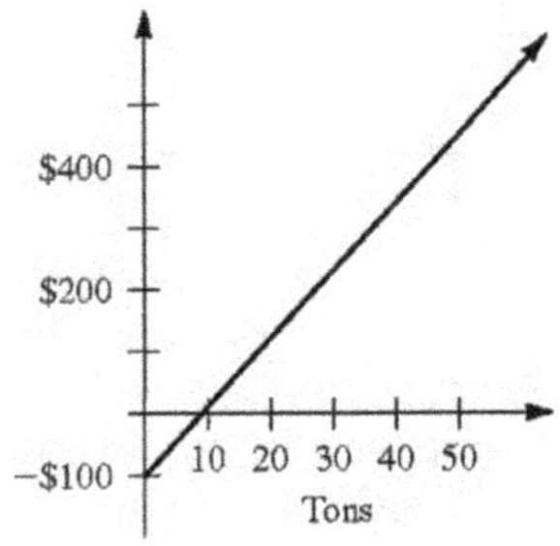

(b) $15x-100=0$

$$15x=100$$

$$x=\frac{100}{15}=\frac{20}{3}\approx 7$$

They need to collect nearly 7 tons.

(c) $15(16)-100=140$

They will earn $140.

(d) $f(x)=17x-125$

51. $C=0.08s+12$

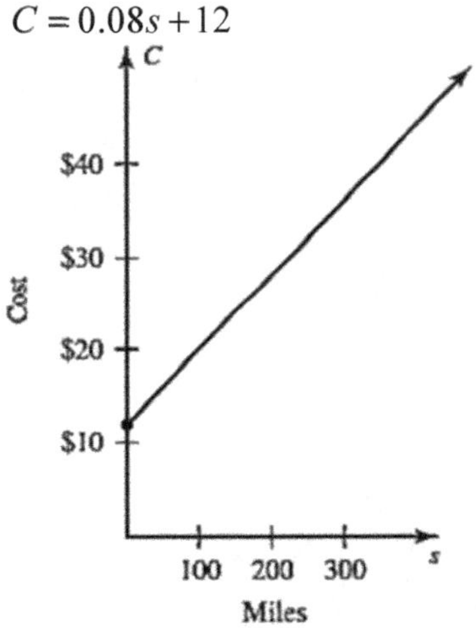

53. $T(h)=35h+75$

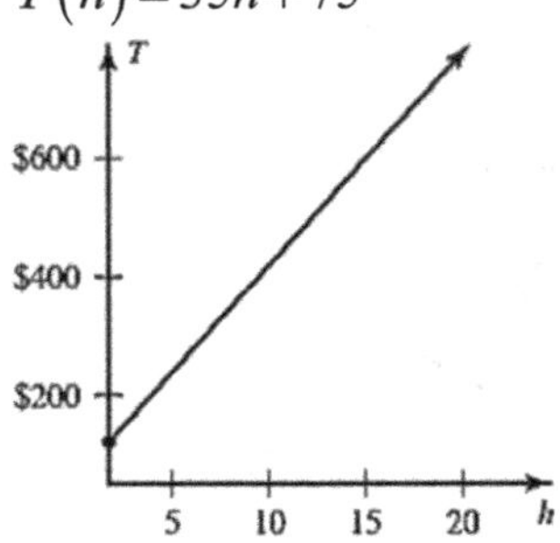

55. always

57. never

59. $y=3x-5$

Enter the right side of the equation into the $\boxed{Y_1 =}$ menu in the Y_1 field, and then press the $\boxed{\text{GRAPH}}$ key.

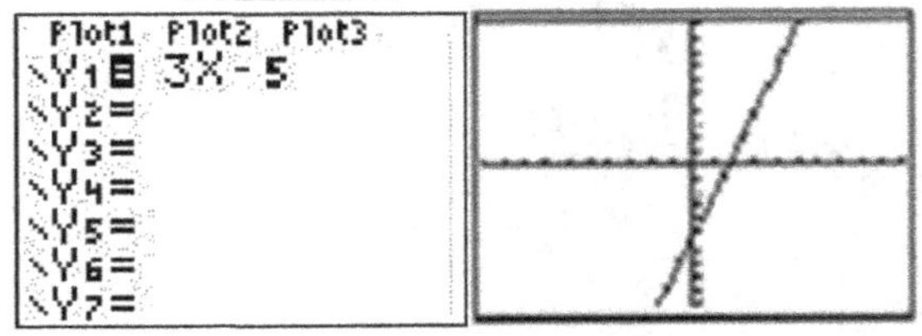

61. $y=-2x+4$

Enter the right side of the equation into the $\boxed{Y_1 =}$ menu in the Y_1 field, and then press the $\boxed{\text{GRAPH}}$ key.

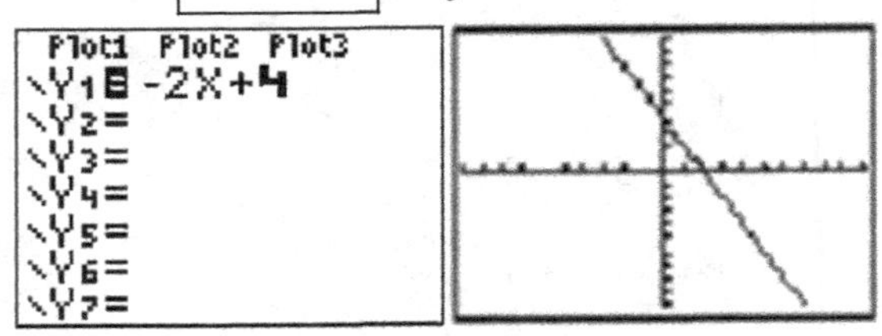

63. $y=\frac{2}{3}x+3$

Enter the right side of the equation into the $\boxed{Y_1 =}$ menu in the Y_1 field, and then press the $\boxed{\text{GRAPH}}$ key.

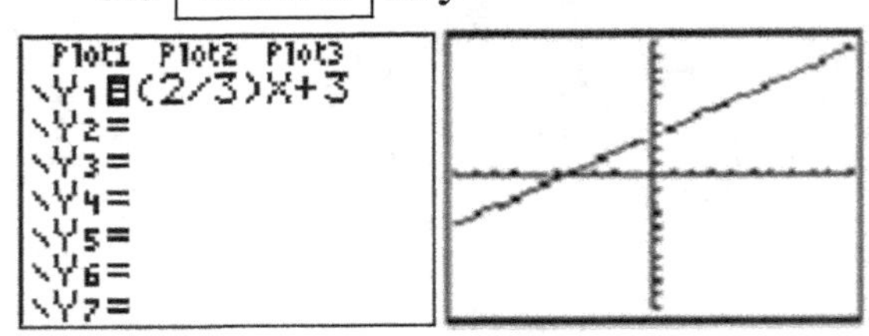

65. $3x+4y=8$

$$4y=-3x+8$$

$$y=-\frac{3}{4}x+2$$

Enter the right side of the equation into the $\boxed{Y_1 =}$ menu in the Y_1 field, and then press the $\boxed{\text{GRAPH}}$ key.

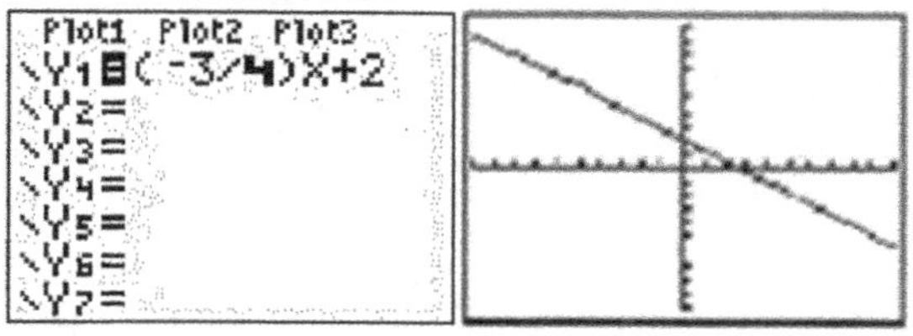

67. Enter $f(x)=17x-125$ as $Y1=17x-125$.

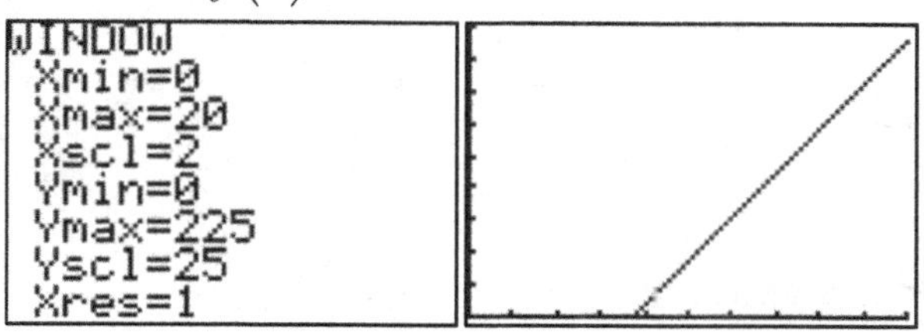

69. Enter $C=0.08s+12$ as $Y1=0.08x+12$.

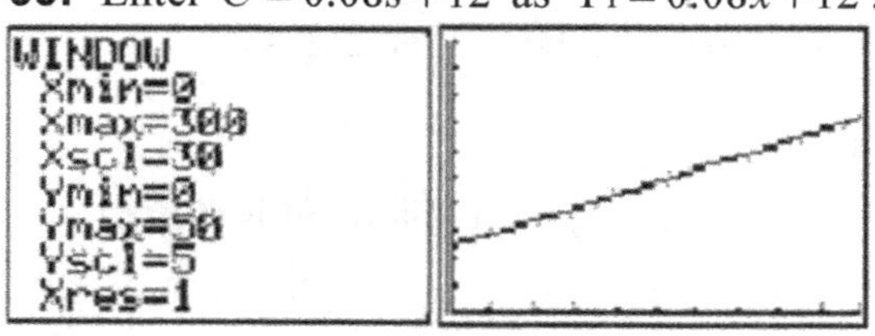

71. $w(d)=-1.75d+25$

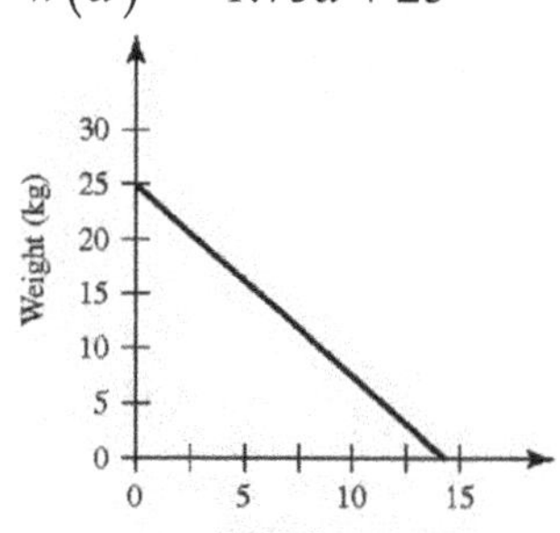

73. $s=\frac{3}{4}L+1$

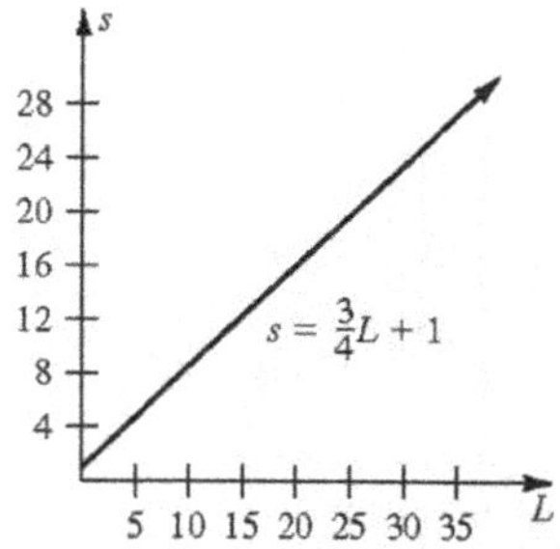

75. $f(x)=2x$

$f(x)=2x+1$

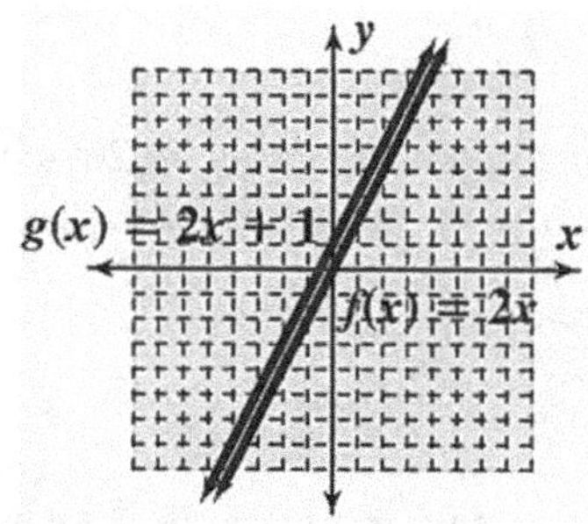

The lines are parallel.

77. $f(x)=2x$

$f(x)=-\frac{1}{2}x$

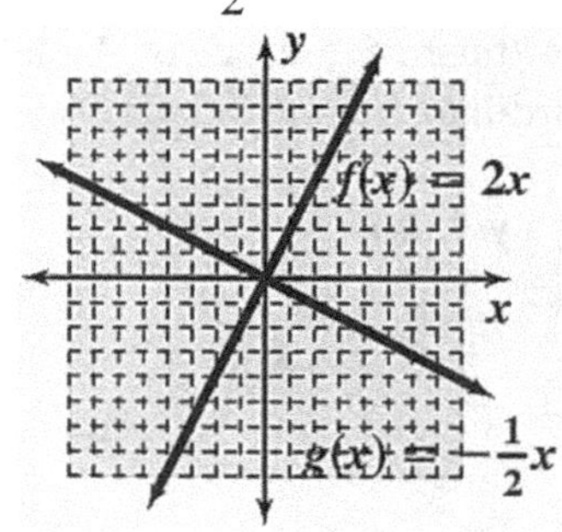

The lines are perpendicular.

79. **(a)** $y = 2x + 3$

Point	x	y
A	5	13
B	6	15
C	7	17
D	8	19
E	9	21

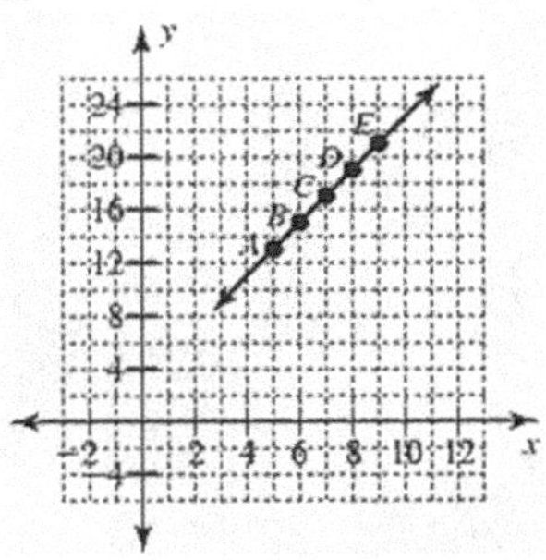

(b) The y-coordinate increases by 2 each time the x-coordinate increases by 1.
(c) Yes.
(d) grows by 2 units

81. $y = 3x - 2$
(b) The y-coordinate increases by 3 each time the x-coordinate increases by 1.
(c) Yes.
(d) grows by 3 units

83. $y = -4x + 50$
(b) The y-coordinate decreases by 4 each time the x-coordinate increases by 1.
(c) Yes.
(d) decreases by 4 units

Exercises 3.2

< Objective 1 >

1. $m = \frac{11-7}{9-5} = \frac{4}{4} = 1$

3. $m = \frac{3-(-1)}{2-(-3)} = \frac{4}{5}$

5. $m = \frac{7-3}{3-(-2)} = \frac{4}{5}$

7. $m = \frac{-8-2}{2-(-3)} = \frac{-10}{5} = -2$

9. $m = \frac{-5-(-2)}{5-3} = \frac{-3}{2} = -\frac{3}{2}$

11. $m = \frac{2-(-4)}{5-5} = \frac{6}{0}$ is undefined.

13. $m = \frac{3-(-2)}{3-(-4)} = \frac{5}{7}$

15. $m = \frac{-6-(-6)}{8-(-2)} = \frac{0}{10} = 0$

17. $m = \frac{3-7}{2-(-1)} = \frac{-4}{3} = -\frac{4}{3}$

< Objective 2 >

19. $y = 3x + 5$
The slope is 3; the y-intercept is (0, 5).

21. $y = -3x - 6$
The slope is –3; the y-intercept is (0, –6).

23. $y = \frac{3}{4}x + 1$
The slope is $\frac{3}{4}$; the y-intercept is (0, 1).

25. $y = \frac{2}{3}x$
The slope is $\frac{2}{3}$; the y-intercept is (0, 0).

27. $4x+3y=12$

$$3y=-4x+12$$

$$y=-\frac{4}{3}x+4$$

$$f(x)=-\frac{4}{3}x+4$$

The slope is $-\frac{4}{3}$; the y-intercept is (0, 4).

29. $f(x)=9$

The slope is 0; the y-intercept is (0, 9).

31. $3x-2y=8$

$$-2y=-3x+8$$

$$y=\frac{3}{2}x-4$$

$$f(x)=\frac{3}{2}x-4$$

The slope is $\frac{3}{2}$; the y-intercept is (0, –4).

< Objective 3 >

33. $y=3x+5$

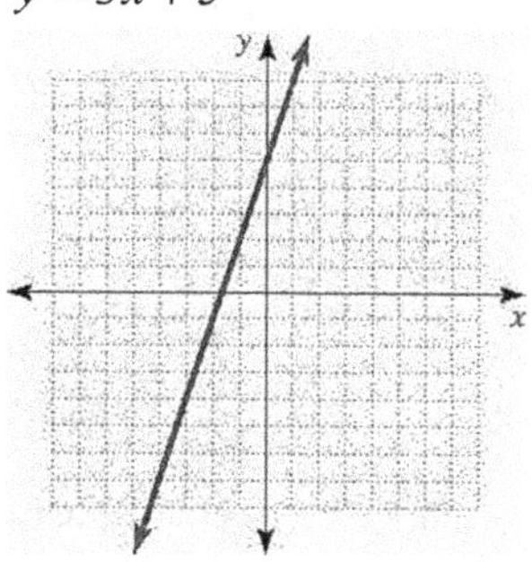

35. $y=-4x+5$

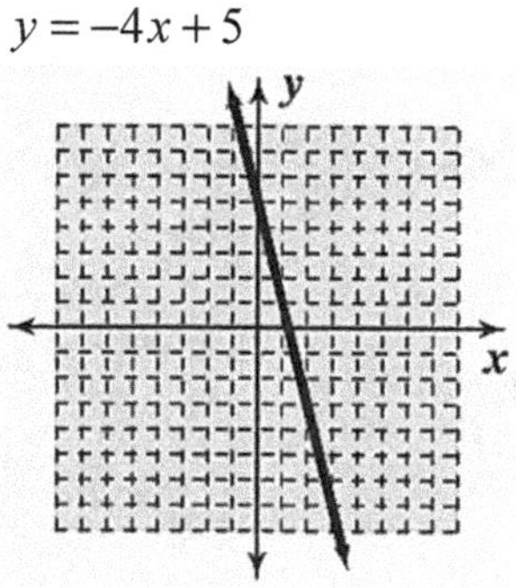

37. $y=\frac{1}{2}x-2$

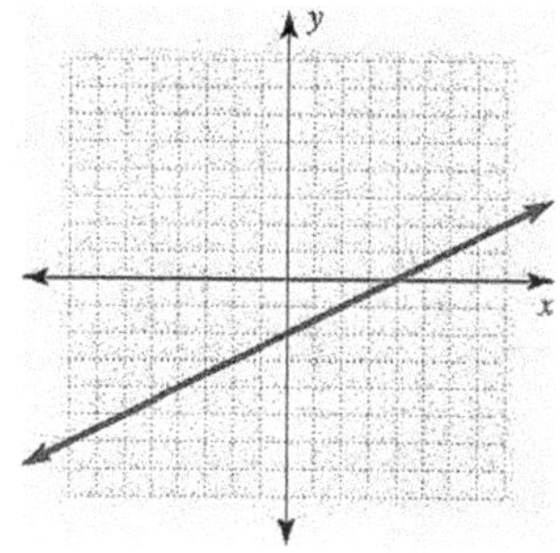

39. $y=\frac{4}{3}x$

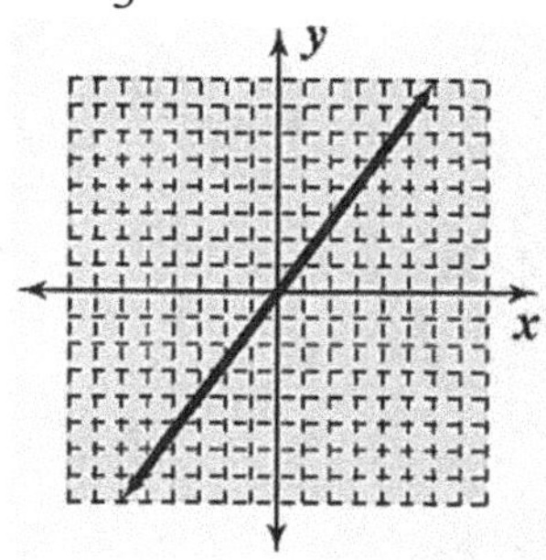

41. $y=\frac{3}{4}x+3$

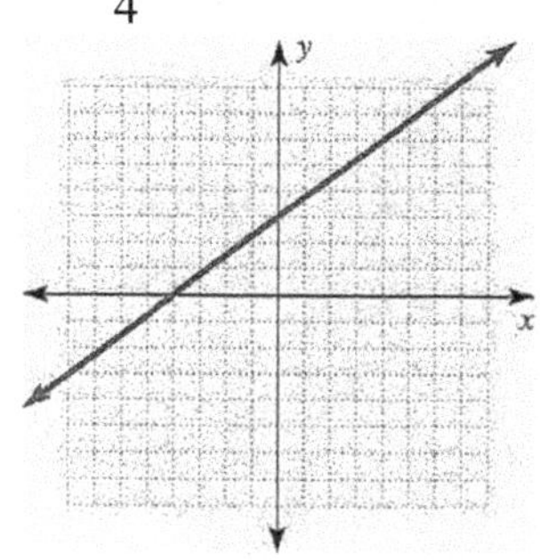

< Objective 4 >

43. $2x+5y=10$

$5y=-2x+10$

$y=-\frac{2}{5}x+2$

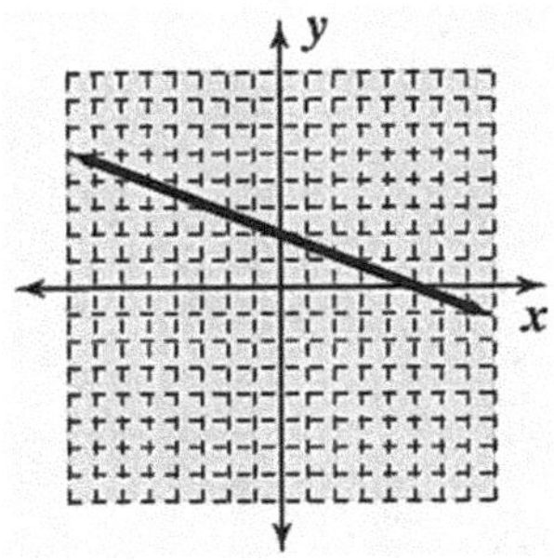

45. $x+7y=14$

$7y=-x+14$

$y=-\frac{1}{7}x+2$

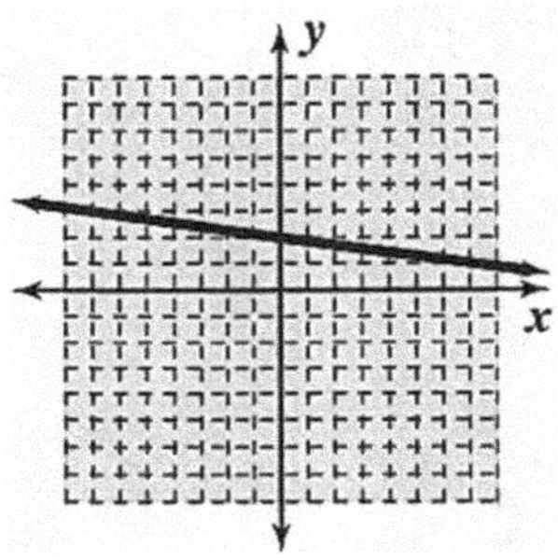

47. The slope is 2.

49. The slope is –2.

51. The slope is 3.

53. The slope is $-\frac{2}{5}$.

55. $y=0.10x+200$
The slope is 0.10; it represents the market price per pound. The y-intercept is (0, 200). The \$200 represents the minimum award.

57. The slope represents the price of newsprint; the y-intercept represents the fixed cost.

59. $m=\frac{-24{,}000 \text{ ft}}{15 \text{ mi}}=\frac{-24{,}000}{15\cdot 5{,}280}=\frac{-24{,}000}{79{,}200}$
≈ -0.30

61. $m=\frac{1.80-0.50}{2002-1960}=\frac{1.3}{42}\approx 0.031$
The cost of the soft drink increases at an average rate of 3.1¢ per year.

63. $y=4x+5$
There are no solutions in quadrant IV.

65. $y=-x+1$
There are no solutions in quadrant III.

67. $y=-2x-5$
There are no solutions in quadrant I.

69. $y=5$
There are no solutions in quadrants III and IV.

71. **(g)** $y=-\frac{3}{4}x+1$

73. **(e)** $y=-3x-2$

75. **(h)** $y=-4x$

77. **(c)** $y=-x+3$

79. sometimes

81. sometimes

83. $d=7.5w$

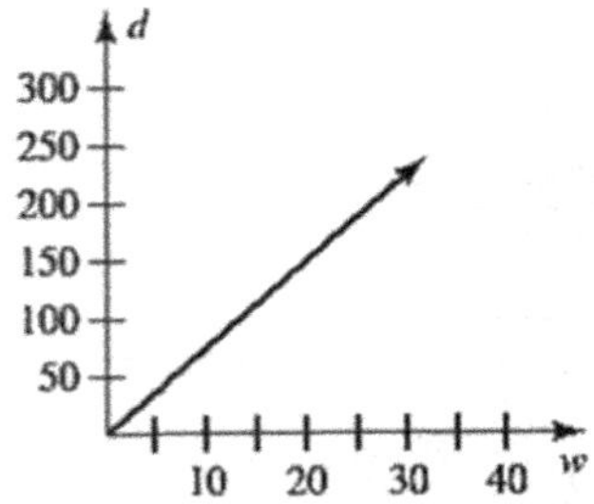

85. $m=\frac{25,000-0}{4-0}=\frac{25,000}{4}=6,250$

For points between 0 and 4 feet from the left end of the wood beam, the bending moment increases at an average rate of 6,250 ft-lb per foot.

87. Above and Beyond

89. Above and Beyond

91. $y=-2x+4$

$y=-2x+1$

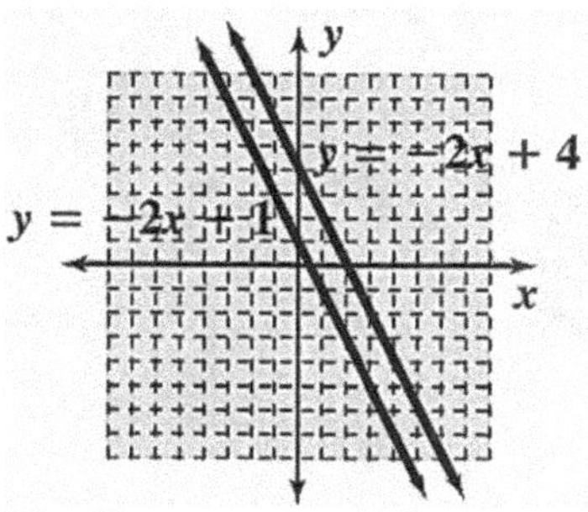

The lines are parallel; they will not intersect.

93. $y=\frac{4}{3}x$

$y=-\frac{3}{4}x$

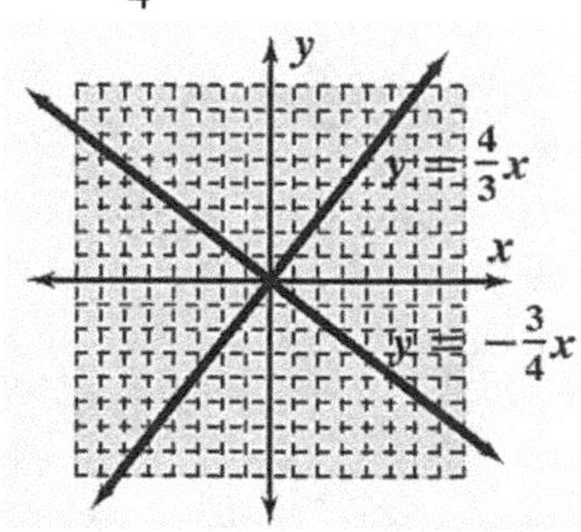

The lines are perpendicular; the product of their slopes is $\frac{4}{3}\left(-\frac{3}{4}\right)=-1$.

Exercises 3.3

< Objective 1 >

1. $m_1=\frac{3-(-3)}{4-(-2)}=\frac{6}{6}=1$

$m_2=\frac{7-5}{5-3}=\frac{2}{2}=1$

The two lines have same slope. Hence, the lines are parallel.

3. $m_1=\frac{-1-4}{5-7}=\frac{-5}{-2}=\frac{5}{2}$

$m_2=\frac{-2-1}{-3-8}=\frac{-3}{-11}=\frac{3}{11}$

The lines are neither parallel nor perpendicular.

5. $x-3y=6$

$-3y=-x+6$

$y=\frac{1}{3}x-2$

$m_1=\frac{1}{3}$

$3x+y=3$

$y=-3x+3$

$m_2=-3$

The product of the two slopes is -1. Hence, lines are perpendicular.

7. $m=\frac{5-3}{4-(-2)}=\frac{2}{6}=\frac{1}{3}$

9. $\frac{y-2}{4-(-1)}=\frac{y-2}{5}$

$m=2$

$\frac{y-2}{5}=2$

$y-2=10$

$y=12$

< Objective 2 >

11. $y=\frac{5}{4}x-5$

13. $y-3=5(x-1)=5x-5$
$y=5x-2$

15. $y+3=-3(x+2)=-3x-6$
$y=-3x-9$

17. $y+3=\frac{2}{5}(x-5)=\frac{2}{5}x-2$
$y=\frac{2}{5}x-5$

19. $x=1$

21. $y-0=-\frac{4}{5}(x-5)$
$y=-\frac{4}{5}x+4$

< Objective 3 >

23. $m=\frac{6-3}{5-2}=1$
$y-3=1(x-2)=x-2$
$y=x+1$

25. $m=\frac{0-(-3)}{2-(-2)}=\frac{3}{4}$
$y+3=\frac{3}{4}(x+2)=\frac{3}{4}x+\frac{3}{2}$
$y=\frac{3}{4}x-\frac{3}{2}$

27. $m=\frac{2-2}{4-(-3)}=0$
$y=2$

29. $m=\frac{-3-0}{0-2}=\frac{3}{2}$
$y=\frac{3}{2}x-3$

31. $m=\frac{-1-4}{-2-0}=\frac{5}{2}$
$y=\frac{5}{2}x+4$

< Objective 4 >

33. $y=4x-2$

35. $m=\frac{2-0}{0-4}=-\frac{1}{2}$
$y=-\frac{1}{2}x+2$

37. $y=4$

39. $y-3=-2(x-2)$
$y=-2x+7$

41. $m=3$
$y=3x+3$

43. $m=-2; m_{\perp}=\frac{1}{2}$
$y=\frac{1}{2}x+4$

45. $m=0$
$y=3$

47. $m=-4$
$y-5=-4(x+4)$
$y=-4x-11$

49. $m=\frac{4}{3}$

$$y-2=\frac{4}{3}(x-3)=\frac{4}{3}x-4$$
$$y=\frac{4}{3}x-2$$

51. $m=-\frac{2}{3};m_{\perp}=\frac{3}{2}$

$$y+1=\frac{3}{2}(x-3)=\frac{3}{2}x-\frac{9}{2}$$
$$y=\frac{3}{2}x-\frac{11}{2}$$

53. $x+2y=4$

$$2y=-x+4$$
$$y=-\frac{1}{2}x+2$$
$$m=-\frac{1}{2}$$
$$y-1=-\frac{1}{2}(x+2)=-\frac{1}{2}x-1$$
$$y=-\frac{1}{2}x$$

55. $(C_1,F_1)=(10,50)$

$$(C_2,F_2)=(40,104)$$
$$m=\frac{104-50}{40-10}=\frac{54}{30}=\frac{9}{5}$$
$$F-50=\frac{9}{5}(C-10)=\frac{9}{5}C-18$$
$$F=\frac{9}{5}C+32$$

57. **(a)** $C=65x+75$

(b) $C=65(3.5)+75=302.5$

The job will cost \$302.50.

(c) $65x+75=247.25$

$$65x=172.25$$
$$x=2.65$$

The job would take 2.65 hr to complete.

59. The slope is 1 and the y-intercept is (0, 3).

61. The slope is 2 and the y-intercept is (0, 1).

63. The slope is –3 and the y-intercept is (0, 1).

65. The slope is –2 and the y-intercept is (0, –3).

67. $m_{\overline{AB}}=\frac{0-0}{2-0}=0$

$$m_{\overline{BC}}=\frac{3-0}{2-2}=\frac{3}{0}\text{ is undefined}$$
$$m_{\overline{CD}}=\frac{3-3}{0-2}=0$$
$$m_{\overline{DA}}=\frac{0-3}{0-0}\text{ is undefined}$$

The opposite sides are parallel and hence the figure is a parallelogram. The adjacent sides are perpendicular and hence the figure is a rectangle.

69. $m_{\overline{AB}}=\frac{0-0}{4-0}=0$

$$m_{\overline{BC}}=\frac{2-0}{5-4}=2$$
$$m_{\overline{CD}}=\frac{2-2}{1-5}=0$$
$$m_{\overline{DA}}=\frac{0-2}{0-1}=2$$

The opposite sides are parallel and hence the figure is a parallelogram. The adjacent sides are not perpendicular and hence the figure is not a rectangle.

71. True

73. always

75. $y = 0.5x + 7$
$y = 0.5x + 3$
$y = 0.5x - 1$
$y = 0.5x - 5$

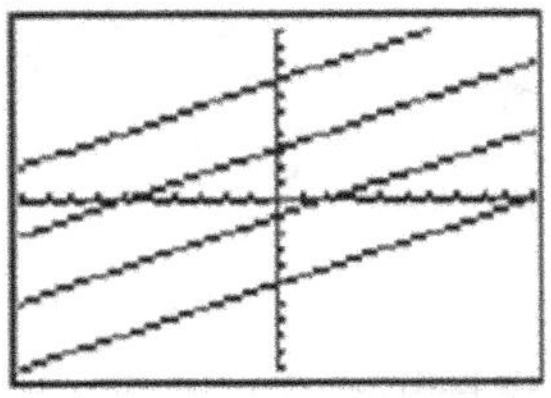

The lines appear to be parallel.

77. **(a)** $m = \dfrac{43,560}{8,000} = 5.445$
To the nearest tenth, the slope is 5.4.
(b) Each additional inch in rainfall yields an additional 5.4 bushels per acre.

79. **(a)** The slope is –4.2.
(b) Each additional year of age reduces the arterial oxygen tension by 0.42 mm Hg.

81. Above and Beyond

Exercises 3.4

1. $y = 3x$
y increases by 3 units when x increases by 1 unit.

3. $y = -x$
y decreases by 1 unit when x increases by 1 unit.

5. $y = \dfrac{3}{2}x - 1$
y increases by 3 units when x increases by 2 units. y increases by $\dfrac{3}{2}$ units when x increases by 1 unit.

7. $y = -0.4x + 11$
y decreases by 0.4 unit when x increases by 1 unit.

< Objective 1 >

9. Let C be the cost of purchasing x pounds of candy.
$C(x) = 1.58x$

11. Let P represent the perimeter of a square with side length s.
$P(s) = 4s$

13. $C(7) = 1.58(7) = \$11.06$
Five pounds of candy will cost \$11.06.

15. $P(18) = 4(18) = 72$ cm
The perimeter of the square with length of a side of 18 cm is 72 cm.

< Objective 3 >

17. We need to find x so that $C(x) = 12$.

$$1.58x = 12$$
$$\frac{1.58x}{1.58} = \frac{12}{1.58}$$
$$x \approx 7.59 \text{ lb}$$

$C(7.59) = 12$
7.59 lb of candy can be purchased for \$12.

19. We need to find s so that $P(s) = 42$.

$$4s = 42$$
$$\frac{4s}{4} = \frac{42}{4}$$
$$s = 10.5 \text{ ft}$$

$P(10.5) = 42$
The length of the square is 10.5 ft if the perimeter is 42 ft.

< Objective 2 >

21. **(a)** Let W represent the weight of the kitten at age t months. The line contains the points (3, 4) and (8, 9).

$$m=\frac{9-4}{8-3}=\frac{5}{5}=1$$

$$y-4=1(x-3)=x-3$$

$$y=x+1$$

$$W(t)=t+1$$

(b) The slope of the function is 1. The kitten gained 1 lb per month.

23. **(a)** Let f represent the size of a song that is t seconds long. The line contains the points (193, 3.0) and (293, 4.2).

$$m=\frac{4.2-3.0}{293-193}=\frac{1.2}{100}=0.012$$

$$y-3.0=0.012(x-193)=0.012x-2.316$$

$$y=0.012x+0.684$$

$$f(t)=0.012t+0.684$$

(b) The slope of the function is 0.012 MB. Each second requires 0.012 MB of space.
(c) A 6:22 song lasts 382 seconds.

$$f(382)=0.012(382)+0.684\approx 5.3 \text{ MB}$$

It takes 5.3 MB to store the song.
(d) We need to find x so that $f(x)=4.6$.

$$0.012x+0.684=4.6$$

$$0.012x=4.6-0.684=3.916$$

$$x=\frac{3.916}{0.012}$$

$$x\approx 326 \text{ s}$$

The song will last 326 s or 5:26.

< Objective 4 >

25. **(a)** $f(x)=-0.036x+50.34$

(b) The slope is −0.36; each additional mi/hr of speed reduces fuel efficiency by 0.36 mi/gal. The y intercept is (0, 50.34); if the car were driven at 0 mi/hr, it would get 50.34 mi/gal.
Note: In this case y-intercept does not apply to any realistic phenomenon.
(c) Use $x=80$.

$$f(80)=-0.36(80)+50.34\approx 21.5$$

If you drove 80 mi/hr, you can expect a fuel efficiency of 21.5 mi/gal.
(d) We need to find x so that $f(x)=25$.

$$-0.36x+50.34=25$$

$$-0.36x=-25.34$$

$$x\approx 70.4$$

For 25 mi/gal, the speed should be 70.4 mi/hr.

27. **(a)** $f(x)=6.8x-49.4$

(b) The slope is 6.8; a person's annual earnings increase $6,800 for each additional year of education. The y-intercept is (0, −49.4); a person with no education is predicted to earn a negative salary.
Note: In this case y-intercept does not apply to any real phenomenon.
(c) Use $x=15$.

$$f(15)=6.8(15)-49.4=52.6$$

A person with 15 years of education would earn $52,600.
(d) We need to find x so that $f(x)=75$.

$$6.8x-49.4=75$$

$$6.8x=124.4$$

$$x\approx 18$$

To earn $75,000, a person needs 18 yr of education.

< Objective 5 >

29. **(a)**

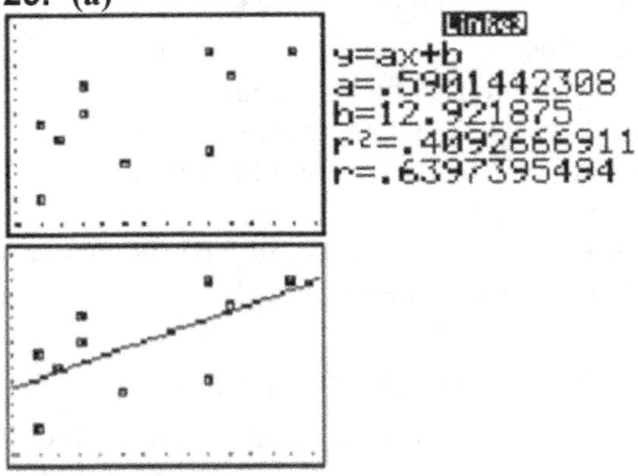

(b) $y = 0.6x + 12.9$

(c) The slope is 0.6. A young girl's weight increases about 0.6 lb for every month she ages.

31. **(a)**

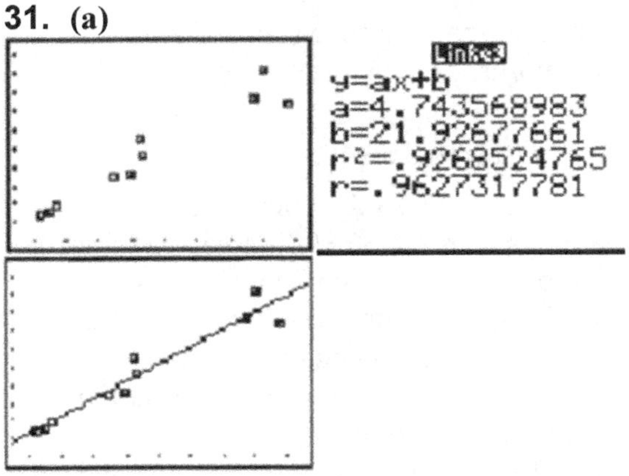

(b) $y = 4.7x + 21.9$

(c) The slope is 4.7. Each additional word leads to approximately 4.7 additional characters in a paragraph.

(d) Use $x = 75$.

$y = 7.7(75) + 21.9 = 374.4$

You would expect 374.4 characters if a paragraph had 75 words.

(e) Substitute $y = 200$ to obtain

$$200 = 4.7x + 21.9$$
$$178.1 = 4.7x$$
$$\frac{178.1}{4.7} = x$$
$$37.9 = x$$

A paragraph with 200 characters will have 38 words.

33. **(a)** The recommended dose *d* mg for an animal weighing *x* kg is given by $d(x) = 4x$.

(b) $d(5) = 4(5) = 20$

A 5-kg cat will require a dose of 20 mg.

(c) $4x = 1{,}450$

$$x = 362.5$$

A 1,450-mg dose of BAL is recommended for an animal weighing 362.5 kg.

35. **(a)** The weight *W* grams of the tumor after *x* days is given by $W(x) = -2.33x + 32$.

(b) $W(5) = -2.33(5) + 32 = 20.35$

After 5 days of treatment, the tumor will weigh 20.35 g.

(c) $-2.33x + 32 = 0$

$$-2.33x = -32$$
$$x \approx 13.7$$

It will take about 14 days for the tumor to be eliminated.

37. **(a)** The input force *F* required to lift an object with weight *x* lb is given by

$F(x) = \frac{1}{2}x + 8$.

(b) $F(300) = \frac{1}{2}(300) + 8 = 158$

The required force to lift a 300-lb object is 158 lb.

(c) $\frac{1}{2}w + 8 = 650$

$$\frac{1}{2}w = 642$$
$$w = 1{,}284$$

A 1,284-lb object can be lifted with an input force of 650 lb.

39. **(a)** The working depth *D* inches of a gear with pitch *p* is given by $D(p) = \frac{2.157}{p}$.

(b) $D(3.5) = \frac{2.157}{3.5} \approx 0.62$

The working depth will be 0.62 inches for a gear with pitch 3.5.

41. $V(0) = 0.28(0) + 2.2 = 2.2$
$V(0°C) = 2.2V$

43. $0.28x + 2.2 = 7.8$
$0.28x = 5.6$
$x = 20$
The temperature will be 20°C if the sensor puts out 7.8 V.

45. $W(7) = 7 + 1 = 8$
The model says that a 7-month-old kitten should weigh 8 lb.

47. Above and Beyond

49. $W(32) = 0.6(40) + 12.9 = 36.9$
The model says that a 40-month-old girl should weigh 36.9 lb.

51. Interpolation: Exercise 48, Extrapolation: Exercises 49-50

Exercises 3.5

< Objective 1 >

1. $x + y < 4$
$y < -x + 4$

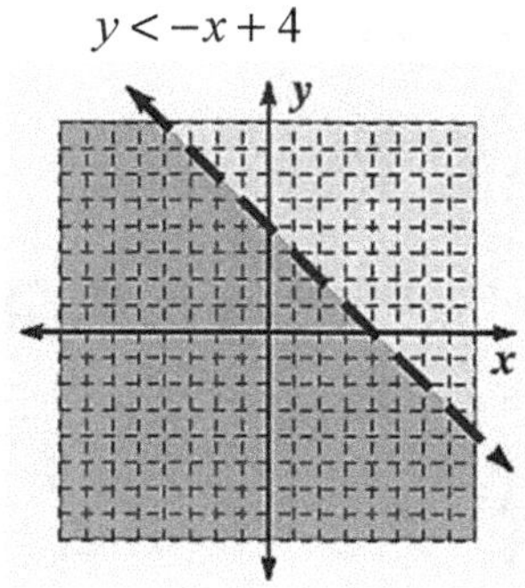

3. $x - y \geq 3$
$-y \geq -x + 3$
$y \leq x - 3$

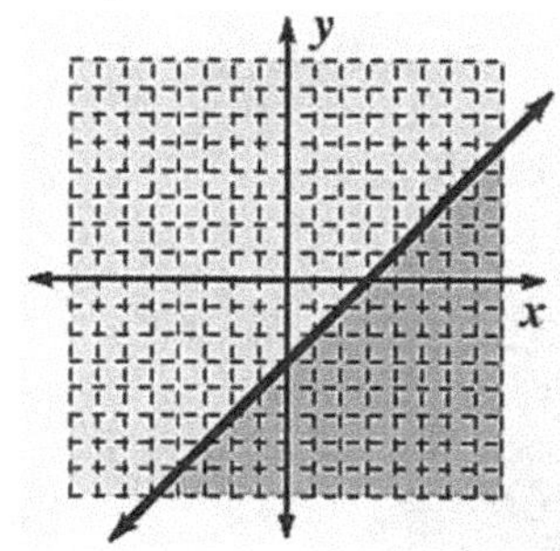

5. $y \geq 2x + 1$

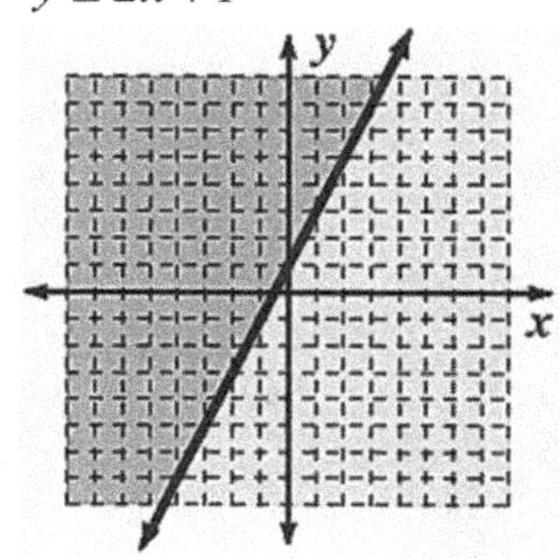

7. $2x + 3y < 6$
$3y < -2x + 6$
$y < -\frac{2}{3}x + 2$

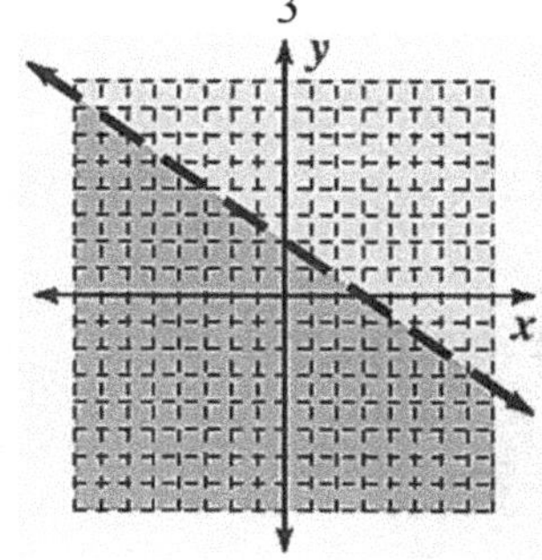

9. $x - 4y > 8$

$-4y > -x + 8$

$y < \dfrac{x}{4} - 2$

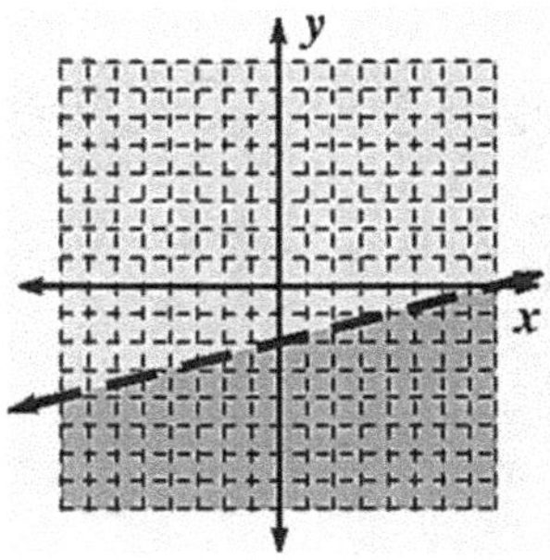

11. $y \geq 3x$

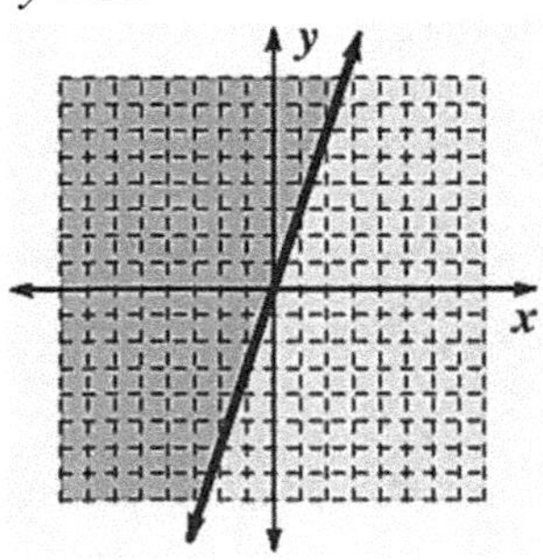

13. $x - 2y > 0$

$-2y > -x$

$y < \dfrac{x}{2}$

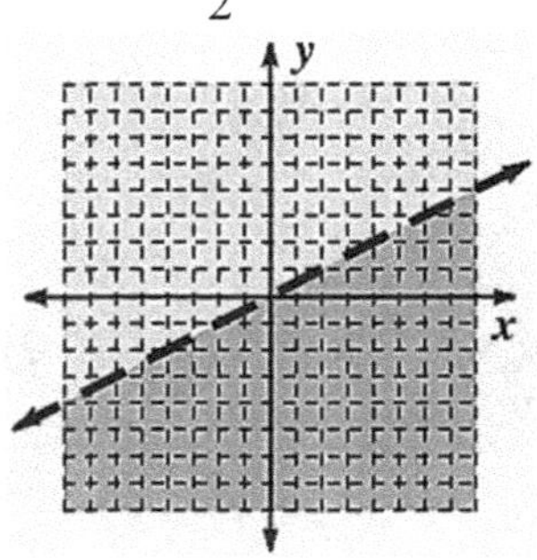

15. $x < 3$

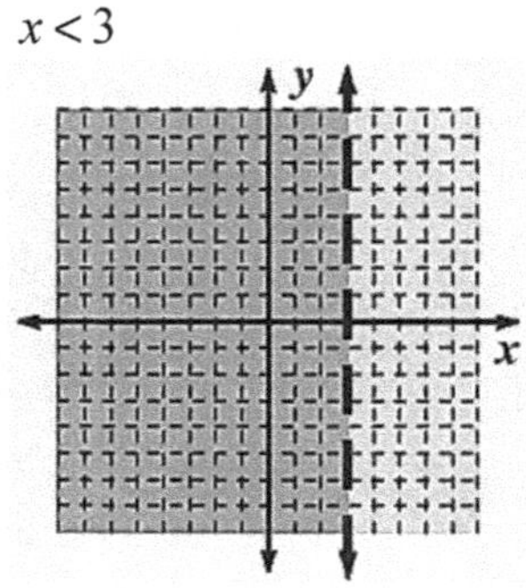

17. $y > 3$

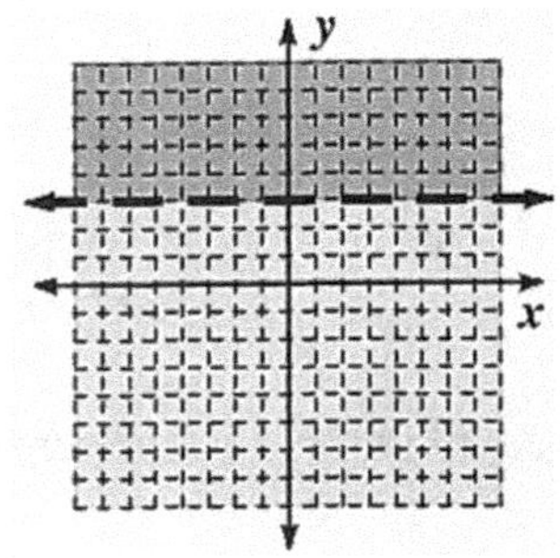

19. $3x - 6 \leq 0$

$3x \leq 6$

$x \leq 2$

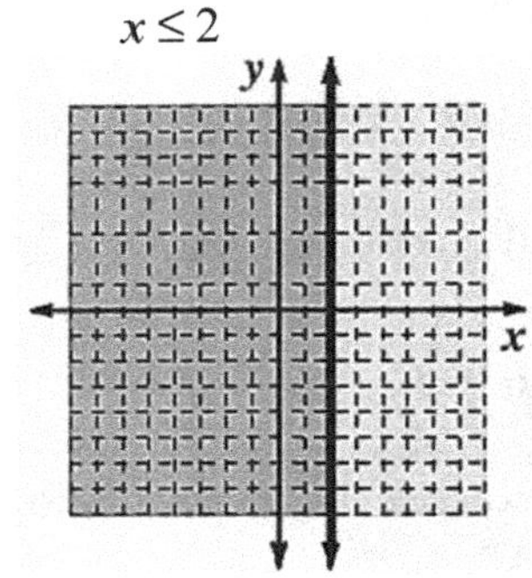

< Objective 2 >

21. $0 < x < 1$

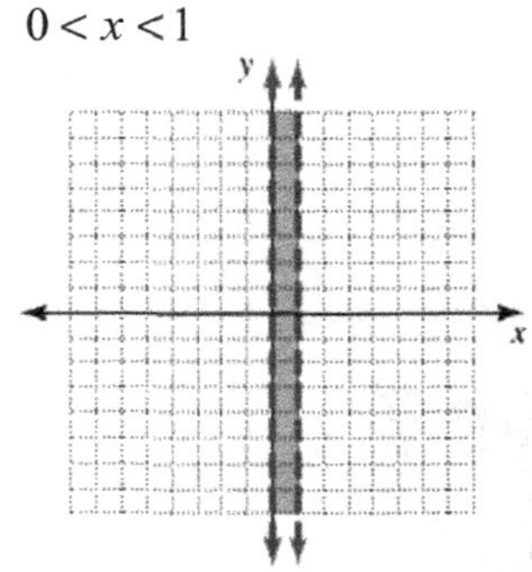

23. $1 \leq x \leq 3$

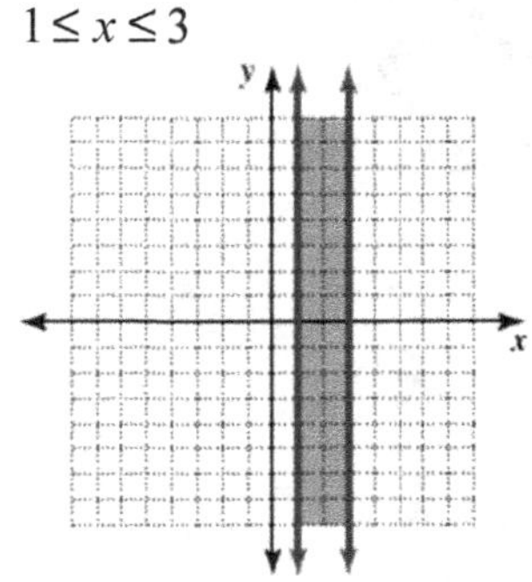

25. $0 \le x \le 3$

$2 \le y \le 4$

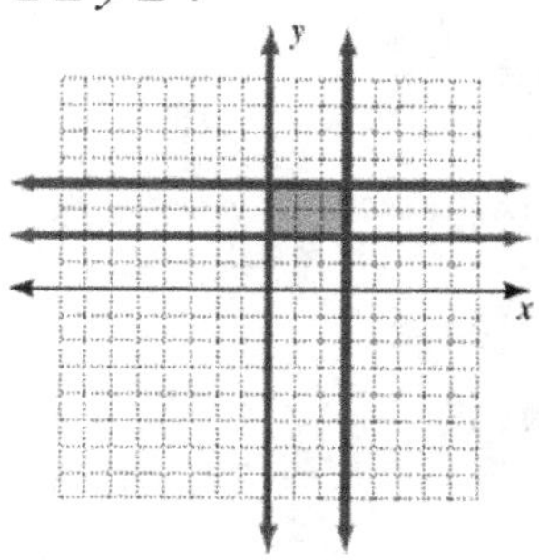

27. $x + 2y \le 4 \rightarrow y \le -\frac{x}{2} + 2$

$x \ge 0$

$y \ge 0$

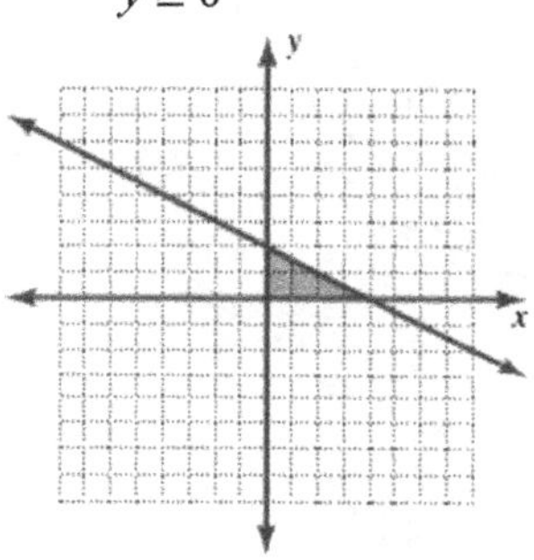

29. $12x + 18y \le 360 \rightarrow y \le -\frac{2}{3}x + 20$

$x \ge 0$

$y \ge 0$

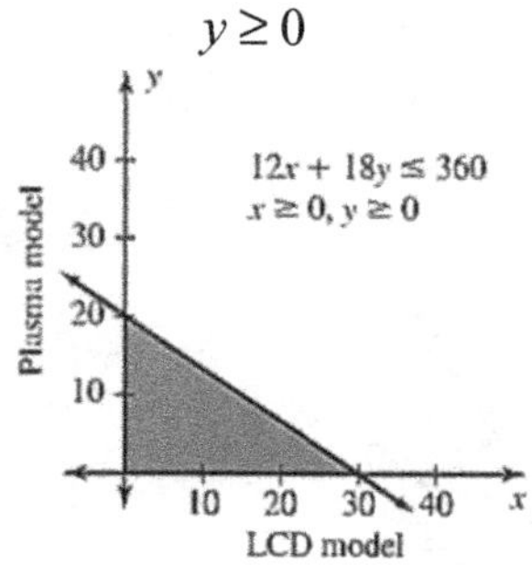

31. $3x + 4y \le 1{,}000 \rightarrow y \le -\frac{3}{4}x + 250$

$x \ge 0$

$y \ge 0$

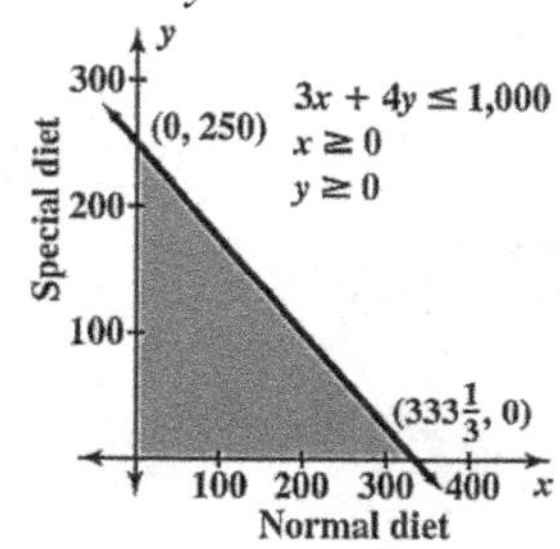

33. $(x_1, y_1) = (0, 4)$

$(x_2, y_2) = (4, 0)$

$$m = \frac{0-4}{4-0} = -1$$

$$y = -x + 4$$

The graphed line is solid and the shading appears above the line.

$y \ge -x + 4$

35. $(x_1, y_1) = (0, -3)$

$(x_2, y_2) = (6, 0)$

$$m = \frac{0-(-3)}{6-0} = \frac{1}{2}$$

$$y = \frac{1}{2}x - 3$$

The graphed line is dashed and the shading appears below the line.

$y < \frac{1}{2}x - 3$

37. True

39. always

41. **(a)** $8x+10y \le 400$

(b)

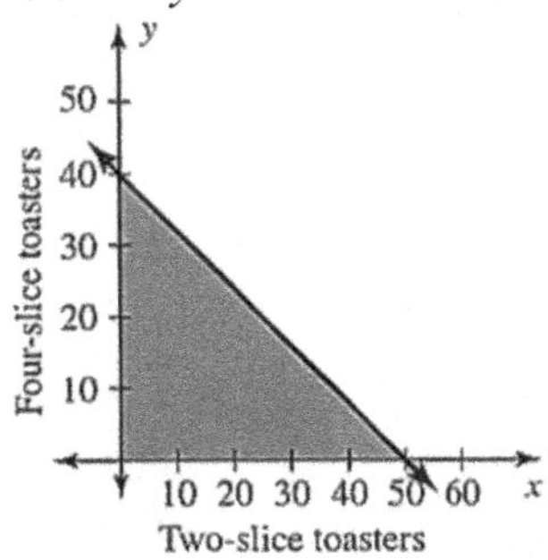

(c) No

43. Above and Beyond

Summary Exercises

1. $x+y=5$

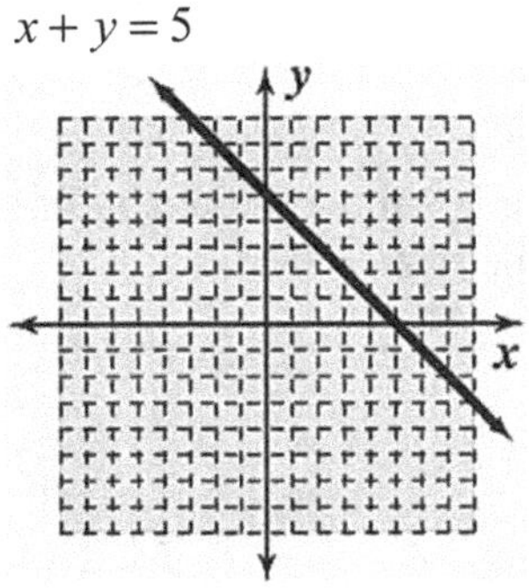

3. $y=5x$

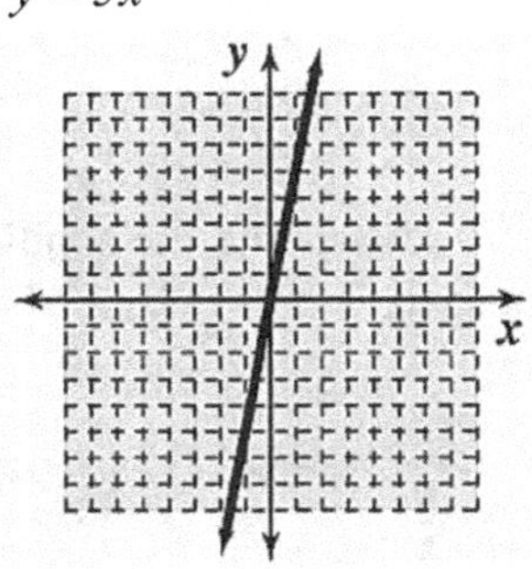

5. $y=\frac{3}{2}x$

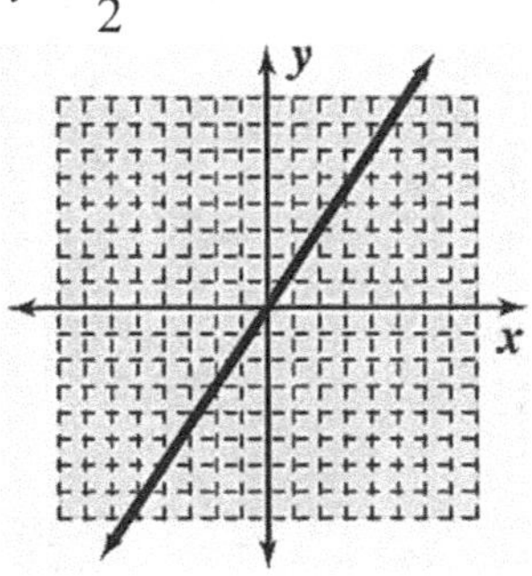

7. $y=-2x+4$

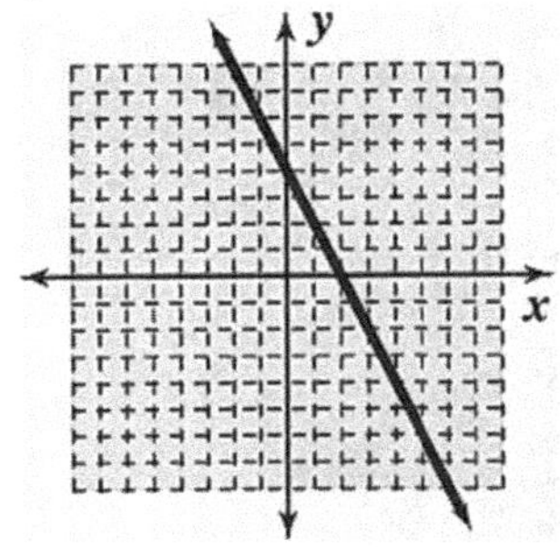

9. $y=\frac{2}{3}x+2$

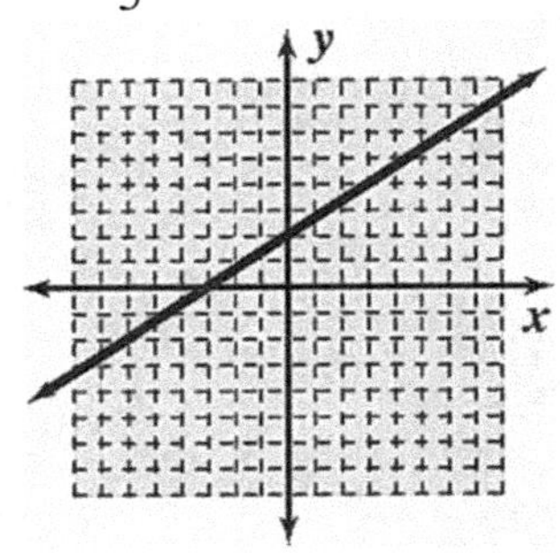

11. $2x+y=6$

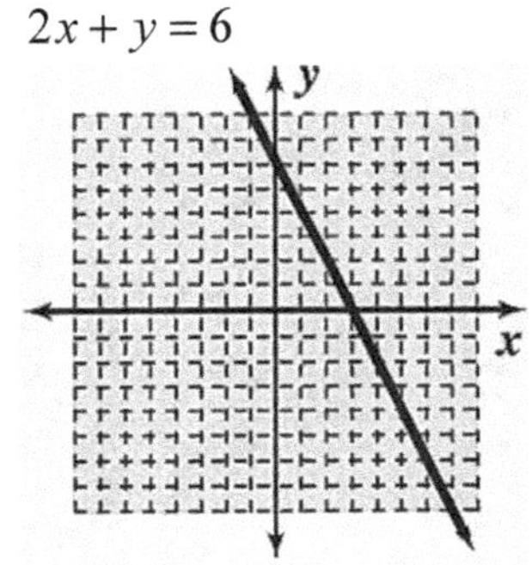

13. $3x-4y=12$

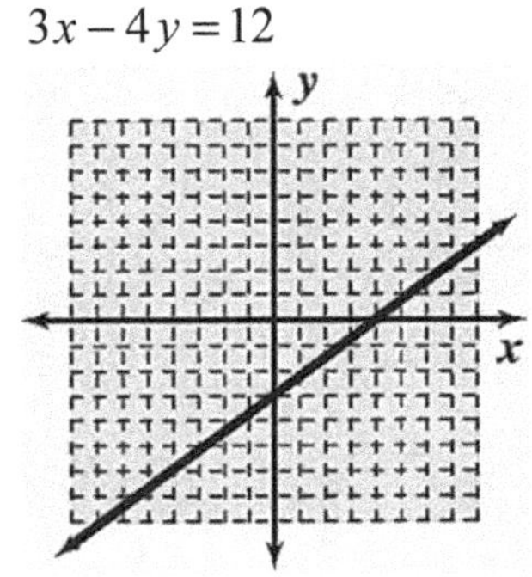

15. $y=-2$

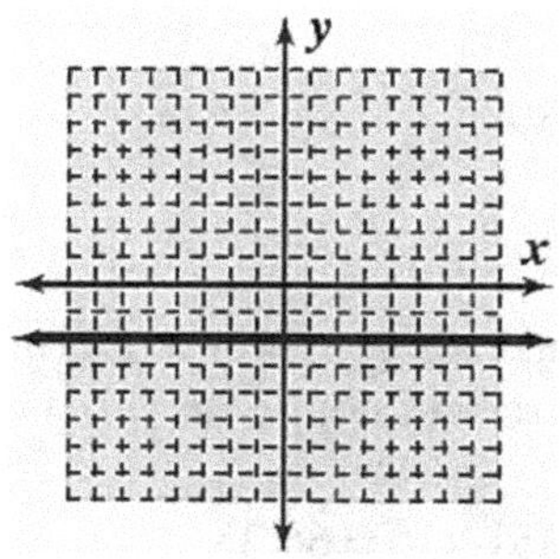

17. $3x+4y=12$

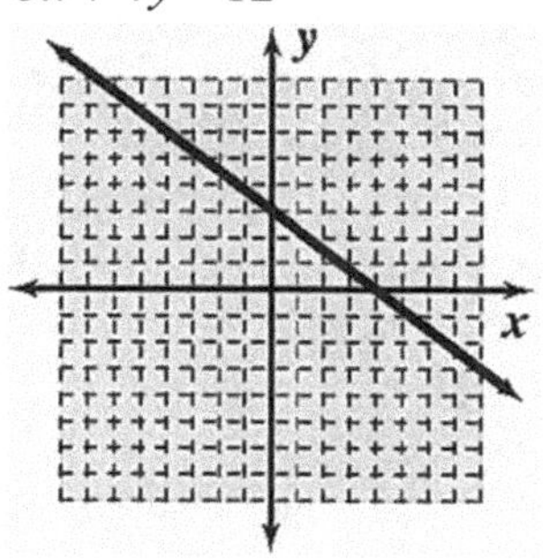

19. $3x+2y=6$

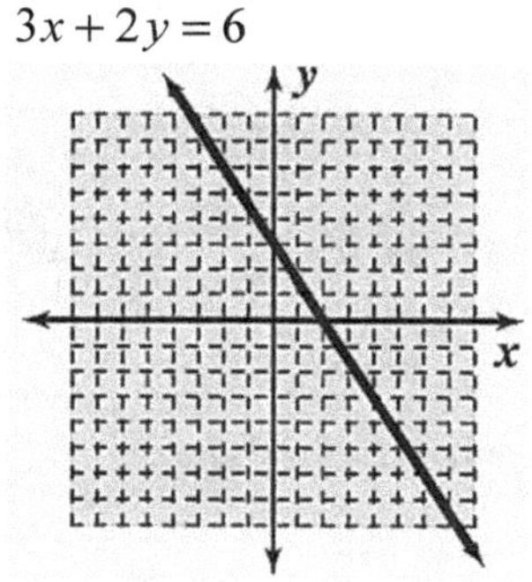

21. $m=\dfrac{8-4}{5-3}=\dfrac{4}{2}=2$

23. $m=\dfrac{3-5}{2-(-2)}=\dfrac{-2}{4}=-\dfrac{1}{2}$

25. $m=\dfrac{6-6}{5-(-2)}=0$

27. $m=\dfrac{-2-(-6)}{5-(-3)}=\dfrac{4}{8}=\dfrac{1}{2}$

29. $y=2x+5$
The slope is 2; the y-intercept is (0, 5).

31. $y=-\dfrac{3}{4}x$

The slope is $-\dfrac{3}{4}$; the y-intercept is (0, 0).

33. $2x+3y=6$

$$3y=-2x+6$$
$$y=-\frac{2}{3}x+2$$

The slope is $-\dfrac{2}{3}$; the y-intercept is (0, 2).

35. $y=-3$
The slope is 0; the y-intercept is (0, –3).

37. $y=2x+3$

39. $y=-\dfrac{2}{3}x+2$

41. $m_1=\dfrac{3-(-2)}{1-(-3)}=\dfrac{5}{4}$

$m_2=\dfrac{8-3}{4-0}=\dfrac{5}{4}$

The lines are parallel.

43. $x+2y=6$

$$2y=-x+6$$
$$y=-\frac{1}{2}x+3$$

$x+3y=9$

$$3y=-x+9$$
$$y=-\frac{1}{3}x+3$$
$$m_1=-\frac{1}{2}; m_2=-\frac{1}{3}$$

The lines are neither parallel nor perpendicular.

45. $y=\dfrac{2}{3}x-5$

47. $y-3=3(x-2)=3x-6$
$y=3x-3$

49. $y+2=\frac{5}{3}(x-3)=\frac{5}{3}x-5$
$y=\frac{5}{3}x-7$

51. $y+4=-\frac{5}{2}(x+2)=-\frac{5}{2}x-5$
$y=-\frac{5}{2}x-9$

53. $y=-5$

55. $m=\frac{3-(-1)}{3-(-3)}=\frac{2}{3}$
$y+1=\frac{2}{3}(x+3)=\frac{2}{3}x+2$
$y=\frac{2}{3}x+1$

57. $y=\frac{3}{4}x+3$

59. $3x-y=6$
$y=3x-6$
$m=3, m_{\parallel}=3$
Thus, $y=3x-6$.

61. $3x-2y=5$
$-2y=-3x+5$
$y=\frac{3}{2}x-\frac{5}{2}$
$m=\frac{3}{2}$
$m_{\perp}=-\frac{2}{3}$
$y+1=-\frac{2}{3}(x-2)=-\frac{2}{3}x+\frac{4}{3}$
$y=-\frac{2}{3}x+\frac{1}{3}$

63. $C(x)=1.75x+30$

65. $150=1.75x+30$
$120=1.75x$
$68.6=x$
The cart can make 68 gyros for \$150 (note that the 69th gyro will not be finished).

67. $P(35)=2.75(35)-30=66.25$
The profit will be \$66.25 for selling 35 gyros in one day.

69. Using the points (63, 2) and (252, 8):
$m=\frac{8-2}{252-63}=\frac{6}{189}=\frac{2}{63}$
$y-2=\frac{2}{63}(x-63)=\frac{2}{63}x-2$
$y=\frac{2}{63}x$
$f(x)=\frac{2}{63}x$

71. Each mile driven requires $\frac{2}{63}\approx 0.03$ gal of gas.

73. Gas is not used when not driving.

75.

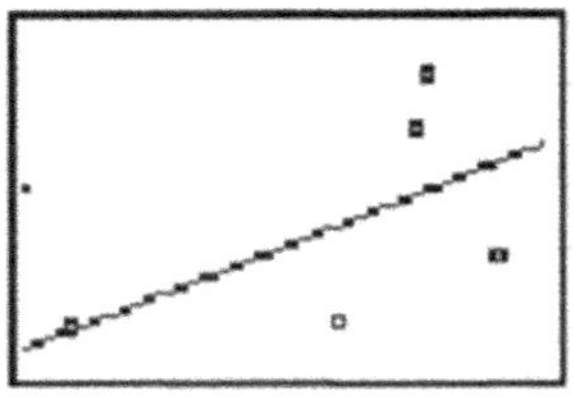

77. The slope of the best-fit line is 5.73.

79. $y=5.73(17)+11.93=109.34$
You would expect 109 books to be acquired if a state's per-student expenditures were \$17.

81. $y < 2x + 1$

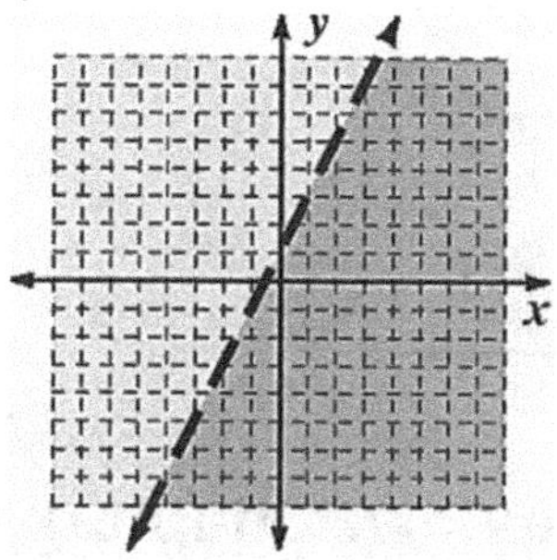

83. $3x + 2y \geq 6 \rightarrow y \geq -\dfrac{3}{2}x + 3$

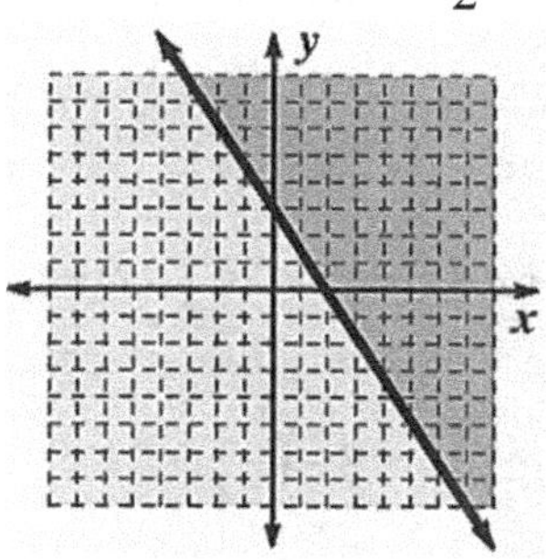

85. $y < -2x$

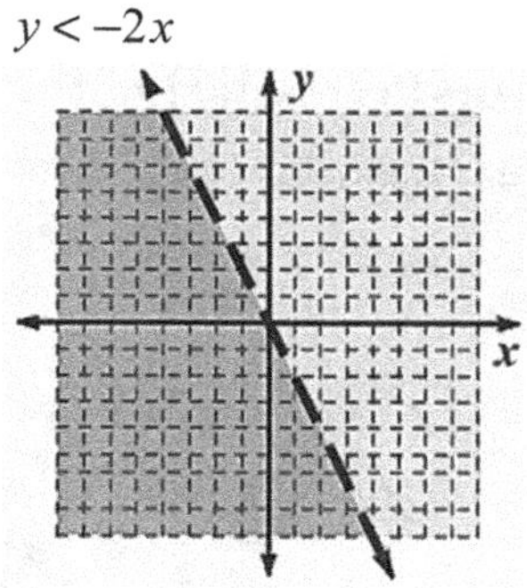

87. $y \geq -3$

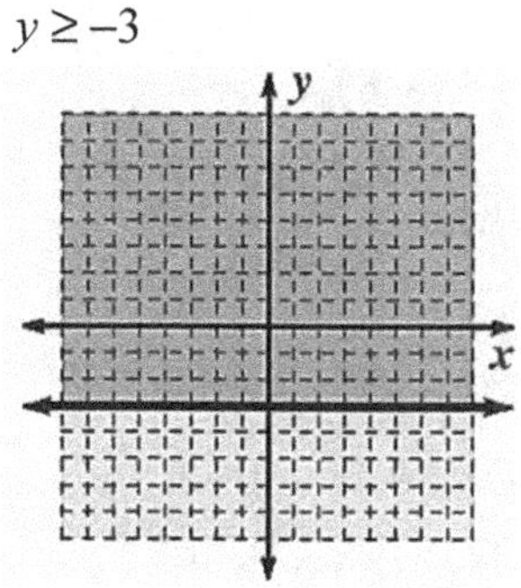

Chapter Test 3

1. $m = \dfrac{10-5}{2-(-3)} = \dfrac{5}{5} = 1$

3. $y = -5x - 9$

The slope is –5; the y-intercept is (0, –9).

5. $y = 5$

The slope is 0; the y-intercept is (0, 5).

7. $y = -3x + 6$

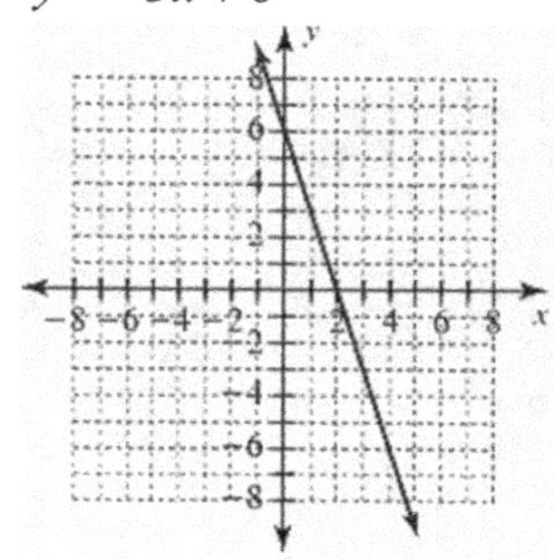

9. $y = 5x - 2$

11. $4x - y = 9$

$$-y = -4x + 9$$
$$y = 4x - 9$$
$$m = 4; m_{\|} = 4$$
$$y = 4x + 3$$

13. $x + y = 4$

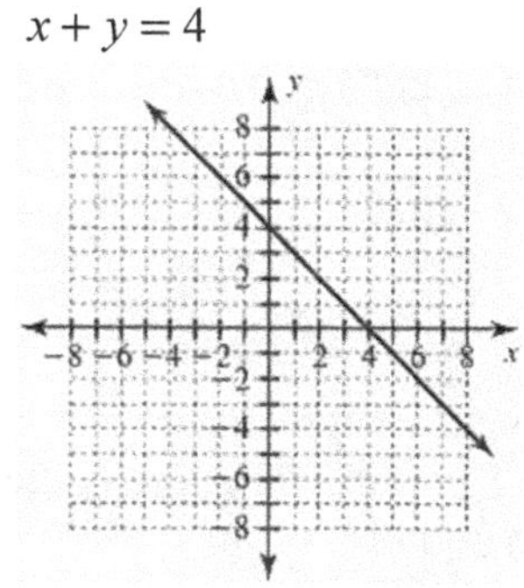

15. $y = \dfrac{3}{4}x - 4$

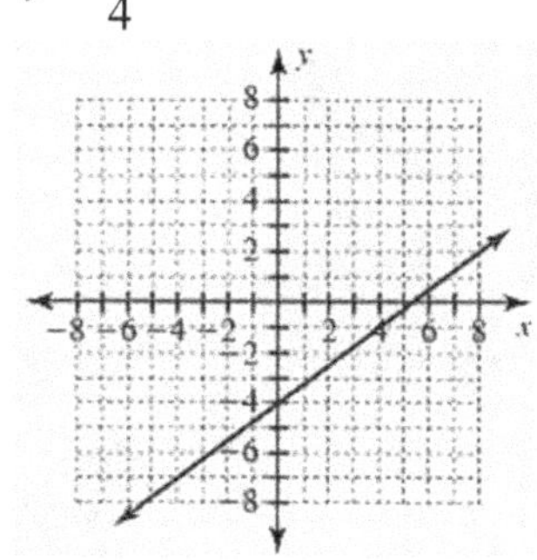

17. $2x+5y=10$

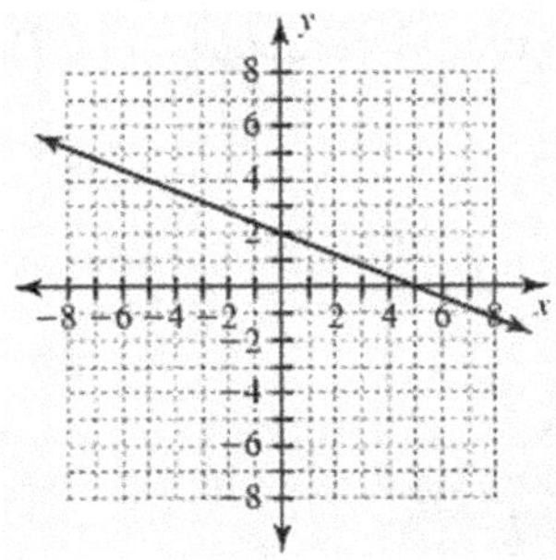

19. $5x+6y\le 30 \rightarrow y\le -\frac{5}{6}x+5$

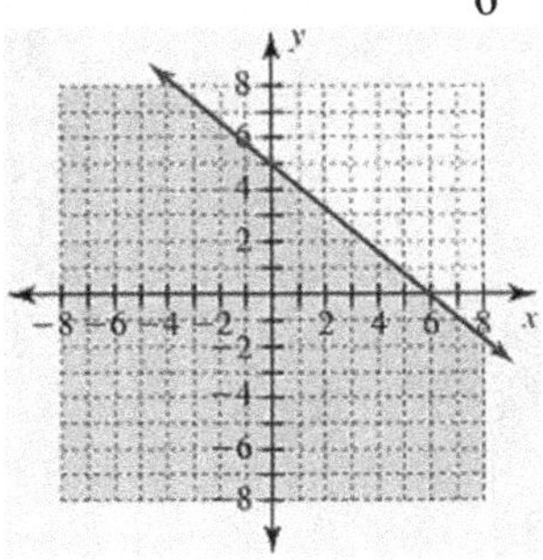

21. $4x-8\le 0 \rightarrow x\le 2$

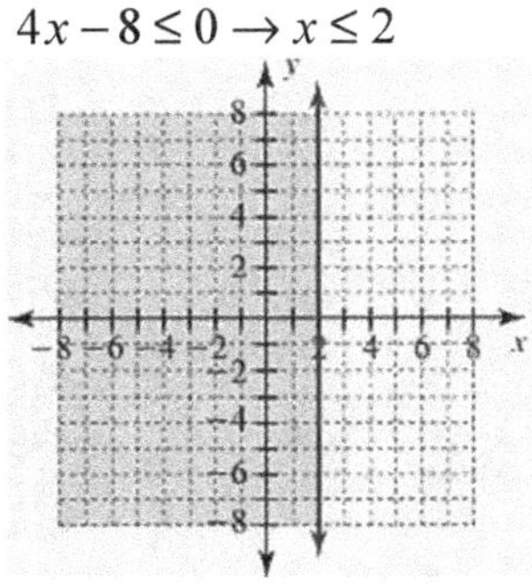

23. Let y represent the time it takes to roast a turkey weighing x pounds. The line contains the points (10, 4) and (18, 6).

$$m=\frac{6-4}{18-10}=\frac{2}{8}=\frac{1}{4}$$

$$y-4=\frac{1}{4}(x-10)=\frac{1}{4}x-\frac{5}{2}$$

$$y=\frac{1}{4}x+\frac{3}{2}$$

25.

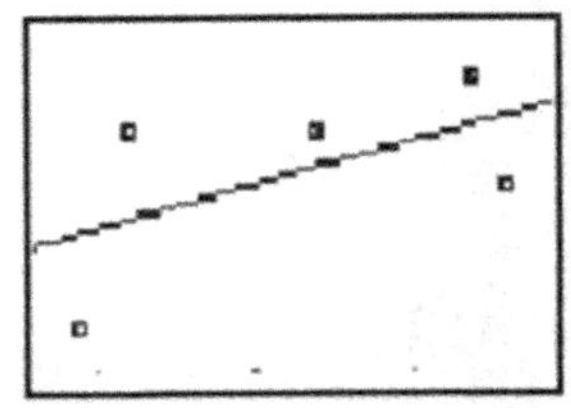

Cumulative Review: Chapters 0–3

1. $$\frac{5}{6}-\left(\frac{2}{3}+\frac{1}{2}\right)=\frac{5}{6}-\left(\frac{4}{6}+\frac{3}{6}\right)=\frac{5}{6}-\left(\frac{7}{6}\right)=-\frac{2}{6}$$
$$=-\frac{1}{3}$$

3. $$\begin{aligned}2^3-|-8|\div(-4)\cdot 2+5&=2^3-8\div(-4)\cdot 2+5\\&=8-8\div(-4)\cdot 2+5\\&=8-(-2)\cdot 2+5\\&=8-(-4)+5=17\end{aligned}$$

5. $$\begin{aligned}-4x^2+3y+2z&=-4(-1)^2+3(3)+2(-2)\\&=-4(1)+9+(-4)\\&=-4+9+(-4)=1\end{aligned}$$

7. $$\begin{aligned}9x-5y-(3x-8y)&=9x-5y-3x+8y\\&=6x+3y\end{aligned}$$

9. $$\begin{aligned}&-4x^2+7x-4-\left(-7x^2+11x\right)\\&\quad-\left(9x^2-5x-6\right)\\&=-4x^2+7x-4+7x^2-11x-9x^2+5x+6\\&=-6x^2+x+2\end{aligned}$$

11. $$\begin{aligned}5x-3(2x-6)+9&=-2(x-5)+6\\5x-6x+18+9&=-2x+10+6\\-x+27&=-2x+16\\x&=-11\end{aligned}$$

13.
$$\frac{x+1}{3}-\frac{2x+3}{4}=\frac{1}{6}$$
$$12\left(\frac{x+1}{3}-\frac{2x+3}{4}\right)=12\bullet\frac{1}{6}$$
$$4(x+1)-3(2x+3)=2$$
$$4x+4-6x-9=2$$
$$-2x-5=2$$
$$-2x=7$$
$$x=-\frac{7}{2}$$
The solution is $-\frac{7}{2}$.

15. $4x-7<9$
$$4x<16$$
$$x<4$$
$$\{x|x<4\}$$

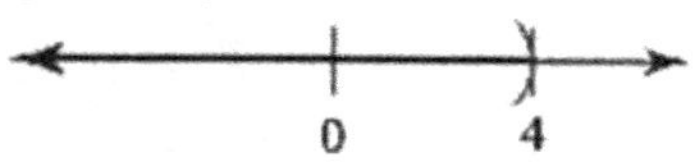

17. $4\le 2x-6\le 12$
$$10\le 2x\le 18$$
$$5\le x\le 9$$
$$\{x|5\le x\le 9\}$$

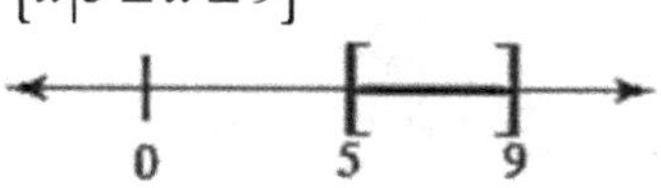

19.
$$F=\frac{9}{5}C+32$$
$$F-32=\frac{9}{5}C$$
$$\frac{5}{9}(F-32)=C$$
$$C=\frac{5}{9}(F-32)$$

21. $m=\frac{12-(-4)}{-2-6}=\frac{16}{-8}=-2$

23. $y=-4x+9$
The slope is –4; the y-intercept is (0, 9).

25. $y=9$
The slope is 0; the y-intercept is (0, 9).

27. $y=5x-6$

29. $2x+3y=6$
$$3y=-2x+6$$
$$y=-\frac{2}{3}x+2$$
$$m=-\frac{2}{3}$$
$$y=-\frac{2}{3}x+6$$

31. $m=\frac{-3-0}{0-2}=\frac{3}{2}$
$$y=\frac{3}{2}x-3$$

33. Let x be the number.
$$\frac{1}{3}x+3x=30$$
$$\frac{10}{3}x=30$$
$$x=9$$
The number is 9.

35. Let x be the cost of a first class ticket. Then $\frac{1}{2}x$ is the cost of a coach ticket.
$$x+\frac{1}{2}x=1350$$
$$\frac{3}{2}x=1350$$
$$x=900$$
$$\frac{1}{2}x=450$$
The first class ticket costs \$900 and the coach ticket costs \$450.

37. $2x+3y<6$

$$3y<-2x+6$$

$$y<-\frac{2}{3}x+2$$

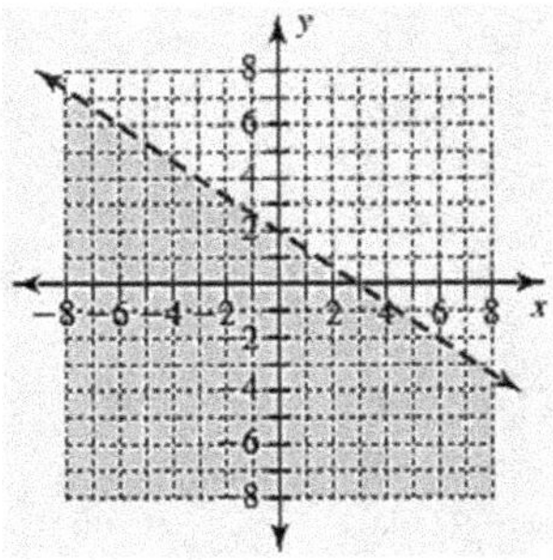

39. $f(12)=3.95(12)+30.77=78.17$

It will take \$78.17 to ship a 12-lb package.

Chapter 4
Systems of Linear Equations

Exercises 4.1

< Objective 1 >

1. $x + y = 6$

$x - y = 4$

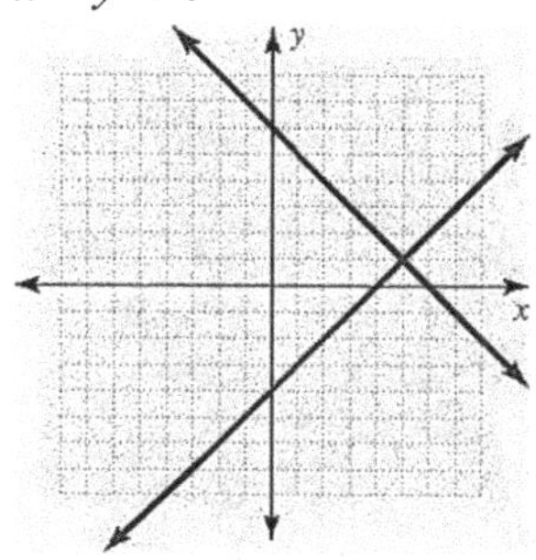

The lines intersect at (5, 1), hence the solution is $\{(5, 1)\}$.

3. $x + y = 5$

$-x + y = 7$

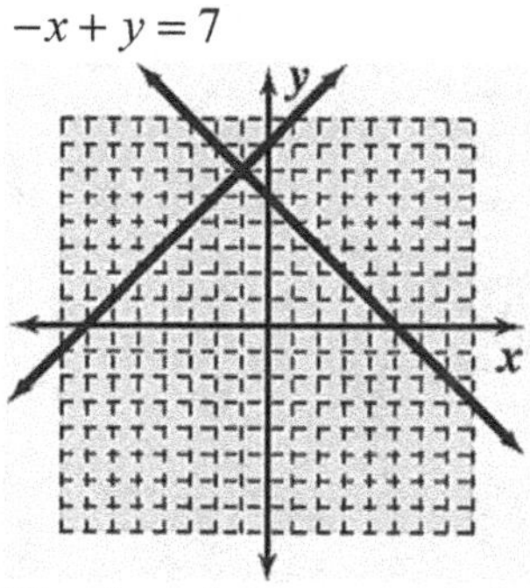

The lines intersect at (−1, 6), hence the solution is $\{(-1, 6)\}$.

5. $x + 2y = 4$

$x - y = 1$

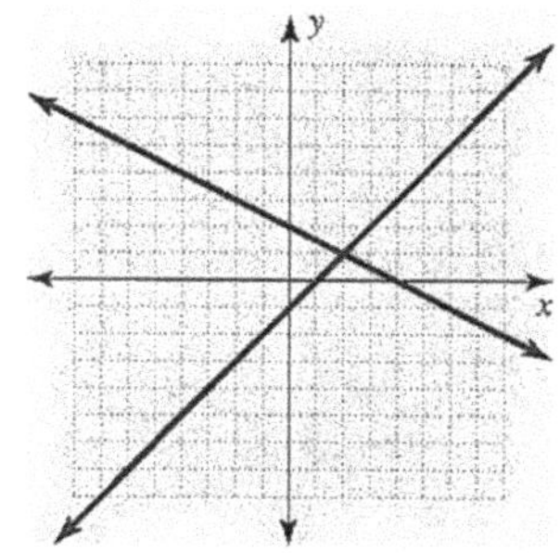

The lines intersect at (2, 1), hence the solution is $\{(2, 1)\}$.

7. $3x - y = 21$

$3x + y = 15$

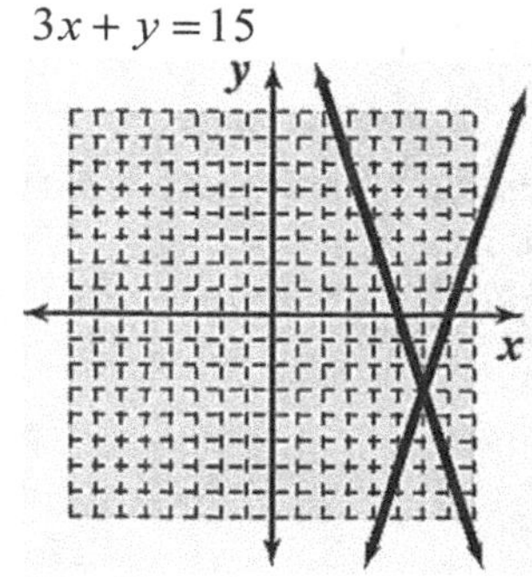

The lines intersect at (6, −3), hence the solution is $\{(6, -3)\}$.

9. $x + 3y = 12$

$2x - 3y = 6$

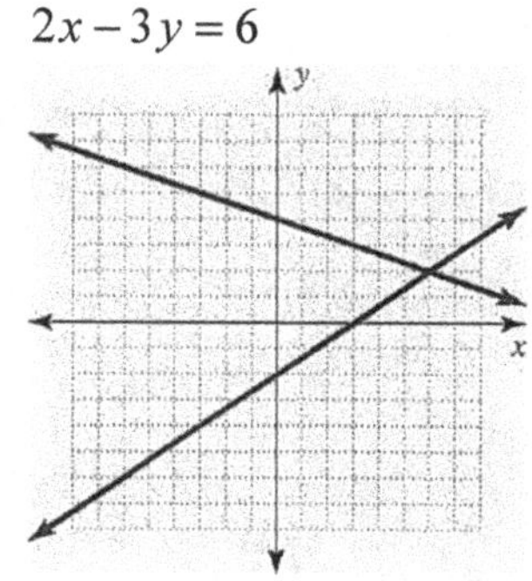

The lines intersect at (6, 2), hence the solution is $\{(6, 2)\}$.

11. $3x + 2y = 12$

$y = 3$

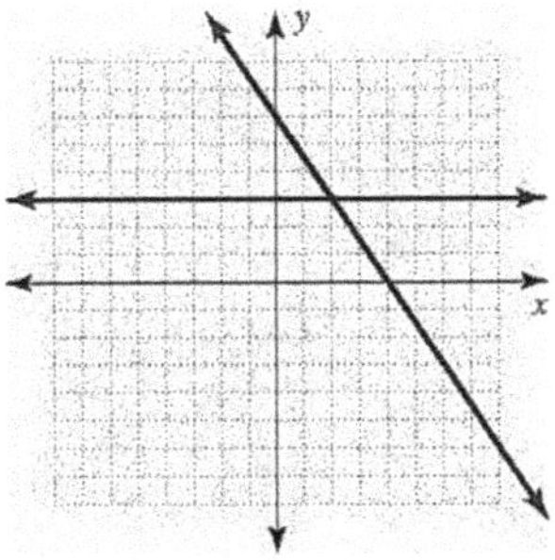

The lines intersect at (2, 3), hence the solution is $\{(2, 3)\}$.

13. $x - y = 4$

$2x - 2y = 8$

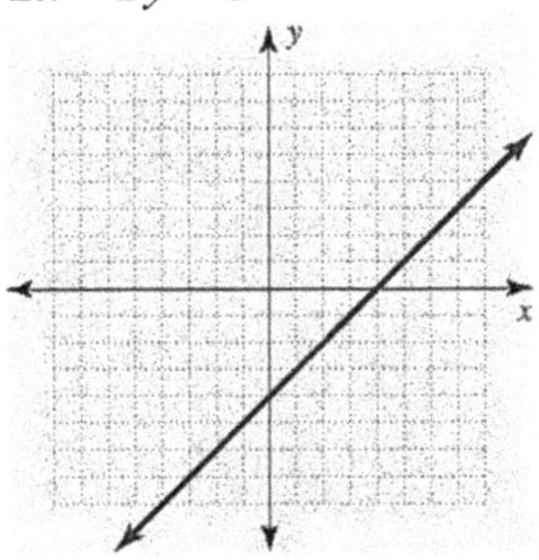

The system is dependent and there is infinite number of solutions.

15. $x - 4y = -4$

$x + 2y = 8$

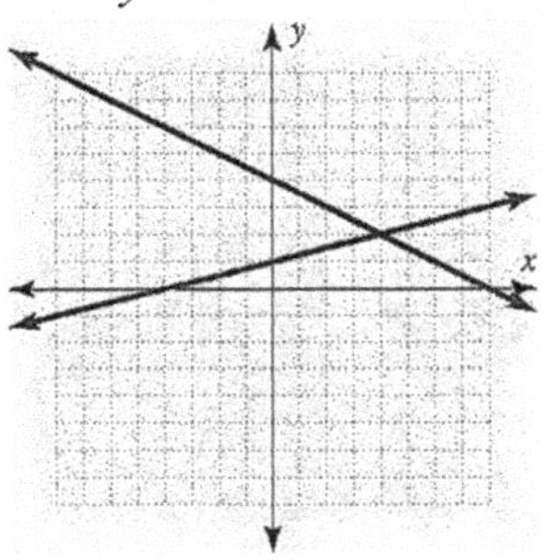

The lines intersect at (4, 2), hence the solution is $\{(4, 2)\}$.

17. $3x - 2y = 6$

$2x - y = 5$

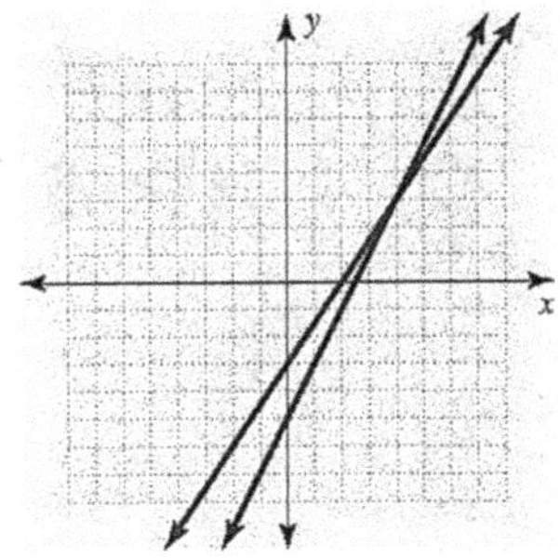

The lines intersect at (4, 3), hence the solution is $\{(4, 3)\}$.

19. $3x - y = 3$

$3x - y = 6$

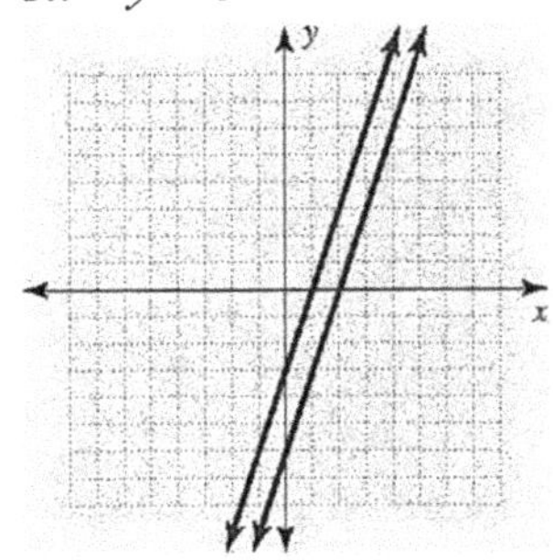

The system is inconsistent, hence there are no solutions.

21. $2y = 3$

$x - 2y = -3$

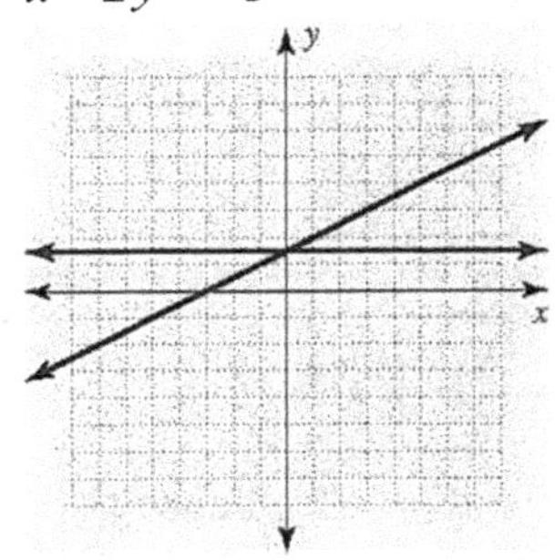

The lines intersect at $\left(0, \frac{3}{2}\right)$, hence the solution is $\left\{\left(0, \frac{3}{2}\right)\right\}$.

23. $x=-5$

$y=3$

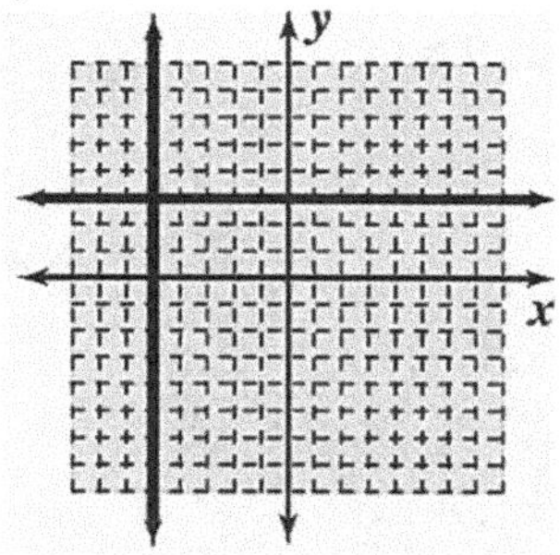

The lines intersect at $(-5, 3)$, hence the solution is $\{(-5, 3)\}$.

< Objective 2 >

25. $y=3x+7$; $m=3$

$y=7x-2$; $m=7$

The system is consistent.

27. $y=7x-1$; $m=7$

$y=7x+8$; $m=7$

The system is inconsistent.

29. $3x+4y=12$; $m=-\dfrac{3}{4}$

$9x-5y=10$; $m=\dfrac{9}{5}$

The system is consistent.

31. $7x-2y=5$; $m=\dfrac{7}{2}$, $b=-\dfrac{5}{2}$

$14x-4y=10$; $m=\dfrac{7}{2}$, $b=-\dfrac{5}{2}$

The system is dependent.

33. $A=15x+5{,}000$

35.

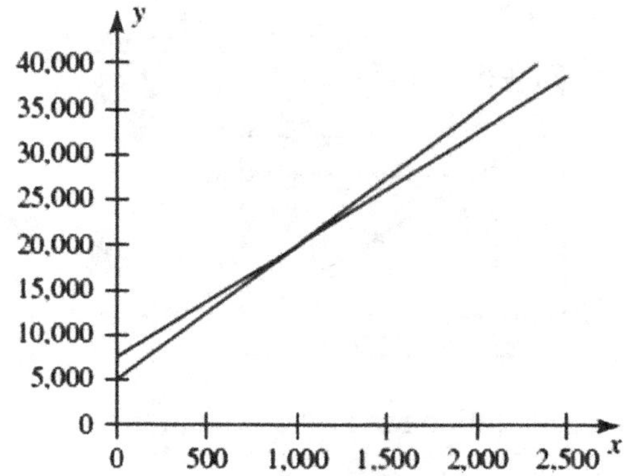

37. To build an 1800 ft^2 shop

$A=15(1800)+5{,}000=32{,}000$

$B=12.5(1800)+7{,}500=30{,}000$

Company B offers the lowest price at \$30,000.

39. sometimes

41. never

43. $88x+57y=1909$

$95x+48y=1674$

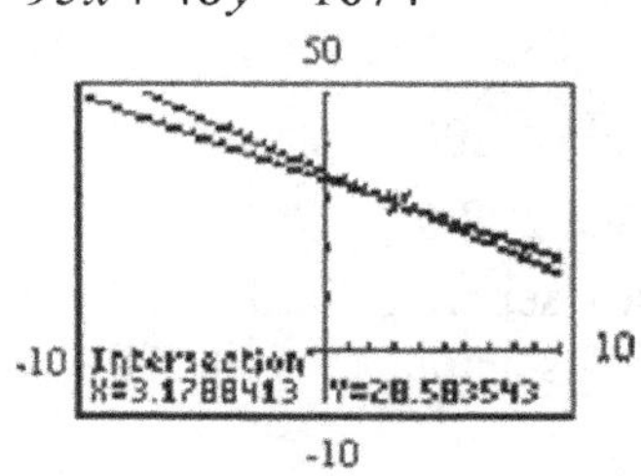

The solution set of the system is $\{(3.18, 28.58)\}$.

45. $25x-65y=5312$

$-21x+32y=1256$

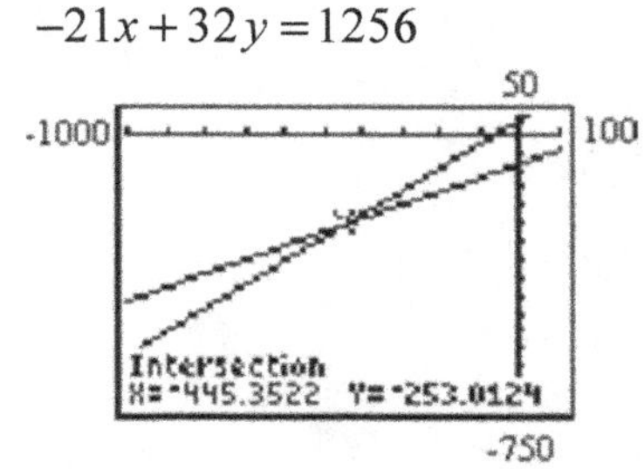

The solution set of the system is $\{(-445.35, -253.01)\}$.

47. $15x+20y=79$

$7x+5y=115$

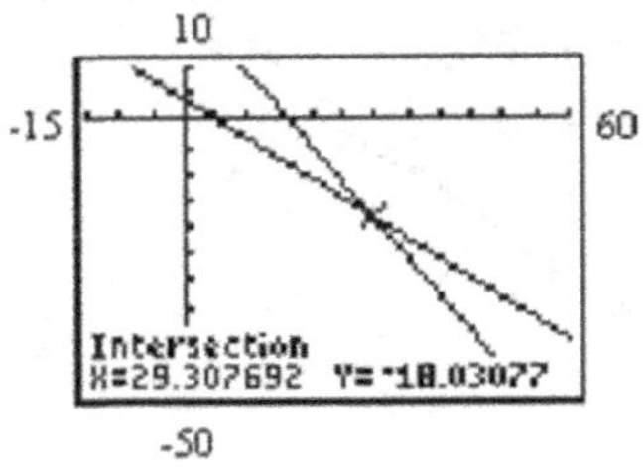

The solution set of the system is $\{(29.31,-18.03)\}$.

49. $x+y=15$

$80x=120y$

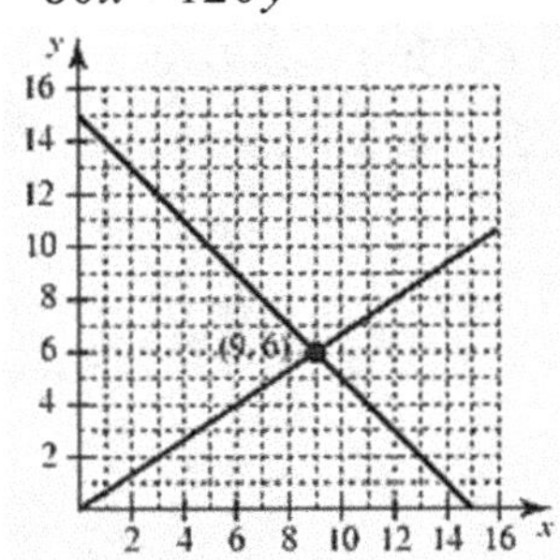

The lines intersect at $\{(9, 6)\}$.
$x=9$ ft, $y=6$ ft

51. $0.78x+.050y=4,200$

$0.22x+0.50y=2,800$

78%/22% alloy separates into 2,500 grams.
50%/50% alloy separates into 4,500 grams.

53. $2a+2b=30$

$\frac{24}{12}a+\frac{6}{12}b=12$

$a=3$; $b=2$

55.

$$ax+3y=8$$
$$a(1)+3(2)=8$$
$$a=2$$
$$3x+4y=b$$
$$-3(1)+4(2)=b$$
$$5=b$$

$a=2$; $b=5$

57. Above and Beyond

59. Above and Beyond

Exercises 4.2

< Objectives 1 and 2 >

1. $2x-8=0$

$f(x)=2x-8$

$g(x)=0$

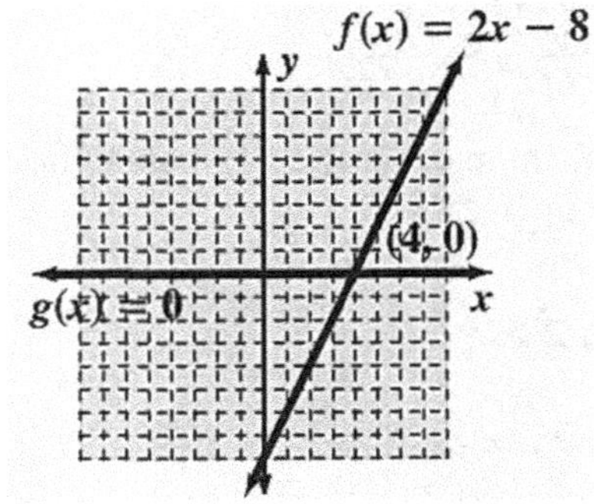

The point of intersection is (4, 0). The solution set is $\{4\}$.

3. $7x-7=0$

$f(x)=7x-7$

$g(x)=0$

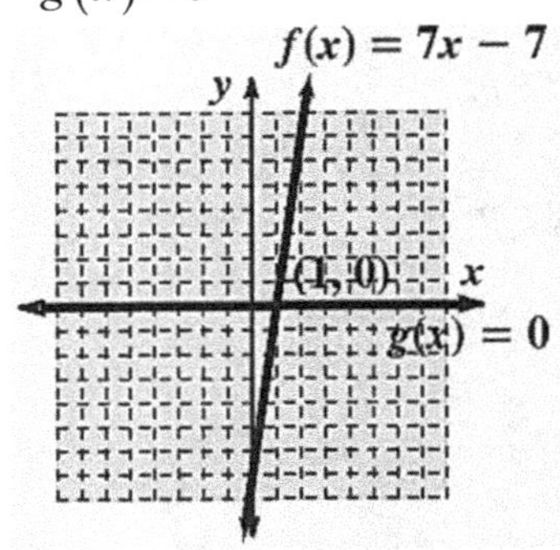

The point of intersection is (1, 0). The solution set is $\{1\}$.

5. $5x-8=2$

$f(x)=5x-8$

$g(x)=2$

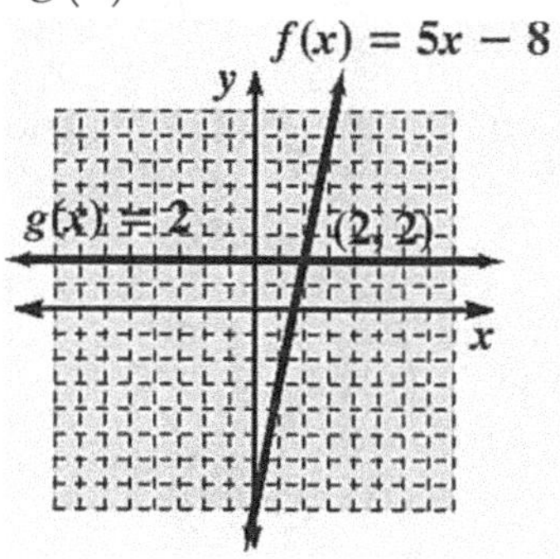

The point of intersection is (2, 2). The solution set is {2}.

7. $2x-3=7$

$f(x)=2x-3$

$g(x)=7$

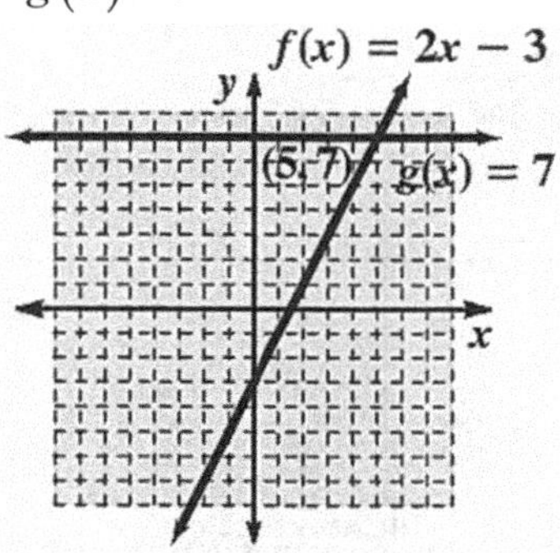

The point of intersection is (5, 7). The solution set is {5}.

9. $4x-2=3x+1$

$f(x)=4x-2$

$g(x)=3x+1$

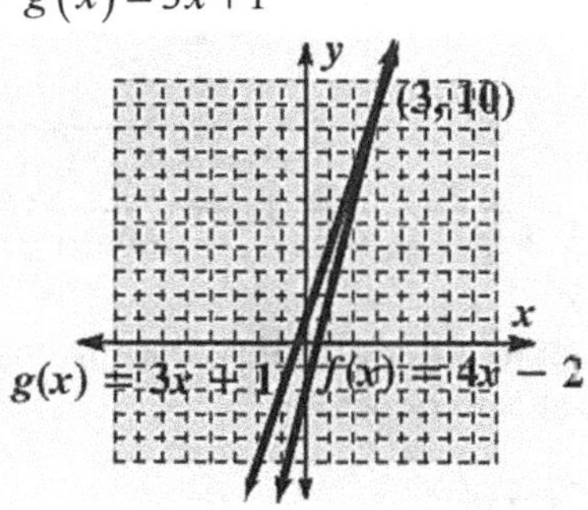

The point of intersection is (3, 10). The solution set is {3}.

11. $\frac{7}{5}x-3=\frac{3}{10}x+\frac{5}{2}$

$f(x)=\frac{7}{5}x-3$

$g(x)=\frac{3}{10}x+\frac{5}{2}$

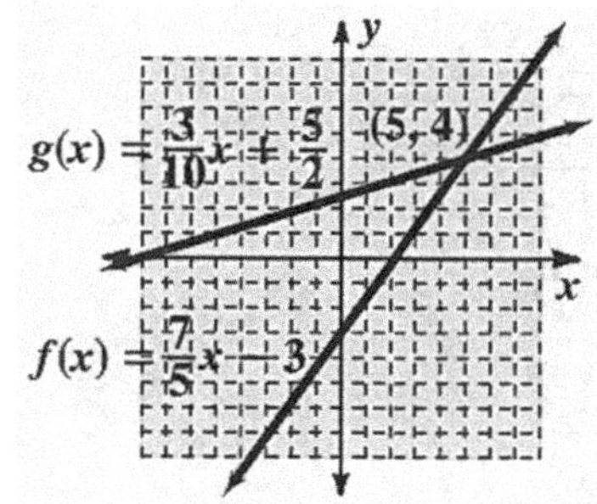

The point of intersection is (5, 4). The solution set is {5}.

13. $3(x-1)=4x-5$

$f(x)=3(x-1)=3x-3$

$g(x)=4x-5$

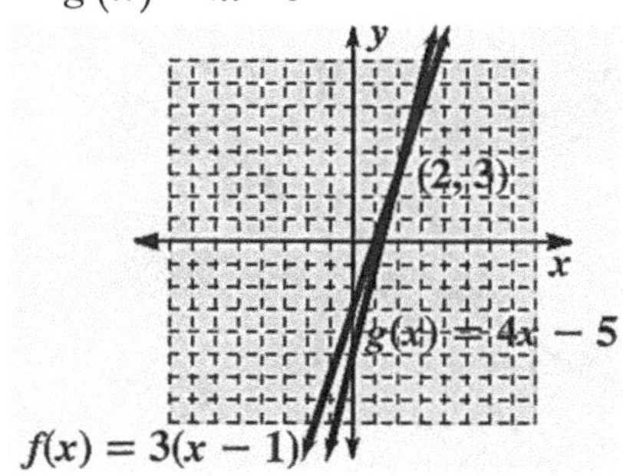

The point of intersection is (2, 3). The solution set is {2}.

15. $7\left(\frac{1}{5}x-\frac{1}{7}\right)=x+1$

$$f(x)=7\left(\frac{1}{5}x-\frac{1}{7}\right)=\frac{7}{5}x-1$$

$$g(x)=x+1$$

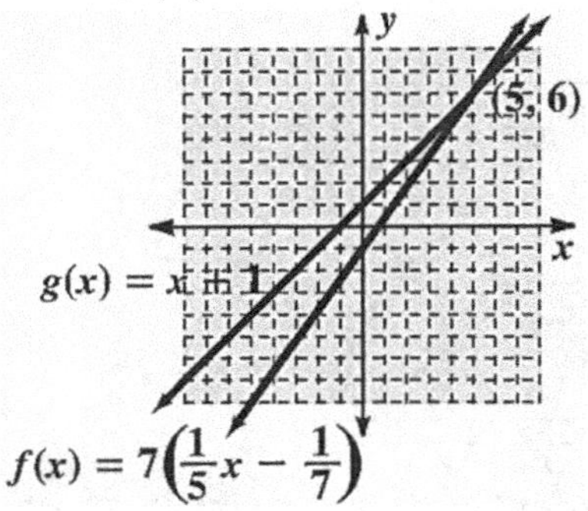

The point of intersection is (5, 6). The solution set is {5}.

17. $C(x)=4.50x+2,400$

$R(x)=7.50x$

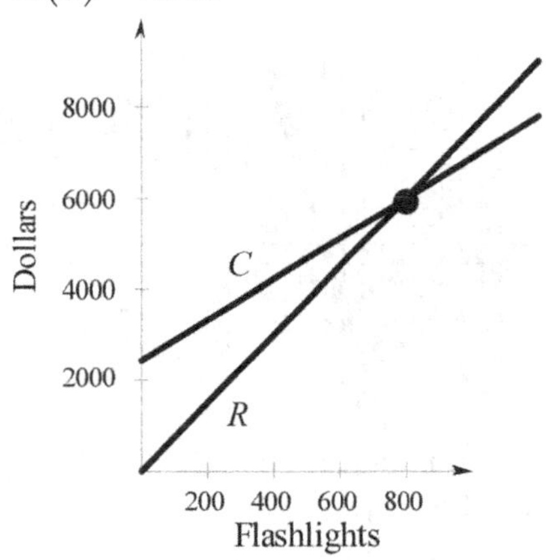

The break-even point is 800 flashlights.

19. True

21. never

23. $4.17(x+3.56)=2.89(x+0.35)$

$$f(x)=4.17(x+3.56)$$

$$g(x)=2.89(x+0.35)$$

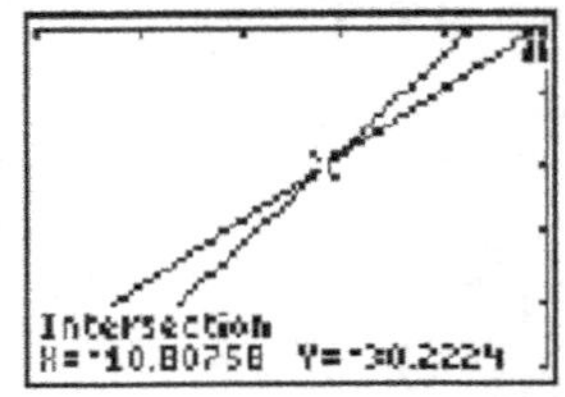

The solution set is {–10.81}.

25. $3.61(x+4.13)=2.31(x-2.59)$

$$f(x)=3.61(x+4.13)$$

$$g(x)=2.31(x-2.59)$$

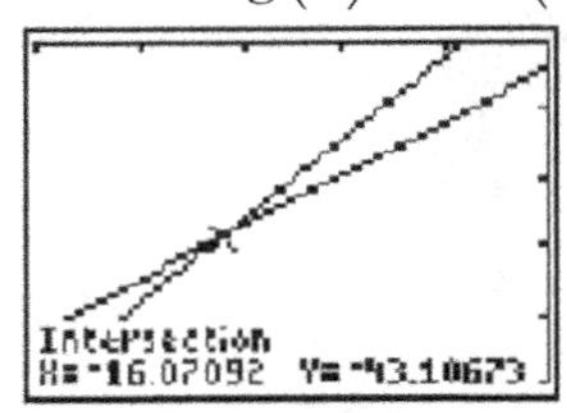

The solution set is {–16.07}

27. $V=0.28C+2.2$

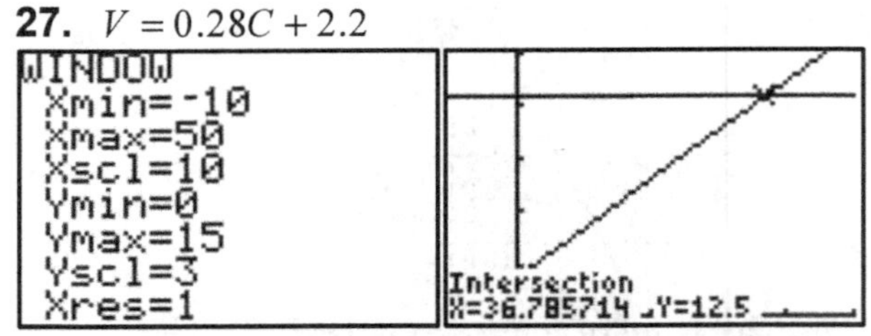

If the sensor outputs 12.5 V, the temperature, to the nearest tenth, is 36.8°C.

29. $d=0.125w$

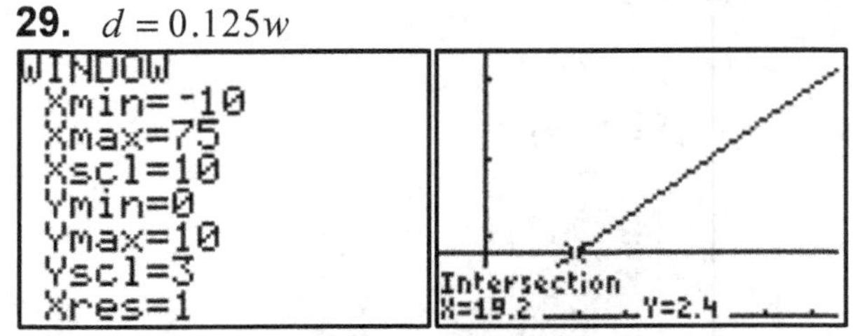

A 19.2-kg fawn requires a 2.4-mg dose.

31. For a given number of miles, the lower graph gives the cheaper cost.

33. Above and Beyond

Exercises 4.3

< Objective 1 >

1. $2x - y = 1$

$-2x + 3y = 5$

$2y = 6$

$y = 3$

$2x - (3) = 1$

$x = 2$

The solution is $\{(2, 3)\}$.

3. $4x + 3y = -5$

$5x + 3y = -8$

Multiply the first equation by –1.

$-4x - 3y = 5$

$5x + 3y = -8$

$x = -3$

$4(-3) + 3y = -5$

$3y = 7$

$y = \frac{7}{3}$

The solution is $\left\{\left(-3, \frac{7}{3}\right)\right\}$.

5. $x + y = 3$

$3x - 2y = 4$

Multiply the first equation by 2.

$2x + 2y = 6$

$3x - 2y = 4$

$5x = 10$

$x = 2$

$(2) + y = 3$

$y = 1$

The solution is $\{(2, 1)\}$.

7. $2x + y = 8$

$-4x - 2y = -16$

Multiply the first equation by 2.

$4x + 2y = 16$

$-4x - 2y = -16$

$0 = 0$

The system is dependent and there is infinite number of solution.

9. $5x - 2y = 31$

$4x + 3y = 11$

Multiply the first equation by 3 and the second by 2.

$15x - 6y = 93$

$8x + 6y = 22$

$23x = 115$

$x = 5$

$5(5) - 2y = 31$

$-2y = 6$

$y = -3$

The solution is $\{(5, -3)\}$.

11. $3x - 2y = 7$

$-6x + 4y = -15$

Multiply the first equation by 2.

$6x - 4y = 14$

$-6x + 4y = -15$

$0 = -1$

The system is inconsistent and there are no solutions.

13. $-2x + 7y = 2$

$3x - 5y = -14$

Multiply the first equation by 3 and the second by 2.

$-6x + 21y = 6$

$6x - 10y = -28$

$11y = -22$

$y = -2$

$-2x + 7(-2) = 2$

$-2x = 16$

$x = -8$

The solution is $\{(-8, -2)\}$.

< Objective 2 >

15. $x-y=7$

$y=2x-12$

Substitute into the first equation.

$x-(2x-12)=7$

$-x+12=7$

$x=5$

$y=2(5)-12=-2$

The solution is $\{(5,-2)\}$.

17. $3x+2y=-18$

$x=3y+5$

Substitute into the first equation.

$3(3y+5)+2y=-18$

$11y+15=-18$

$11y=-33$

$y=-3$

$x=3(-3)+5=-4$

The solution is $\{(-4,-3)\}$.

19. $10x-2y=4$

$y=5x-2$

Substitute into the first equation.

$10x-2(5x-2)=4$

$4=4$

The system is dependent and there is infinite number of solutions.

21. $3x+4y=9$

$y=3x+1$

$3x+4(3x+1)=9$

$15x+4=9$

$x=\frac{1}{3}$

$y=3\left(\frac{1}{3}\right)+1=2$

The solution is $\left\{\left(\frac{1}{3},2\right)\right\}$.

23. $x-7y=3$

$2x-5y=15$

Solve the first equation for x.

$x=7y+3$

$2(7y+3)-5y=15$

$9y+6=15$

$y=1$

$x=7(1)+3=10$

The solution is $\{(10,1)\}$.

25. $3x-9y=7$

$-x+3y=2$

Solve the second equation for x.

$x=3y-2$

$3(3y-2)-9y=7$

$-6=7$

The system is inconsistent and there are no solutions.

27. $2x-3y=4$

$x=3y+6$

Substitute into the first equation.

$2(3y+6)-3y=4$

$3y+12=4$

$y=-\frac{8}{3}$

$x=3\left(-\frac{8}{3}\right)+6=-2$

The solution is $\left\{\left(-2,-\frac{8}{3}\right)\right\}$.

29. $4x-3y=0$

$5x+2y=23$

Use the addition method.

$8x-6y=0$

$15x+6y=69$

$23x=69$

$x=3$

$4(3)-3y=0$

$y=4$

The solution is $\{(3,4)\}$.

31.
$$3x - y = 17 \rightarrow y = 3x - 17$$
$$5x + 3y = 5$$
$$5x + 3(3x - 17) = 5$$
$$14x - 51 = 4$$
$$x = 4$$
$$3(4) - y = 17$$
$$y = -5$$
The solution is {(4, –5)}.

33.
$$\frac{1}{4}x - \frac{1}{6}y = 2 \rightarrow 3x - 2y = 24$$
$$\frac{3}{4}x + y = 3 \rightarrow 3x + 4y = 12 \rightarrow -3x - 4y = -12$$
$$3x - 2y = 24$$
$$-3x - 4y = -12$$
$$-6y = 12$$
$$y = -2$$
$$\frac{3}{4}x + (-2) = 3$$
$$x = \frac{20}{3}$$
The solution is $\left\{\left(\frac{20}{3}, -2\right)\right\}$.

35.
$$\frac{2}{3}x + \frac{3}{5}y = -3 \rightarrow 10x + 9y = -45$$
$$\frac{1}{3}x + \frac{2}{5}y = -3 \rightarrow 5x + 6y = -45$$
$$\rightarrow -10x - 12y = 90$$
$$10x + 9y = -45$$
$$-10x - 12y = 90$$
$$-3y = 45$$
$$y = -15$$
$$\frac{1}{3}x + \frac{2}{5}(-15) = -3$$
$$\frac{1}{3}x = 3$$
$$x = 9$$
The solution is {(9, –15)}.

< Objective 3 >

37. **(d)**
$$x + y = 36$$
$$y = 3x - 4$$

39. **(g)**
$$L = 2W + 3$$
$$2L + 2W = 36$$

41. **(h)**
$$x + y = 120$$
$$2.20x + 5.40y = 360$$

43. **(c)**
$$x + y = 200$$
$$0.20x + 0.60y = 90$$

45. Let *a* be the number of adult tickets and *s* the number of student tickets sold.
$$a + s = 750$$
$$8a + 4.50s = 5{,}300$$
$$a + s = 750 \rightarrow a = 750 - s$$
$$8(750 - s) + 4.50s = 5{,}300$$
$$-3.5s = -700$$
$$s = 200$$
$$a = 750 - (200) = 550$$
550 adult and 200 student tickets were sold.

47. Let *L* and *W* be the length and width of the rectangle, respectively.
$$L = 2W - 3$$
$$2L + 2W = 84 \rightarrow L + W = 42$$
$$(2W - 3) + W = 42$$
$$3W = 45$$
$$W = 15$$
$$L = 2(15) - 3 = 27$$
The rectangle is 15 in. by 27 in.

49. Let m be the cost of a bag of mulch and f be the cost of a bag of fertilizer.

$$8m+3f=96$$
$$5m+5f=100 \rightarrow m+f=20$$
$$\rightarrow f=20-m$$
$$8m+3(20-m)=96$$
$$5m=36$$
$$m=\frac{36}{5}=7.20$$
$$f=20-(7.20)=12.80$$

Mulch costs \$7.20 per bag; fertilizer costs \$12.80 per bag.

51. Let x be the less expensive and y be the more expensive bean (in pounds).

$$x+y=150 \rightarrow y=150-x$$
$$4x+6.50y=4.75 \bullet 150$$
$$\rightarrow 4x+6.5y=712.5$$
$$4x+6.5(150-x)=712.5$$
$$-2.5x=-262.5$$
$$x=105$$
$$y=45$$

The retailer should use 105 lb of the \$4 bean and 45 lb of the \$6.50 bean.

53. Let d be the amount invested in a time deposit and b be the amount invested in a bond.

$$d+b=12{,}000 \rightarrow b=12{,}000-d$$
$$0.04d+0.06b=640 \rightarrow 2d+3b=32{,}000$$
$$2d+3(12{,}000-d)=32{,}000$$
$$-d=-4{,}000$$
$$d=4{,}000$$
$$b=8{,}000$$

Cheryl invested \$4,000 in a time deposit and \$8,000 in a bond.

55. Let x be the amount of 10% acid solution and y be the amount of 50% acid solution (in ml).

$$x+y=400 \rightarrow x=400-y$$
$$0.1x+0.5y=0.4 \bullet 400 \rightarrow x+5y=1{,}600$$
$$(400-y)+5y=1{,}600$$
$$4y=1{,}200$$
$$y=300$$
$$x=100$$

The chemist should use 100 mL of 10% solution and 300 mL of 50% solution.

57. Let x be the rate of the boat in still water and y be the rate of the current (in mph).

	d	***r***	***t***
Downstream	36	$x+y$	2
Upstream	36	$x-y$	3

$$2(x+y)=36 \rightarrow x+y=18$$
$$3(x-y)=36 \rightarrow x-y=12$$

Use the addition method.

$$2x=30$$
$$x=15$$
$$y=3$$

The boat's rate in still water is 15 mi/hr and the rate of the river's current is 3 mi/hr.

59. Let t be the tens digit and u be the units digit.

$$t+u=8$$
$$10u+t=10t+u+36 \rightarrow u-t=4$$

Use the addition method.

$$2u=12$$
$$u=6$$
$$t=2$$

The original number is 26.

61. Let x be the number of battery-powered models produced and y be the number of solar models produced.

$10x+15y=450 \rightarrow 2x+3y=90$

$8x+8y=280 \rightarrow x+y=35 \rightarrow y=35-x$

Use the substitution method.

$$2x+3(35-x)=90$$
$$-x=-15$$
$$x=15$$
$$y=20$$

The manufacturer should produce 15 battery-powered calculators and 20 solar models.

63.
$$S=D$$
$$10p=210-4p$$
$$14p=210$$
$$p=15$$
$$S=150=D$$

The equilibrium point is (15, 150).

65. Let m be the number of miles driven. Then $48(2)+0.22m$ gives the cost of 2 day rental from Company A; and $46(2)+0.26m$ gives the cost of 2 day rental from Company B.

$$48(2)+0.22m=46(2)+0.26m$$
$$96+0.22m=92+0.26m$$
$$.04m=4$$
$$m=100$$

The total charges will be the same if 100 mi are driven.

67. sometimes

69. always

71.
$$y=2x-3$$
$$2x+3y=1 \rightarrow y=-\frac{2}{3}x+\frac{1}{3}$$

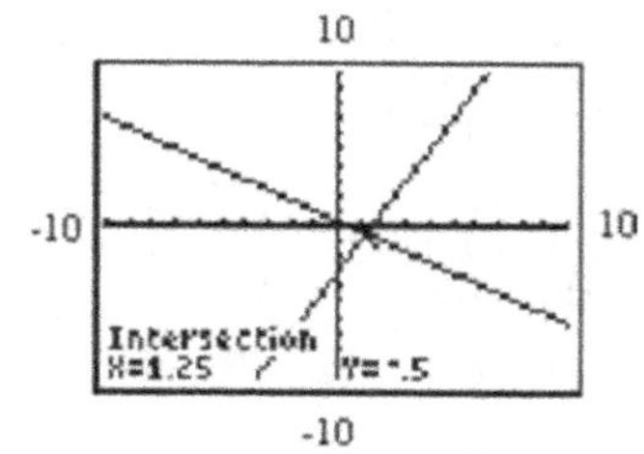

The solution set of the system is {(1.3, –0.5)}.

73. $5x-12y=8 \rightarrow y=\frac{5}{12}x-\frac{2}{3}$

$$7x+2y=44 \rightarrow y=-\frac{7}{2}x+22$$

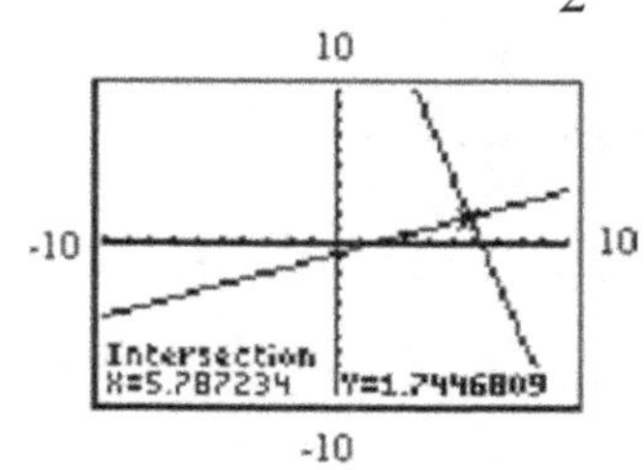

The solution set of the system is {(5.8, 1.7)}.

75. **(a)** $x+y=24$

$90x=280y$

(b) $x+y=24 \rightarrow y=24-x$

$90x=280y \rightarrow 9x=28y$

Use the substitution method.

$$9x=28(24-x)=672-28x$$
$$37x=672$$
$$x\approx 18.16$$
$$y\approx 24-18.16=5.84$$
$$x=18.16,\ y=5.84$$

77. **(a)**
$$x + y = 75$$
$$0.09x + 0.02y = 3$$

(b) Use the substitution method.
$$x + y = 75 \rightarrow y = 75 - x$$
$$0.09x + 0.02(75 - x) = 3$$
$$0.09x + 1.5 - .02x = 3$$
$$0.07x = 1.5$$
$$x \approx 21.4$$
$$y \approx 75 - 21.4 = 53.6$$
The technician should use 21.4 mL of 9% solution and 53.6 mL of 2% solution.

79. $\frac{1}{x} + \frac{1}{y} = 4 \rightarrow u + v = 4$

$\frac{1}{x} - \frac{3}{y} = -6 \rightarrow u - 3v = -6$

Multiply the second equation by −1.
$$u + v = 4$$
$$-u + 3v = 6$$
$$4v = 10$$
$$v = \frac{5}{2} \rightarrow y = \frac{2}{5}$$
$$u = \frac{3}{2} \rightarrow x = \frac{2}{3}$$
The solution set for the system is $\left\{\left(\frac{2}{3}, \frac{2}{5}\right)\right\}$.

81. $\frac{2}{x} + \frac{3}{y} = 4 \rightarrow 2u + 3v = 4$

$\frac{2}{x} - \frac{6}{y} = 10 \rightarrow 2u - 6v = 10$

Multiply the second equation by −1.
$$2u + 3v = 4$$
$$-2u + 6v = -10$$
$$9v = -6$$
$$v = -\frac{2}{3} \rightarrow y = -\frac{3}{2}$$
$$2u + 3\left(-\frac{2}{3}\right) = 4$$
$$2u = 6$$
$$u = 3 \rightarrow x = \frac{1}{3}$$
The solution set for the system is $\left\{\left(\frac{1}{3}, -\frac{3}{2}\right)\right\}$.

83. $1 = 2m + b \rightarrow -1 = -2m - b$

$4 = 4m + b$

Use the addition method.
$$3 = 2m$$
$$m = \frac{3}{2}$$
$$1 = 2\left(\frac{3}{2}\right) + b$$
$$b = -2$$
$$y = \frac{3}{2}x - 2$$

85. Above and Beyond

87. **(a)** $Ax+By=C \rightarrow -ADx-BDy=-CD$

$Dx+Ey=F \rightarrow ADx+AEy=AF$

$$-ADx-BDy=-CD$$
$$ADx+AEy=AF$$
$$(AE-BD)y=AF-CD$$
$$y=\frac{AF-CD}{AE-BD}$$

Because we do not allow zero denominators to be equal to zero, we have the condition $AE-BD\neq 0$.

(b) $Ax+By=C \rightarrow AEx+BEy=CE$

$Dx+Ey=F \rightarrow -BDx-BEy=-BF$

$$(AE-BD)x=CE-BF$$
$$x=\frac{CE-BF}{AE-BD}$$

This gives the condition $AE-BD\neq 0$.

Exercises 4.4

< Objective 1 >

1. $x-y+z=3$

$2x+y+z=8$

$3x+y-z=1$

Pair the first and third equations; and the second and third equations; eliminate z.

$$x-y+z=3$$
$$3x+y-z=1$$
$$4x=4$$
$$x=1$$

$$2x+y+z=8$$
$$3x+y-z=1$$
$$5x+2y=9$$

Solve for y: $5(1)+2y=9$

$y=2$

Substitute (1, 2) into the first equation to determine z: $(1)-(2)+z=3$

$z=4$

The solution set for the system is $\{(1, 2, 4)\}$.

3. $x+y+z=1$

$2x-y+2z=-1$

$-x-3y+z=1$

Pair the first and second equations; and the first and third equations (multiply the first by 3); eliminate y.

$x+y+z=1$ and $3x+3y+3z=3$

$2x-y+2z=-1$ $\quad -x-3y+z=1$

$3x+3z=0$ $\quad 2x+4z=4$

$x+z=0$ $\quad x+2z=2$

Solve for x and z: $x+z=0 \rightarrow x=-z$

$$x+2z=2$$
$$(-z)+2z=2$$
$$z=2$$
$$x=-2$$

Substitute $x=-2$ and $z=2$ into the first equation to determine y: $(-2)+y+(2)=1$

$y=1$

The solution set for the system is $\{(-2, 1, 2)\}$.

5. $x+y+z=1$

$-2x+2y+3z=20$

$2x-2y-z=-16$

Pair twice the first with the second equation; and the second and third equations; eliminate x.

$2x+2y+2z=2$ and $-2x+2y+3z=20$

$-2x+2y+3z=20$ $\quad 2x-2y-z=-16$

$4y+5z=22$ $\quad z=2$

Solve for y: $4y+5(2)=22$

$y=3$

Substitute $y=3$ and $z=2$ into the first equation to determine x: $x+(3)+(2)=1$

$x=-4$

The solution set for the system is $\{(-4, 3, 2)\}$.

7. $2x+y-z=2$

$-x-3y+z=-1$

$-4x+3y+z=-4$

Pair the first and second equations; and the first and third equations; eliminate z.

$2x+y-z=2$ and $2x+y-z=2$

$-x-3y+z=-1$ $\quad -4x+3y+z=-4$

$x-2y=1$ $\quad -2x+4y=-2$

$x-2y=1$

Both pairs give the same equation. There is infinite number of solutions. The system is dependent.

9. $3x-y+z=5$

$x+3y+3z=-6$

$x+4y-2z=12$

Pair three times the first with the second equation; and four times the first with the third equation; eliminate y.

$9x-3y+3z=15$ and $12x-4y+4z=20$

$x+3y+3z=-6$ $\quad x+4y-2z=12$

$10x+6z=9$ $\quad 13x+2z=32$

Solve for x and z: $10x+6z=9$

$13x+2z=32$

Multiply the second equation by –3 and use the method of addition.

$10x+6z=9$ and $10(3)+6z=9$

$-39x-6z=-96$ $\quad 6z=-21$

$-29x=-87$ $\quad z=-\frac{21}{6}=-\frac{7}{2}$

$x=\frac{87}{29}=3$

Substitute $x=3$ and $z=-\frac{7}{2}$ into the first equation to determine y:

$3(3)-y+\left(-\frac{7}{2}\right)=5$

$y=\frac{1}{2}$

The solution set for the system is $\left\{\left(3,\frac{1}{2},-\frac{7}{2}\right)\right\}$.

11. $x+2y+z=2$

$2x+3y+3z=-3$

$2x+3y+2z=2$

Pair –2 times the first with the second; and –2 times the first with the third; eliminate x.

$-2x-4y-2z=-4$ and $-2x-4y-2z=-4$

$2x+3y+3z=-3$ $\quad 2x+3y+2z=2$

$-y+z=-7$ $\quad -y=-2$

$x+z=1$ $\quad y=2$

Solve for z: $-(2)+z=-7$

$z=-5$

Use one of the original equations to determine x: $x+2(2)+(-5)=2$

$x=3$

The solution set for the system is {(3, 2, –5)}.

13. $x+3y-2z=8$

$3x+2y-3z=15$

$4x+2y+3z=-1$

Pair –3 times the first with the second; and –4 times the first with the third; eliminate x.

$-3x-9y+6z=-24$ and $-4x-12y+8z=-32$

$3x+2y-3z=15$ $\quad 4x+2y+3z=-1$

$-7y+3z=-9$ $\quad -10y+11z=-33$

Solve the resulting system.

$-7y+3z=-9$

$-10y+11z=-33$

Multiply the first equation by –10, the second equation by 7, and use the addition method to solve for z.

$70y-30z=90$

$-70y+77z=-231$

$47z=-141$

$z=-3$

Solve for y: $-7y+3(-3)=-9$

$y=0$

Use one of the original equations to determine x: $x+3(0)-2(-3)=8$

$x=2$

The solution set for the system is {(2, 0, –3)}.

15. $x+y-z=2$

$x-2z=1$

$2x-3y-z=8$

Pair 3 times the first with the third; and use the second equation; eliminate y.

$3x+3y-3z=6$

$2x-3y-z=8$

$5x-4z=14$

Solve the resulting system for z.

$5x-4z=14$

$x-2z=1 \to x=2z+1$

$5(2z+1)-4z=14$

$6z=9$

$z=\frac{3}{2}$

Solve for x: $x=2\left(\frac{3}{2}\right)+1=4$

Use one of the original equations to determine y: $(4)+y-\left(\frac{3}{2}\right)=2$

$y=-\frac{1}{2}$

The solution set for the system is $\left\{\left(4,-\frac{1}{2},\frac{3}{2}\right)\right\}$.

17. $x-3y+2z=1$

$16y-9z=5$

$4x+4y-z=8$

Pair –4 times the first with the third; and use the second equation; eliminate x.

$-4x+12y-8z=-4$

$4x+4y-z=8$

$16y-9z=4$

Solve the resulting system: $16y-9z=4$

$16y-9z=5$

The system is inconsistent and there are no solutions.

19. $x+2y-4z=13$

$3x+4y-2z=19$

$3x+2z=3$

Pair –2 times the first with the second; and use the third equation; eliminate y.

$-2x-4y+8z=-26$

$3x+4y-2z=19$

$x+6z=-7$

Solve the resulting system for z.

$x+6z=-7 \to x=-7-6z$

$3x+2z=3$

$3(-7-6z)+2z=3$

$-16z=24$

$z=-\frac{3}{2}$

Solve for x.

$x=-7-6\left(-\frac{3}{2}\right)=2$ and $(-1)+3z=-5$

$z=-2$

Use one of the original equations to determine y: $(2)+2y-4\left(-\frac{3}{2}\right)=13$

$2y=5$

$y=\frac{5}{2}$

The solution set for the system is $\left\{\left(2,\frac{5}{2},-\frac{3}{2}\right)\right\}$.

< Objective 2 >

21. Let x, y, and z be the three numbers (smallest to largest).

$x+y+z=16$

$z=x+y \rightarrow x+y-z=0$

$3x=z+1 \rightarrow 3x-z=1$

Pair –1 times the first with the second; and use the third equation; eliminate y.

$-x-y-z=-16$

$x+y-z=0$

$-2z=-16$

$z=8$

Solve the (original) third equation for x.

$3x-(8)=1$

$x=3$

Use one of the original equations to determine y: $(3)+y+(8)=16$

$y=5$

The numbers are 3, 5, and 8.

23. Let n, d, and q be the number of nickels, dimes, and quarters, respectively.

$n+d+q=25$

$0.05n+0.10d+0.25q=4.90$

$d=2n-1$

The second equation is equivalent to

$5n+10d+25q=490$

$n+2d+5q=98$

Pair –5 times the first with this (new) second equation; and use the third equation; eliminate q.

$-5n-5d-5q=-125$

$n+2d+5q=98$

$-4n-3d=-27$

$4n+3d=27$

Solve the resulting system.

$d=2n-1$

$4n+3d=27$

$4n+3(2n-1)=27$

$10n=30$

$n=3$

$d=5$

$q=17$

The cashier has 3 nickels, 5 dimes, and 17 quarters.

25. Let x, y, and z be the lengths of the three sides (smallest to largest).

$x+y+z=19$

$z=2x$

$z=x+y-3 \rightarrow -x-y+z=-3$

Pair the first and third equations to eliminate y: $x+y+z=19$

$-x-y+z=-3$

$2z=16$

$z=8$

The second equation gives $x=4$. From the first equation, $y=7$. The sides of the triangle are 4 cm, 7 cm, and 8 cm.

27. Let s, b, and m be the amount Jovita has invested in a savings account, bond, and money market fund, respectively.

$$s+b+m=12{,}000$$
$$0.04s+0.06b+0.09m=860$$
$$b=2s$$

The second equation is equivalent to $4s+6b+9m=86{,}000$.

Pair –9 times the first equation with this (new) second equation to eliminate m.

$$-9s-9b-9m=-108{,}000$$
$$4s+6b+9m=86{,}000$$
$$-5s-3b=-22{,}000$$
$$5s+3b=22{,}000$$

Combine this last equation with the (original) third equation and solve the resulting system.

$$b=2s$$
$$5s+3b=22{,}000$$
$$5s+3(2s)=22{,}000$$
$$s=2{,}000$$
$$b=4{,}000$$

From the first equation, we have $m=6000$. Jovita invested \$2,000 in a savings account, \$4,000 in a bond, and \$6,000 in a money market fund.

29. Let h, t, and u be the hundreds, tens, and units digits, respectively, of the number.

$$h+t+u=9$$
$$t=2h$$
$$100h+10t+u=100u+10t+h-99$$
$$99h-99u=-99$$
$$h-u=-1$$
$$u=h+1$$

Use substitution: $h+t+u=9$

$$h+(2h)+(h+1)=9$$
$$h=2$$
$$t=4$$
$$u=3$$

The original number is 243.

31. Let r, s, and j be the amount driven each day by Roy, Sally, and Jeff, respectively.

$$r+s+j=50$$
$$s=2r$$
$$j=s+10$$

Substitute from the second to the third.

$$j=(2r)+10$$

Substitute back into the original equation.

$$r+(2r)+(2r+10)=50$$
$$r=8$$
$$s=16$$
$$j=26$$

Roy drives 8 mi, Sally drives 16 mi, and Jeff drives 26 mi each day.

33. True

35. $x+2y+3z+w=0$
$-x-y-3z+w=-2$
$x-3y+2z+2w=-11$
$-x+y-2z+w=1$

Eliminate the x-variable by pairing the first and second equations, the second and third equations, and the third and fourth equations.

$x+2y+3z+w=0$ and $-x-y-3z+w=-2$
$-x-y-3z+w=-2$ $\quad x-3y+2z+2w=-11$
$y+2w=-2$ $\quad -4y-z+3w=-13$

and $x-3y+2z+2w=-11$
$-x+y-2z+w=1$
$-2y+3w=-10$

Solve the resulting system.

$y+2w=-2$ and $y+2w=-2$
$-4y-z+3w=-13$ $\quad y=-2w-2$
$-2y+3w=-10$ $\quad -2y+3w=-10$
$-2(-2w-2)+3w=-10$

Giving $7w=-14$
$w=-2$
$y=2$

Use the middle equation from the 3-by-3 system to solve for z.

$-4(2)-z+3(-2)=-13$
$z=-1$

Choose any equation from the original 4-by-4 system to determine x.

$x+2(2)+3(-1)+(-2)=0$
$x=1$

The solution set for the system is $\{(1, 2, -1, -2)\}$.

37. $x-y=5 \rightarrow y=x-5$
$2x+3y=20$
$4x+5y=38$

Use the substitution method with the first two equations.

$y=x-5$
$2x+3y=20$
$2x+3(x-5)=20$
$5x=35$
$x=7$
$y=2$

Substitute these values into the third equation to check: $4(7)+5(2)\stackrel{?}{=}38$
$38=38$

The solution set for the system is $\{(7, 2)\}$.

39. Rewrite the system after multiplying the second and third equations by 10.

$T+C+B=85$
$15T+11C+4B=735$
$20T+15C+18B=1,495$

Eliminate B.

$-4T-4C-4B=-340$ and $-18T-18C-18B=-1,530$
$15T+11C+4B=735$ $\quad 20T+15C+18B=1,495$
$11T+7C=395$ $\quad 2T-3C=-35$

Solve the resulting 2-by-2 system for T.

$11T+7C=395 \rightarrow 33T+21C=1,185$
$2T-3C=-35 \rightarrow 14T-21C=-245$
$47T-940$
$T=20$

Then solve it for C: $2(20)-3C=-35$
$3C=75$
$C=25$

Finally, solve for B: $T+C+B=85$
$(20)+(25)+B=85$
$B=40$

The car drove 25 mi, the truck drove 20 mi, and the bus drove 40 mi.

41. **(a)** $a+b+c=5$

$4a+2b+c=10$

$9a+3b+c=19$

Eliminate c.

$$\begin{array}{rcr} -a-b-c=-5 & \text{and} & -a-b-c=-5 \\ 4a+2b+c=10 & & 9a+3b+c=19 \\ 3a+b=5 & & 8a+2b=14 \\ & & 4a+b=7 \end{array}$$

Solve the resulting system.

$3a+b=5 \rightarrow b=5-3a$

$4a+(5-3a)=7$

$a=2$

$b=-1$

Then determine c: $(2)+(-1)+c=5$

$c=5$

The solution set of the system is $\{(a,b,c)\}=\{(2,-1,4)\}$.

The function is $f(x)=2x^2-x+4$.

(b) $a+b+c=2$

$4a+2b+c=9$

$9a+3b+c=22$

Eliminate c.

$$\begin{array}{rcr} -a-b-c=-2 & \text{and} & -a-b-c=-2 \\ 4a+2b+c=9 & & 9a+3b+c=22 \\ 3a+b=7 & & 8a+2b=20 \\ & & 4a+b=10 \end{array}$$

Solve the resulting system.

$3a+b=7 \rightarrow b=7-3a$

$4a+(7-3a)=10$

$a=3$

$b=-2$

Then determine c: $(3)+(-2)+c=2$

$c=1$

The solution set of the system is $\{(a,b,c)\}=\{(3,-2,1)\}$.

The function is $f(x)=3x^2-2x+1$.

Exercises 4.5

< Objective 1 >

1. $x+2y\le 4 \rightarrow y\le -\dfrac{x}{2}+2$

$x-y\ge 1 \rightarrow y\le x-1$

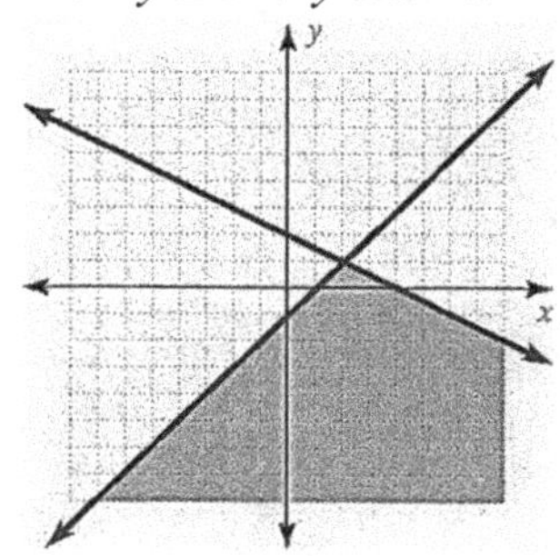

3. $3x+y<6 \rightarrow y<-3x+6$

$x+y>4 \rightarrow y>-x+4$

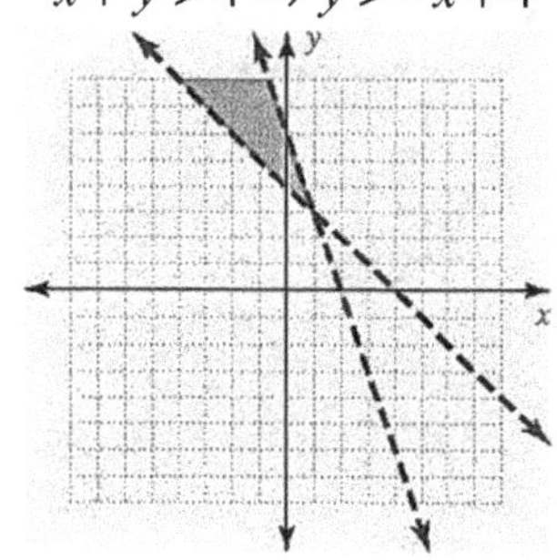

5. $x+3y\le 12 \rightarrow y\le -\dfrac{x}{3}+4$

$2x-3y\le 6 \rightarrow y\ge \dfrac{2}{3}x-2$

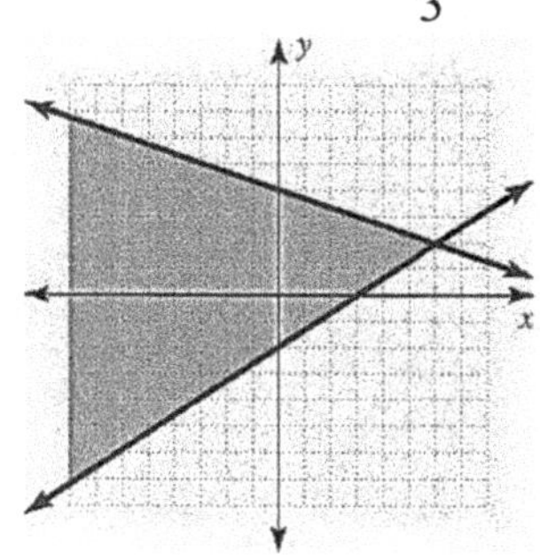

7. $3x+2y\le 12 \rightarrow y\le -\dfrac{3}{2}x+6$

$x\ge 2$

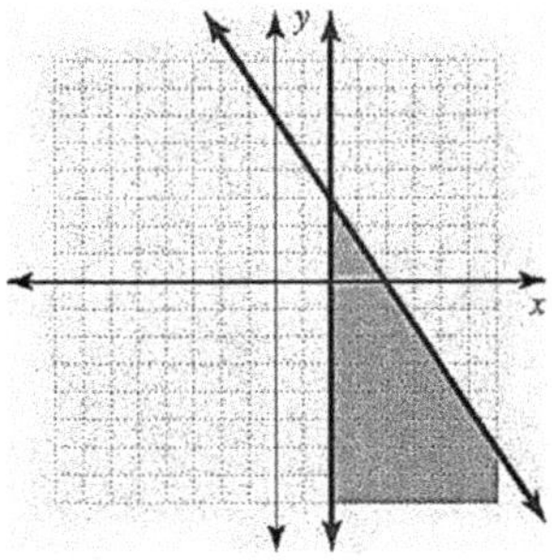

9. $2x+y\le 8 \rightarrow y\le -2x+8$

$x>1$

$y>2$

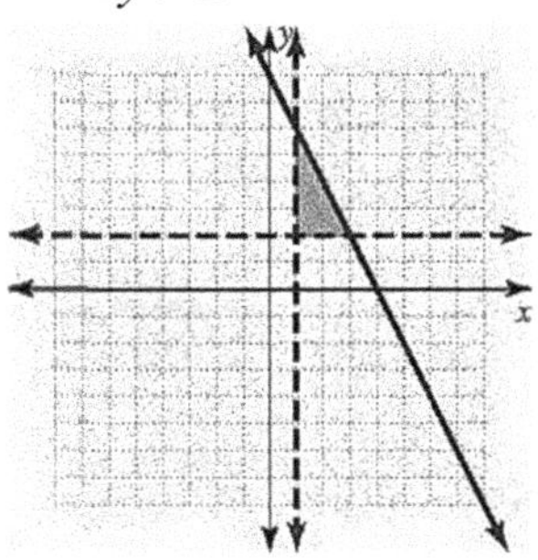

11. $x+2y\le 8 \rightarrow y\le -\dfrac{x}{2}+4$

$2\le x\le 6$

$y\ge 0$

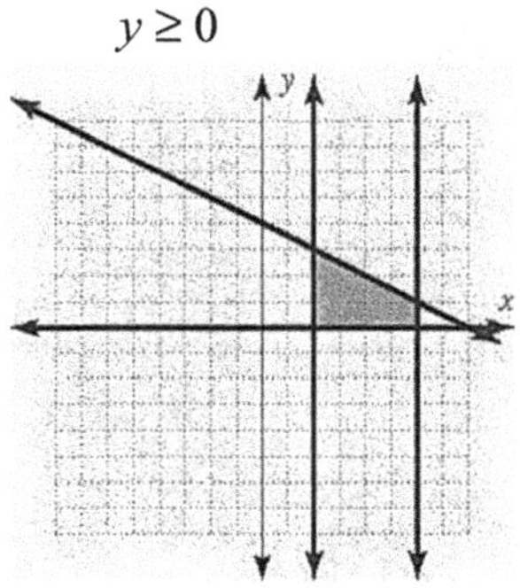

13. $3x+y\le 6 \rightarrow y\le -3x+6$

$x+y\le 4 \rightarrow y\le -x+4$

$x\ge 0$

$y\ge 0$

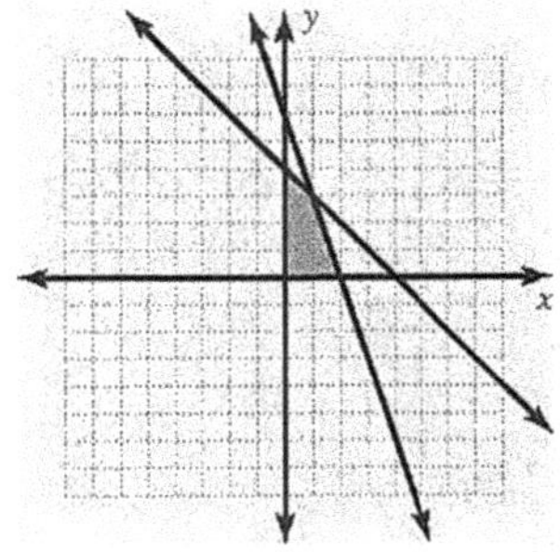

15. $4x+3y\le 12 \rightarrow y\le -\dfrac{4}{3}x+4$

$x+4y\le 8 \rightarrow y\le -\dfrac{x}{4}+2$

$x\ge 0$

$y\ge 0$

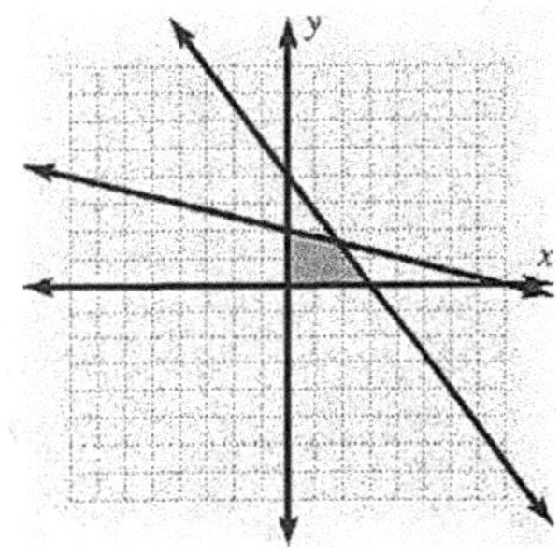

17. $x-4y\le -4 \rightarrow y\ge \dfrac{x}{4}+1$

$x+2y\le 8 \rightarrow y\le -\dfrac{x}{2}+4$

$x\ge 2$

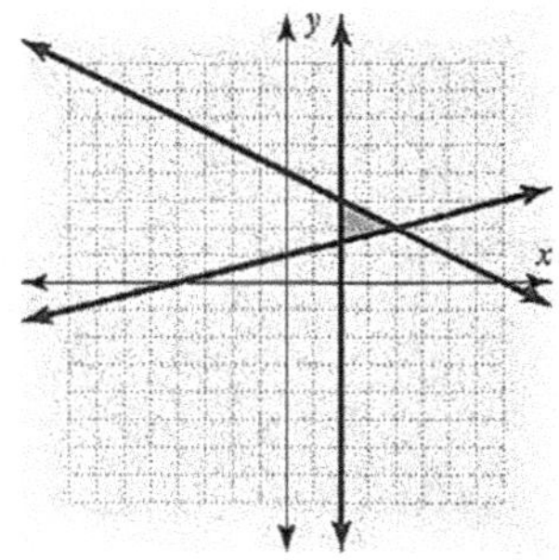

< Objective 2 >

19. $6x+10y\le 300 \rightarrow y\le -\frac{3}{5}x+30$

$x+y\le 40 \rightarrow y\le -x+40$

$x\ge 0$

$y\ge 0$

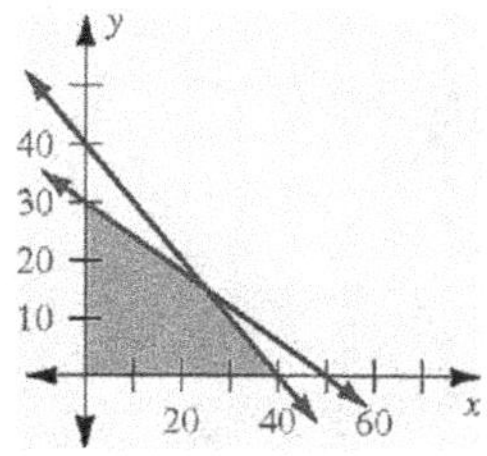

21. True

23. sometimes

25. Above and Beyond

27. Find two points on each line.
L_1: (0, –1), (7, –8)
L_2: (1, 5), (2, 7)
L_3: (0, 5), (2, –1)
Then, determine the equation for each line.

L_1: $m=\frac{-8-(-1)}{7-0}=-1$

$y=-x-1$

L_2: $m=\frac{7-5}{2-1}=2$

$y-5=2(x-1)$

$y=2x+3$

L_3: $m=\frac{-1-5}{2-1}=-3$

$y=-3x+5$

The shading is above L_1, and below L_2 and L_3. All three border lines are solid.

$y\ge -x-1$

$y\le 2x+3$

$y\le -3x+5$

Summary Exercises

1. $x+y=8 \rightarrow y=-x+8$

$x-y=4 \rightarrow y=x-4$

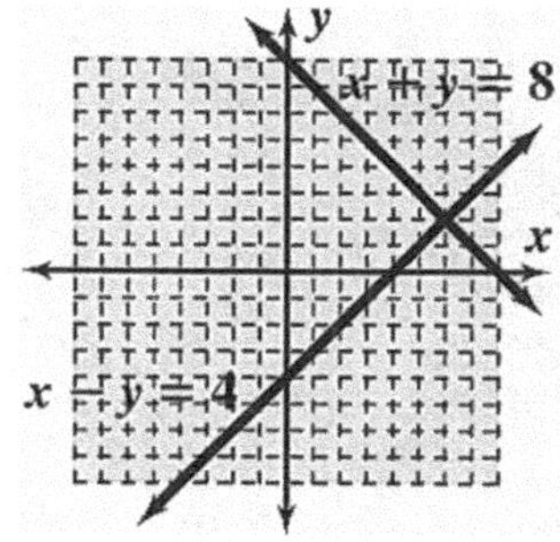

The lines intersect at (6, 2), hence the solution is {(6, 2)}.

3. $2x+3y=12 \rightarrow y=-\frac{2}{3}x+4$

$2x+8=8 \rightarrow y=-2x+8$

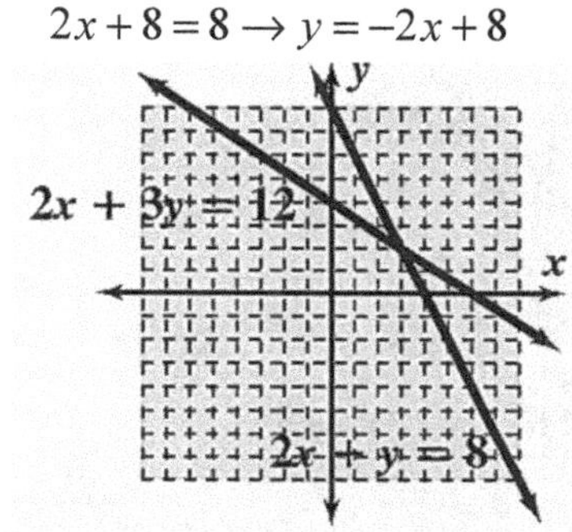

The lines intersect at (3, 2), hence the solution is {(3, 2)}.

5. $44x+35y=1{,}115 \rightarrow y=\frac{1{,}115-44x}{35}$

$11x-27y=850 \rightarrow y=\frac{11x-850}{27}$

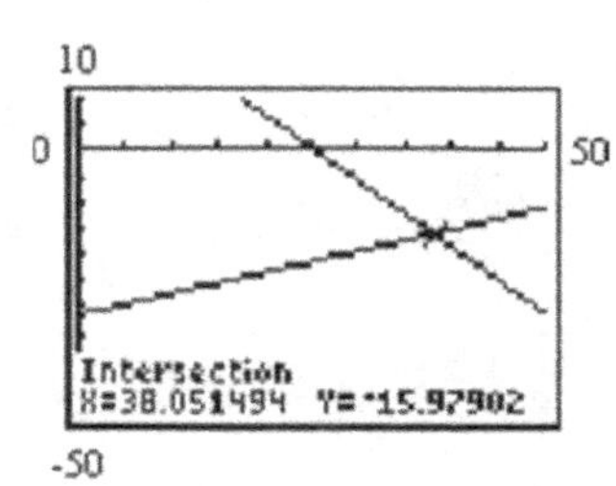

The solution is {(38.05, –15.98)}.

7. $3x-6=0$

$f(x)=3x-6$

$g(x)=0$

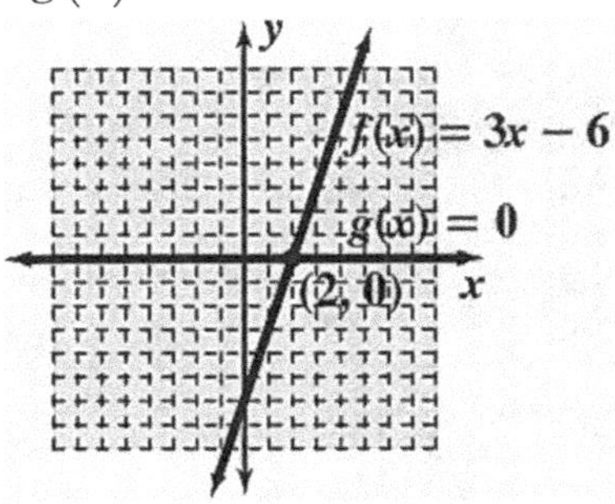

The point of intersection in (2, 0).
The solution set is {2}.

9. $3x+5=x+7$

$f(x)=3x+5$

$g(x)=x+7$

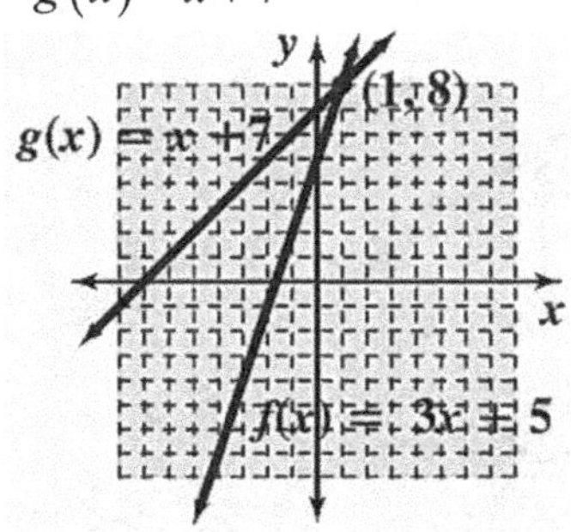

The point of intersection is (1, 8). The solution set is {1}.

11. $\dfrac{6x-1}{2}=2(x-1)$

$f(x)=\dfrac{6x-1}{2}=3x-\dfrac{1}{2}$

$g(x)=2(x-1)=2x-2$

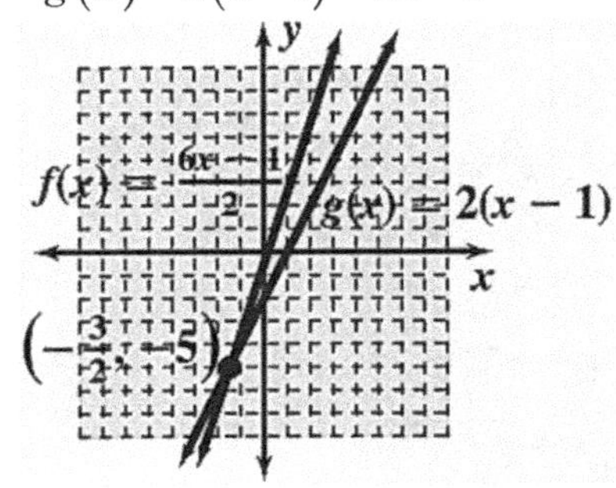

The point of intersection is $\left(-\dfrac{3}{2},-5\right)$. The solution set is $\left\{-\dfrac{3}{2}\right\}$.

13. $3(x-2)=2(x-1)$

$f(x)=3x-6$

$g(x)=2x-2$

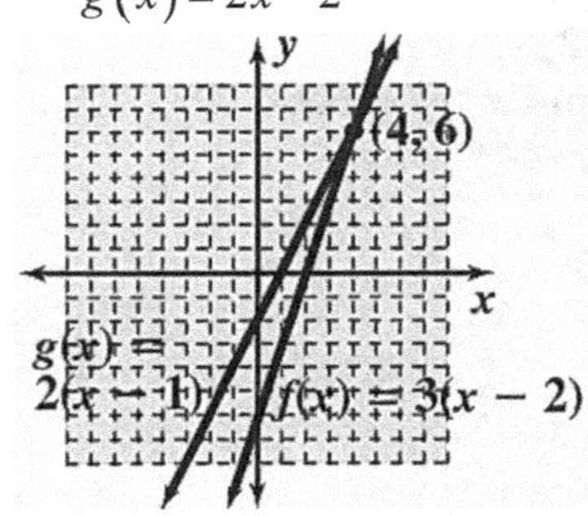

The point of intersection is (4, 6). The solution set is {4}.

15. $x+2y=7$

$x-y=1$

Multiply the second equation by 2, then add.

$$\begin{aligned} x+2y&=7 \\ 2x-2y&=2 \\ 3x&=9 \\ x&=3 \end{aligned} \qquad \text{and} \qquad \begin{aligned} x-y&=1 \\ (3)-y&=1 \\ y&=2 \end{aligned}$$

The solution set for the system is {(3, 2)}.

17. $3x-5y=5$

$-x+y=-1$

Multiply the second equation by 3, then add.

$$\begin{aligned} 3x-5y&=5 \\ -3x+3y&=-3 \\ -2y&=2 \\ y&=-1 \end{aligned} \quad \text{and} \quad \begin{aligned} -x+y&=-1 \\ -x+(-1)&=-1 \\ x&=0 \end{aligned}$$

The solution set for the system is $\{(0, -1)\}$.

19. $6x+5y=-9$

$-5x+4y=32$

Multiply the first equation by 5 and the second by 6, then add.

$$\begin{aligned} 30x+25y&=-45 \\ -30x+24y&=192 \\ 49y&=147 \\ y&=3 \end{aligned} \quad \text{and} \quad \begin{aligned} 6x+5y&=-9 \\ 6x+5(3)&=-9 \\ x&=-4 \end{aligned}$$

The solution set for the system is $\{(-4, 3)\}$.

21. $5x-y=-17$

$4x+3y=-6$

Multiply the first equation by 3, then add.

$$\begin{aligned} 15x-3y&=-51 \\ 4x+3y&=-6 \\ 19x&=-57 \\ x&=-3 \end{aligned} \quad \text{and} \quad \begin{aligned} 5x-y&=-17 \\ 5(-3)-y&=-17 \\ y&=2 \end{aligned}$$

The solution set for the system is $\{(-3, 2)\}$.

23. $x-\frac{1}{2}y=8$

$\frac{2}{3}x+\frac{3}{2}y=-2$

Multiply the first equation by 3, then add.

$$\begin{aligned} 3x-\frac{3}{2}y&=24 \\ \frac{2}{3}x+\frac{3}{2}y&=-2 \\ \frac{11}{3}x&=22 \\ x&=6 \end{aligned} \quad \text{and} \quad \begin{aligned} x-\frac{1}{2}y&=8 \\ (6)-\frac{1}{2}y&=8 \\ y&=-4 \end{aligned}$$

The solution set for the system is $\{(6, -4)\}$.

25. $2x+y=23$

$x=y+4$

Substitute for x in the first equation.

$$\begin{aligned} 2(y+4)+y&=23 \\ 3y+8&=23 \\ y&=5 \end{aligned} \quad \text{and} \quad \begin{aligned} x&=y+4 \\ x&=(5)+4 \\ &=9 \end{aligned}$$

The solution set for the system is $\{(9, 5)\}$.

27. $3x+y=7$

$y=-3x+5$

Substitute for y in the first equation.

$$\begin{aligned} 3x+(-3x+5)&=7 \\ 5&=7 \end{aligned}$$

The system is inconsistent and there are no solutions.

29. $5x-3y=13$

$x-y=3 \rightarrow x=y+3$

Substitute for x in the first equation.

$$\begin{aligned} 5(y+3)-3y&=13 \\ 2y-15&=13 \\ y&=-1 \end{aligned} \quad \text{and} \quad \begin{aligned} x&=y+3 \\ &=(-1)+3=2 \end{aligned}$$

The solution set for the system is $\{(2, -1)\}$.

31. $3x-2y=-12$

$6x+y=1 \rightarrow y=-6x+1$

Substitute for y in the first equation.

$$\begin{aligned} 3x-2(-6x+1)&=-12 \\ 15x-2&=-12 \\ x&=-\frac{2}{3} \end{aligned} \quad \text{and} \quad \begin{aligned} y&=-6x+1 \\ &=-6\left(-\frac{2}{3}\right)+1=5 \end{aligned}$$

The solution set for the system is $\left\{\left(-\frac{2}{3}, 5\right)\right\}$.

33. Let x and y be the numbers.

$$y = 3x + 2$$
$$x + y = 30$$

Substitute for y in the second equation.

$$x + (3x + 2) = 30$$
$$4x + 2 = 30$$
$$x = 7$$
$$3x + 2 = 23$$

The numbers are 7 and 23.

35. Let x and y be the number of adult and student tickets sold, respectively.

$$x + y = 1{,}200 \to x = 1{,}200 - y$$
$$7x + 4.50y = 7{,}400$$

Substitute for x in the second equation.

$$7(1{,}200 - y) + 4.50y = 7{,}400$$
$$8{,}400 - 2.5y = 7{,}400$$
$$-2.5y = -1{,}000$$
$$y = 400$$

Solve for x.

$$x = 1{,}200 - y = 1{,}200 - (400) = 800$$

800 adult and 400 student tickets were sold.

37. Let L and W be the length and the width of the rectangle.

$$L = 2W - 4$$
$$2L + 2W = 64 \to L + W = 32$$

Substitute for L in the second equation.

$$(2W - 4) + W = 32 \quad \text{and} \quad L = 2W - 4$$
$$3W - 4 = 32 \qquad\qquad = 2(12) - 4 = 20$$
$$W = 12$$

The rectangle is 12 cm by 20 cm.

39. Let x be the amount invested in a savings account. Let y be the amount in a time deposit.

$$x + y = 17{,}000 \to y = 17{,}000 - x$$
$$0.06x + 0.08y = 1{,}200$$

Multiply the second equation by 100, simplify, then substitute for y.

$$6x + 8y = 120{,}000 \quad \text{and} \quad y = 17{,}000 - x$$
$$3x + 4y = 60{,}000 \qquad\qquad = 17{,}000 - (8{,}000)$$
$$3x + 4(17{,}000 - x) = 60{,}000 \qquad\qquad = 9{,}000$$
$$-x + 68{,}000 = 60{,}000$$
$$x = 8{,}000$$

Reggie has \$8,000 in a savings account and \$9,000 in a time deposit.

41. Let x be the plane's rate in still air and y the rate of the wind.

	d	*r*	*t*
With wind	2,200	$x + y$	4
Against wind	1,800	$x - y$	4

Which gives the system

$$4(x + y) = 2{,}200 \to 4x + 4y = 2{,}200$$
$$4(x - y) = 1{,}800 \to 4x - 4y = 1{,}800$$

Add the two equations.

$$4x + 4y = 2{,}200 \quad \text{and} \quad 4x + 4y = 2{,}200$$
$$4x - 4y = 1{,}800 \qquad\qquad 4(500) + 4y = 2{,}200$$
$$8x = 4{,}000 \qquad\qquad y = 50$$
$$x = 500$$

The plane's speed is 500 mi/hr in still air and the rate of the wind is 50 mi/hr.

43.

$$D = S$$
$$270 - 5p = 13p$$
$$270 = 18p$$
$$15 = p$$
$$13p = 13(15) = 195$$

The equilibrium point is (15, 195).

45. $x-y+z=0$

$x+4y-z=14$

$x+y-z=6$

Eliminate z by pairing the first two equations, and by pairing the first and third equation.

$$\begin{aligned} x-y+z&=0 \\ x+4y-z&=14 \\ 2x+3y&=14 \end{aligned} \quad \text{and} \quad \begin{aligned} x-y+z&=0 \\ x+y-z&=6 \\ 2x&=6 \\ x&=3 \end{aligned}$$

Solve the resulting system for y.

$$\begin{aligned} 2x+3y&=14 \\ x&=3 \\ 2(3)+3y&=14 \\ 3y&=8 \\ y&=\frac{8}{3} \end{aligned}$$

Use one of the original equations to find z.

$$\begin{aligned} x-y+z&=0 \\ (3)-\left(\frac{8}{3}\right)+z&=0 \\ z&=-\frac{1}{3} \end{aligned}$$

The solution set for the system is $\left\{\left(3,\frac{8}{3},-\frac{1}{3}\right)\right\}$.

47. $x-y-z=2$

$-2x+2y+z=-5$

$-3x+3y+z=-10$

Eliminate z by pairing the first and second equations, and by pairing the first and third equations.

$$\begin{aligned} x-y-z&=2 \\ -2x+2y+z&=-5 \\ -x+y&=-3 \end{aligned} \quad \text{and} \quad \begin{aligned} x-y-z&=2 \\ -3x+3y+z&=-10 \\ -2x+2y&=-8 \end{aligned}$$

Giving the system $\begin{aligned} -x+y&=-3 \\ -2x+2y&=-8 \end{aligned}$

Multiply the first equation by –2 and add.

$$\begin{aligned} 2x-2y&=6 \\ -2x+2y&=-8 \\ 0&=-2 \end{aligned}$$

The system is inconsistent.

49. $x+y-z=-1$

$x-y-2z=2$

$-5x-y-z=-1$

Eliminate y by pairing the first two equations, and by pairing the first and third equation.

$$\begin{aligned} x+y-z&=-1 \\ x-y-2z&=2 \\ 2x-3z&=1 \end{aligned} \quad \text{and} \quad \begin{aligned} x+y-z&=-1 \\ -5x-y-z&=-1 \\ -4x-2z&=-2 \\ -2x-z&=-1 \end{aligned}$$

Giving the system $\begin{aligned} 2x-3z&=1 \\ -2x-z&=-1. \end{aligned}$

Use the addition method.

$$\begin{aligned} 2x-3z&=1 \\ -2x-z&=-1 \\ -4z&=0 \\ z&=0 \end{aligned} \quad \text{and} \quad \begin{aligned} 2x-3z&=1 \\ 2x-3(0)&=1 \\ x&=\frac{1}{2} \end{aligned}$$

And finally,
$$\begin{aligned} x+y-z&=-1 \\ \left(\frac{1}{2}\right)+y-(0)&=-1 \\ y&=-\frac{3}{2} \end{aligned}$$

The solution set for the system is $\left\{\left(\frac{1}{2},-\frac{3}{2},0\right)\right\}$

51. Let x, y, and z be the three numbers (smallest to largest).

$x+y+z=15$

$z=4x=x+y+1$

Substitute for z in the third equation, solve for y and substitute for y and z in the first equation.

$$\begin{aligned} 4x&=x+y+1 \\ y&=3x-1 \end{aligned} \quad \text{and} \quad \begin{aligned} x+(3x-1)+(4x)&=15 \\ 8x-1&=15 \\ x&=2 \end{aligned}$$

Back substitute to find y and z.

$y=3x-1=3(2)-1=5$

$z=4x=4(2)=8$

The numbers are 2, 5, and 8.

53. Let x, y, and z be the number of orchestra, box, and balcony tickets sold, respectively.

$$x+y+z=360$$
$$10x+7y+5z=3{,}040$$
$$x=y+z+40\rightarrow x-y-z=40$$

Eliminate z by pairing –5 times the first equation with the second equation, and pairing the first and third equation.

$$\begin{aligned}-5x-5y-5z&=-1{,}800\\10x+7y+5z&=3{,}040\\5x+2y&=1{,}240\end{aligned}\quad\text{and}\quad\begin{aligned}x+y+z&=360\\x-y-z&=40\\2x&=400\\x&=200\end{aligned}$$

Back-substitute to find y and z.

$$\begin{aligned}5x+2y&=1{,}240\\5(200)+2y&=1{,}240\\y&=120\end{aligned}\quad\text{and}\quad\begin{aligned}x+y+z&=360\\(200)+(120)+z&=360\\z&=40\end{aligned}$$

The theater sold 200 orchestra, 120 box, and 40 balcony tickets.

55. $x-y<7\rightarrow y>x-7$

$x+y=3\rightarrow y=-x+3$

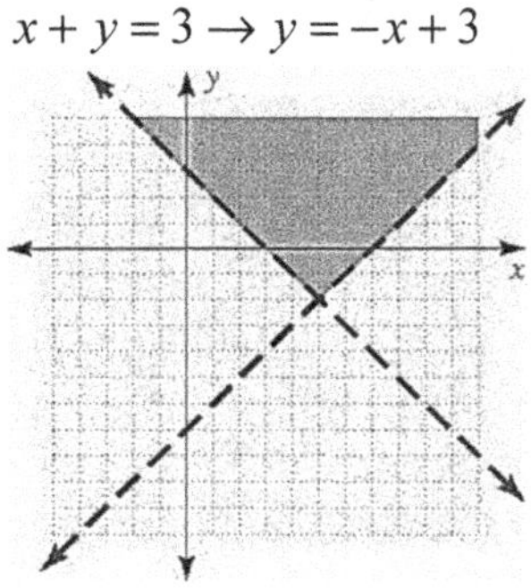

57. $x-6y<6\rightarrow y>\dfrac{x}{6}-1$

$-x+y<4\rightarrow y<x+4$

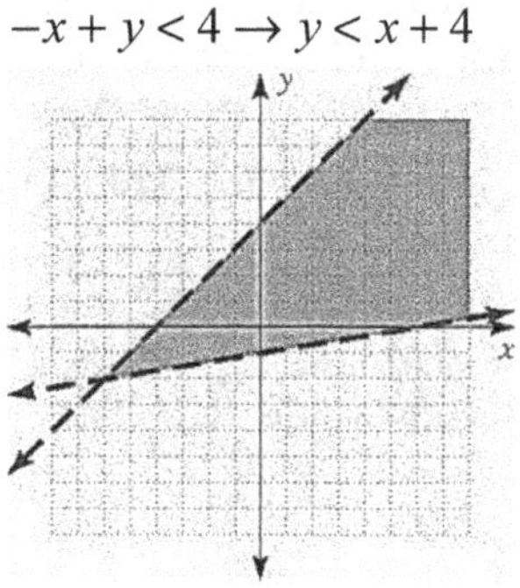

59. $2x+y\le 6\rightarrow y\le -2x+6$

$x\ge 1$

$y\ge 0$

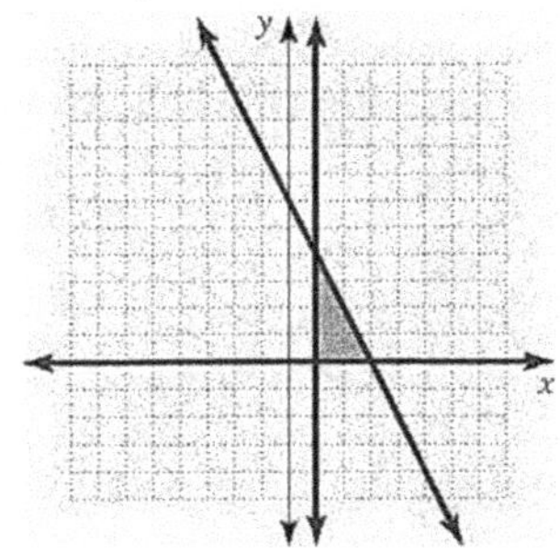

61. $4x+2y\le 8\rightarrow y\le -2x+4$

$x+y\le 3\rightarrow y\le -x+3$

$x\ge 0$

$y\ge 0$

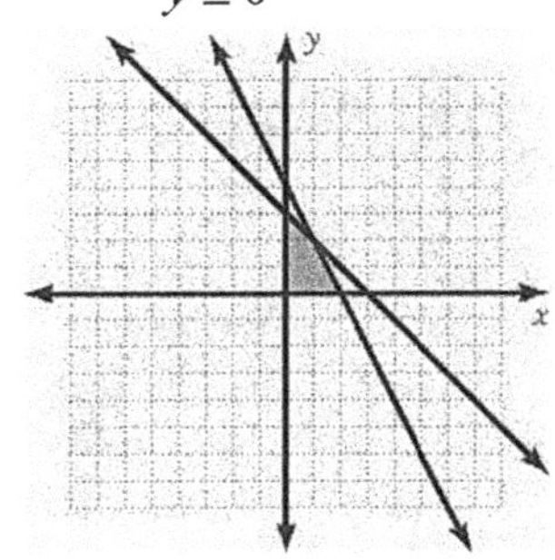

Chapter Test 4

1. $4x-7=5$

$f(x)=4x-7$

$g(x)=5$

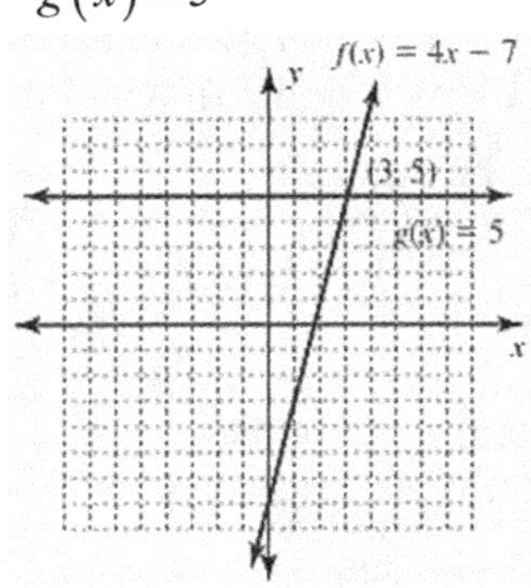

The point of intersection is (3, 5). The solution set is $\{3\}$.

3. $-8x+11=2x-9$

$f(x)=-8x+11$

$g(x)=2x-9$

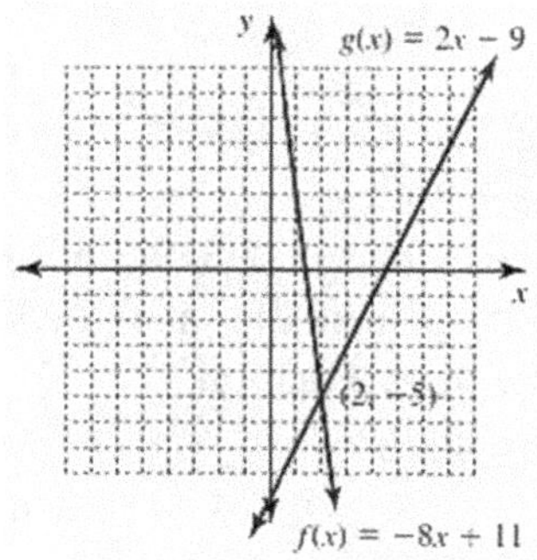

The point of intersection is (2, –5). The solution set is {2}.

5. $3x+y=-5 \rightarrow y=-3x-5$

$5x-2y=-23$

Substitute for y in the second equation.

$5x-2(-3x-5)=-23$ and $y=-3x-5$

$11x+10=-23$ $\quad =-3(-3)-5$

$x=-3$ $\quad =4$

The solution set for the system is {(–3, 4)}.

7. $9x-3y=4$

$-3x+y=-1 \rightarrow y=3x-1$

Substitute for y in the first equation.

$9x-3(3x-1)=4$

$3=4$

There are no solutions. The system is inconsistent.

9. $x-2y=5 \rightarrow x=2y+5$

$2x+5y=10$

Substitute for x in the second equation.

$2(2y+5)+5y=10$ and $x=2y+5$

$9y+10=10$ $\quad =2(0)+5=5$

$y=0$

$x=3$

The solution set for the system is {(5, 0)}.

11. $x-y+z=1$

$-2x+y+z=8$

$x+5z=19$

Pair the first two equations to eliminate y.

$x-y+z=1$

$-2x+y+z=8$

$-x+2z=9$

Giving the system $x+5z=19$

$-x+2z=9$

Add these two equations to solve the system.

$7z=28$ and $x+5z=19$

$z=4$ $\quad x+5(4)=19$

$x=-1$

And finally, $x-y+z=1$

$(-1)-y+(4)=1$

$y=2$

The solution set for the system is {(–1, 2, 4)}.

13. $x-2y<6 \rightarrow y>\frac{x}{2}-3$

$x+y<3 \rightarrow y<-x+3$

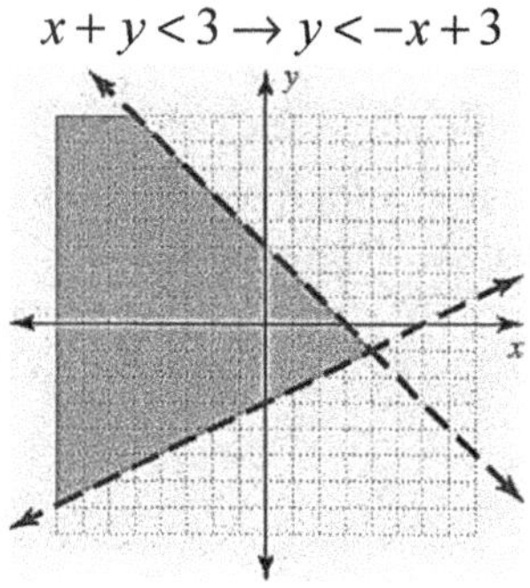

15. $x+2y\le 8 \rightarrow y\le -\frac{x}{2}+4$

$x+y\le 6 \rightarrow y\le -x+6$

$x\ge 0$

$y\ge 0$

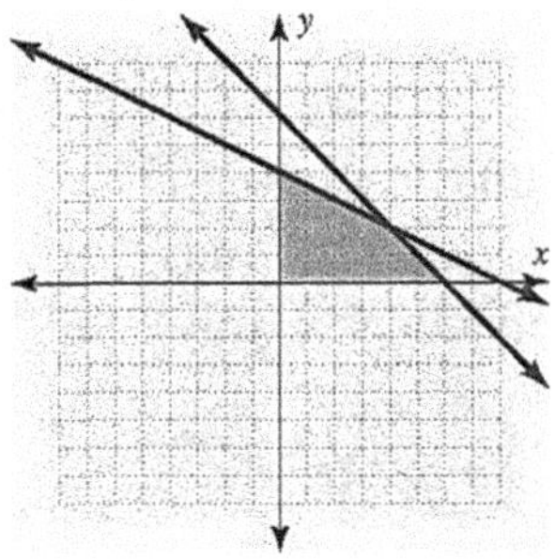

17. Let x be the amount of jawbreakers and y the amount of licorice (in pounds) that should be used.

$x+y=100 \rightarrow y=-x+100$

$2.40x+3.90y=3\bullet 100 \rightarrow 24x+39y=3{,}000$

$\rightarrow 8x+13y=1{,}000$

Substitute for y in the second equation.

$8x+13(-x+100)=1{,}000$ and $y=-x+100$

$-5x+1{,}300=1{,}000$ $\quad =-(60)+100$

$x=60$ $\quad =40$

The dealer should use 60 lb of jawbreakers and 40 lb of licorice.

19. Let x, y, and z be the amount invested in the savings account, bond, and mutual fund, respectively.

$x+y+z=16{,}000$

$0.03x+0.05y+0.07z=900$

$\rightarrow 3x+5y+7z=90{,}000$

$z=x+y \rightarrow -x-y+z=0$

Eliminate x by pairing the first and third equations, and by pairing the second equation with 3 times the third equation.

$x+y+z=16{,}000$ and $3x+5y+7z=90{,}000$

$-x-y+z=0$ $\quad -3x-3y+3z=0$

$2z=16{,}000$ $\quad 2y+10z=90{,}000$

$z=8{,}000$ $\quad y+5z=45{,}000$

Giving $y+5z=45{,}000$

$y+5(8{,}000)=45{,}000$

$y=5{,}000$

And finally, $x+y+z=16{,}000$

$x+(5{,}000)+(8{,}000)=16{,}000$

$x=3{,}000$

Hans invested \$3000 in a savings account, \$5000 in a bond, and \$8000 in a mutual fund.

Cumulative Review: Chapters 0–4

1. $3x-2(x+5)=12-3x$

$3x-2x-10=12-3x$

$4x=22$

$x=\frac{22}{4}=\frac{11}{2}; \left\{\frac{11}{2}\right\}$

3. $x+8\ge 4x-3$

$-3x\ge -11$

$x\le \frac{11}{3}$

$\left\{x \middle| x\le \frac{11}{3}\right\}$

5. $5x+7y=35 \rightarrow y=-\frac{5}{7}x+5$

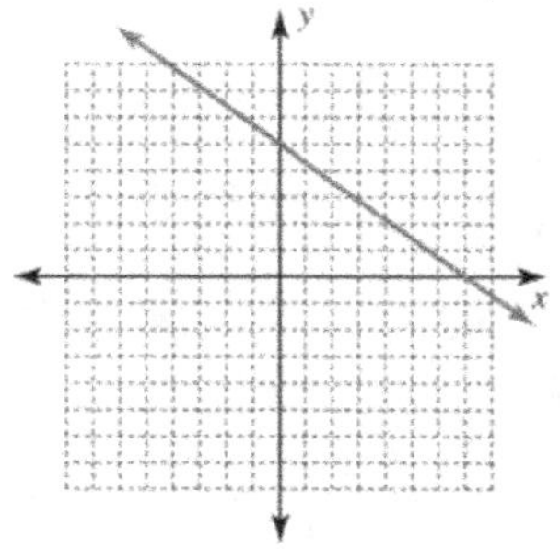

7.
$$P=P_0+IRT$$
$$P-P_0=IRT$$
$$R=\frac{P-P_0}{IT}$$

9.
$$m=\frac{-2-4}{5-(-1)}-1$$
$$y-4=-(x+1)$$
$$y=-x+3$$

11. $f(x)=3x^2-4x-5$

$$f(-5)=3(-5)^2-4(-5)-5=90$$

13.
$$x+y+z=3$$
$$2x-y+2z=0$$
$$-x-3y+z=-9$$

Eliminate y by pairing the first two equations, and by pairing three times the first with the third.

$$\begin{aligned} x+y+z&=3 \\ 2x-y+2z&=0 \\ 3x+3z&=3 \\ x+z&=1 \end{aligned} \quad \text{and} \quad \begin{aligned} 3x+3y+3z&=9 \\ -x-3y+z&=-9 \\ 2x+4z&=0 \\ x+2z&=0 \end{aligned}$$

Giving the system $x+z=1 \rightarrow x=1-z$

$$x+2z=0$$

Substitute for x in the second equation.

$(1-z)+2z=0$ and $x=1-z$

$z=-1$ $\quad =1-(-1)=2$

And finally, $x+y+z=3$

$$(2)+y+(-1)=3$$
$$y=2$$

The solution set of the system is $\{(2, 2, -1)\}$.

15. Let t and u represent the tens and units digits of the number.

$$t+u=10$$
$$10u+t=10t+u-36$$

Simplify this last equation, and add.

$$9t-9u=36 \rightarrow t-u=4$$

$$t+u=10$$
$$t-u=4$$
$$2t=14$$
$$t=7$$
$$u=3$$

Thus, $10t+u=10(7)+3=73$.

The original number is 73.

Chapter 5
Exponents and Polynomials

Exercises 5.1

< Objectives 1 and 2 >

1. $x^4 \bullet x^5 = x^{4+5} = x^9$

3. $x^5 \bullet x^3 \bullet x^2 = x^{5+3+2} = x^{10}$

5. $3^5 \bullet 3^2 = 3^{5+2} = 3^7$

7. $(-2)^3(-2)^5 = (-2)^{3+5} = (-2)^8 = 2^8$

9. $4 \bullet x^2 \bullet x^4 \bullet x^7 = 4x^{2+4+7} = 4x^{13}$

11. $\left(\frac{1}{2}\right)^2\left(\frac{1}{2}\right)^3\left(\frac{1}{2}\right) = \left(\frac{1}{2}\right)^{2+3+1} = \left(\frac{1}{2}\right)^6$

13. $(-2)^2(-2)^3(x^4)(x^5) = (-2)^{2+3}(x^{4+5})$
$= (-2)^5 x^9$

15. $(2x)^2(2x)^3(2x)^4 = (2x)^{2+3+4} = (2x)^9$

17. $(x^2y^3)(x^4y^2) = (x^2x^4)(y^3y^2) = x^6y^5$

19. $(x^3y^2)(x^4y^2)(x^2y^3) = (x^3x^4x^2)(y^2y^2y^3)$
$= x^9y^7$

21. $(2x^4)(3x^3)(-4x^3) = [2 \bullet 3 \bullet (-4)](x^4x^3x^3)$
$= -24x^{10}$

23. $(5x^2)(3x^3)(x)(-2x^3)$
$= [5 \bullet 3 \bullet 1 \bullet (-2)](x^2 \bullet x^3 \bullet x \bullet x^3) = -30x^9$

25. $(5xy^3)(2x^2y)(3xy)$
$= (5 \bullet 2 \bullet 3)(x \bullet x^2 \bullet x)(y^3 \bullet y \bullet y) = 30x^4y^5$

27. $(x^2yz)(x^3y^5z)(x^4yz)$
$= (x^2 \bullet x^3 \bullet x^4)(y \bullet y^5 \bullet y)(z \bullet z \bullet z) = x^9y^7z^3$

29. $\frac{x^{10}}{x^7} = x^{10-7} = x^3$

31. $\frac{x^7y^{11}}{x^4y^3} = x^{7-4}y^{11-3} = x^3y^8$

33. $\frac{x^5y^4z^2}{xy^2z} = x^{5-1}y^{4-2}z^{2-1} = x^4y^2z$

35. $\frac{21x^4y^5}{7xy^2} = \frac{21}{7}x^{4-1}y^{5-2} = 3x^3y^3$

37. $(-3x)(5x^5) = (-3 \bullet 5)(x \bullet x^5) = -15x^6$

39. $(2x)^3 = 2^3x^3 = 8x^3$

41. $(x^3)^7 = x^{3 \cdot 7} = x^{21}$

43. $(3x)(-2x)^3 = (3x)(-2)^3x^3 = 3(-8)(x \bullet x^3)$
$= -24x^4$

45. $(2x^3)^5 = 2^5x^{3 \cdot 5} = 32x^{15}$

47. $(-2x^2)^3(3x^2)^3 = (-2)^3x^{2 \cdot 3}(3)^3x^{2 \cdot 3}$
$= (-8)(27)(x^6 \bullet x^6)$
$= -216x^{12}$

49. $(3x^3)^2(x^2)^4 = 3^2x^{3\cdot 2}x^{2\cdot 4} = 9x^6x^8 = 9x^{14}$

51. $\left(\frac{3}{4}\right)^2 = \frac{3^2}{4^2} = \frac{9}{16}$

53. $\left(\frac{x}{5}\right)^3 = \frac{x^3}{5^3} = \frac{x^3}{125}$

55. $\left(\frac{m^3}{n^2}\right)^3 = \frac{(m^3)^3}{(n^2)^3} = \frac{m^9}{n^6}$

57. $\left(\frac{a^3b^2}{c^4}\right)^2 = \frac{(a^3b^2)^2}{(c^4)^2} = \frac{(a^3)^2(b^2)^2}{c^8} = \frac{a^6b^4}{c^8}$

59. $\left(\frac{2x^5}{y^3}\right)^2 = \frac{(2x^5)^2}{(y^3)^2} = \frac{2^2(x^5)^2}{y^6} = \frac{4x^{10}}{y^6}$

61. $(-8x^2y)(-3x^4y^5)^4$
$= (-8x^2y)\left[(-3)^4(x^4)^4(y^5)^4\right]$
$= (-8x^2y)(81x^{16}y^{20})$
$= (-8\cdot 81)(x^2\cdot x^{16})(y\cdot y^{20}) = -648x^{18}y^{21}$

63. $\left(\frac{3x^4y^9}{2x^2y^7}\right)\left(\frac{x^6y^3}{x^3y^2}\right)^2$
$= \left(\frac{3}{2}x^{4-2}y^{9-7}\right)\left(x^{6-3}y^{3-2}\right)^2$
$= \left(\frac{3}{2}x^2y^2\right)(x^3y)^2 = \left(\frac{3}{2}x^2y^2\right)(x^6y^2)$
$= \frac{3x^8y^4}{2}$

65. False

67. False

69. With a scientific calculator:
4 [x^y] 3 [=]
With a graphing calculator:
4 [∧] 3 [ENTER]
Display: 64
The result is 64.

71. With a scientific calculator:
[(] 3 [+/−] [)] [x^y] 4 [=]
With a graphing calculator:
[(] [(−)] 3 [)] [∧] 4 [ENTER]
Display: 81
The result is 81.

73. With a scientific calculator:
2 [x^y] 3 [×] 3 [x^y] 5 [=]
With a graphing calculator:
[(] 2 [∧] 3 [)] [×] [(] 2 [∧] 5 [)] [ENTER]
Display: 256
The result is 256.

75. $(3x^2)(2x^4) = \left[3(2)^2\right]\left[2(2)^4\right] = 384$

77. $(2x^4)(4x^2) = \left[2(-2)^4\right]\left[4(-2)^2\right] = 512$

79. $(-2x^3)(-3x^5) = \left[-2(2)^3\right]\left[-3(2)^5\right] = 1{,}536$

81. If $P = \$2{,}000$, $r = 0.05$, and $t = 8$ then
$A = P(1+r)^t = (2{,}000)(1+0.05)^8$
$= 2{,}000\cdot 1.05^8 = 2{,}954.91$
There will be $2,954.91 in the account.

83. **(a)** If $m = 12$ and $v = 4.9t^2$, then

$$KE = \frac{1}{2}mv^2 = \frac{1}{2}(12)(4.9t^2)^2 = \frac{1}{2}(12)(24.01t^4)$$
$$= 144.06t^4$$

The kinetic energy of a 12-kg falling object is given by $KE = 144.06t^4$.

(b) $KE = 144.06t^4 = 144.06(4)^4$

$$= 144.06(256) = 36{,}879.36 \text{ joules}$$

The kinetic energy of a 12-kg object 4s after it is dropped is 36,879.36 joules.

85. Let D be the depth of the beam; L be the length of the beam; and M be the moment of inertia of the beam.

Then $D = 4L^3$ and $M = \frac{1}{9}D^3$

So, $M = \frac{1}{9}D^3 = \frac{1}{9}(4L^3)^3 = \frac{1}{9}(64L^9) = \frac{64}{9}L^9$

The moment of inertia of a 4-in-wide beam is $M = \frac{64}{9}L^9$.

87. Above and Beyond

89. $x^{12} = x^{2\cdot 6} = (x^2)^6$

91. $a^{16} = a^{2\cdot 8} = (a^2)^8$

93. $2^{12} = 2^{3\cdot 4} = (2^3)^4 = 8^4$

$2^{18} = 2^{3\cdot 6} = (2^3)^6 = 8^6$

$(2^5)^3 = 2^{15} = 2^{3\cdot 5} = (2^3)^5 = 8^5$

$(2^7)^6 = 2^{42} = 2^{3\cdot 14} = (2^3)^{14} = 8^{14}$

95. $-8x^6y^9z^{15} = (-2)^3 x^{2\cdot 3}y^{3\cdot 3}z^{5\cdot 3} = (-2x^2y^3z^5)^3$

$-2x^2y^3z^5$ raised to the third power is $-8x^6y^9z^{15}$.

97. **(a)** If a country has a 2% growth rate for 35 years, then it takes that long for its population to double. Therefore, over a period of 105 years, $\frac{105}{35}$ doublings will occur. This means there will be 3 doublings.

$$G = (1+R)^y = (1+0.02)^{105} = (1.02)^{105} = 7.998 \approx 8$$

The country's population will be 8 times as large.

(b) The total population of a less developed country in 2010 is 5.9 billion. If the average growth rate remains unchanged in the next 105 years, the population will be 8 times as large. Therefore the country's population after 105 years is: $5.9 \times 8 = 47.2$ billion.

99. Above and Beyond

Exercises 5.2

< Objectives 1 and 2 >

1. $x^{-5} = \frac{1}{x^5}$

3. $5^{-2} = \frac{1}{5^2} = \frac{1}{25}$

5. $(-5)^{-2} = \frac{1}{(-5)^2} = \frac{1}{25}$

7. $(-2)^{-3} = \frac{1}{(-2)^3} = \frac{1}{-8} = -\frac{1}{8}$

9. $\left(\frac{2}{3}\right)^{-3} = \left(\frac{3}{2}\right)^3 = \frac{3^3}{2^3} = \frac{27}{8}$

11. $3x^{-2} = 3 \cdot \frac{1}{x^2} = \frac{3}{x^2}$

13. $-5x^{-4} = -5 \cdot \frac{1}{x^4} = -\frac{5}{x^4}$

15. $(-3x)^{-2} = \dfrac{1}{(-3x)^2} = \dfrac{1}{9x^2}$

17. $\dfrac{1}{x^{-3}} = x^3$

19. $\dfrac{2}{5x^{-3}} = \dfrac{2}{5} \bullet \dfrac{1}{x^{-3}} = \dfrac{2}{5}x^3$

21. $\dfrac{x^{-3}}{y^{-4}} = \dfrac{y^4}{x^3}$

23. $x^5 \bullet x^{-3} = x^{5+(-3)} = x^2$

25. $a^{-9} \bullet a^6 = a^{-9+6} = a^{-3} = \dfrac{1}{a^3}$

27. $z^{-2} \bullet z^{-8} = z^{-2+(-8)} = z^{-10} = \dfrac{1}{z^{10}}$

29. $a^{-5} \bullet a^5 = a^{-5+5} = a^0 = 1$

31. $\dfrac{x^{-5}}{x^{-2}} = x^{-5-(-2)} = x^{-3} = \dfrac{1}{x^3}$

33. $(x^5)^3 = x^{5\bullet 3} = x^{15}$

35. $(2x^{-3})(x^2)^4 = 2 \bullet x^{-3} \bullet x^{2\bullet 4} = 2 \bullet x^{-3+8} = 2x^5$

37. $(3a^{-4})(a^3)(a^2) = 3 \bullet a^{-4+3+2} = 3 \bullet a^1 = 3a$

39. $(x^4y)(x^2)^3(y^3)^0 = (x^4y)x^{2\bullet 3} \bullet y^{3\bullet 0}$

$= (x^4y) \bullet x^6 \bullet y^0$

$= (x^4 \bullet x^6)(y \bullet y^0) = x^{10}y$

41. $(ab^2c)(a^4)^4(b^2)^3(c^3)^4$

$= a^{1+4\bullet 4} \bullet b^{2+2\bullet 3} \bullet c^{1+3\bullet 4} = a^{17}b^8c^{13}$

43. $(x^5)^{-3} = x^{5(-3)} = x^{-15} = \dfrac{1}{x^{15}}$

45. $(b^{-4})^{-2} = b^{(-4)(-2)} = b^8$

47. $(x^5y^{-3})^2 = (x^5)^2(y^{-3})^2 = x^{5\cdot 2} \bullet y^{-3\bullet 2}$

$= x^{10} \bullet y^{-6} = \dfrac{x^{10}}{y^6}$

49. $(x^{-4}y^{-2})^{-3} = x^{(-4)(-3)} \bullet y^{(-2)(-3)} = x^{12}y^6$

51. $(2x^{-3}y^0)^{-5} = 2^{-5} \bullet x^{(-3)(-5)} \bullet y^{0(-5)} = \dfrac{x^{15}y^0}{2^5}$

$= \dfrac{x^{15}}{32}$

53. $\dfrac{x^{-2}}{y^{-4}} = \dfrac{y^4}{x^2}$

55. $\dfrac{x^{-4}}{y^{-2}} = \dfrac{y^2}{x^4}$

57. $(4x^{-2})^2(3x^{-4}) = 4^2 \bullet 3 \bullet x^{-2\bullet 2+(-4)} = 16 \bullet 3x^{-8}$

$= \dfrac{48}{x^8}$

59. $(2x^5)^4(x^3)^2 = 2^4(x^5)^4x^{3\bullet 2} = 16 \bullet x^{5\bullet 4+6}$

$= 16x^{26}$

61. $(2x^{-3})^3(3x^3)^2 = 2^3(x^{-3})^3 \bullet 3^2(x^3)^2$

$= 8 \bullet 9 \bullet x^{-3\bullet 3} \bullet x^{3\bullet 2} = 72 \bullet x^{-9+6}$

$= 72 \bullet x^{-3} = \dfrac{72}{x^3}$

63. $(xy^5z)^4(xyz^2)^8(x^6yz)^5$

$= (x^4y^{5\bullet 4}z^4)(x^8y^8z^{2\bullet 8})(x^{6\bullet 5}y^5z^5)$

$= x^{4+8+30} \bullet y^{20+8+5} \bullet z^{4+16+5} = x^{42}y^{33}z^{25}$

65. $(3x^{-2})(5x^2)^2 = (3x^{-2})(25x^4) = 75 \bullet x^{-2+4}$

$= 75x^2$

67. $\left(2w^3\right)^4\left(3w^{-5}\right)^2 = 2^4 \bullet w^{3\bullet 4} \bullet 3^2 \bullet w^{-5\bullet 2}$
$= 16 \bullet 9 \bullet w^{12-10} = 144w^2$

69. $\dfrac{3x^6}{2y^9} \bullet \dfrac{y^5}{x^3} = \dfrac{3}{2} \bullet x^{6-3} \bullet y^{5-9} = \dfrac{3}{2} \bullet x^3 \bullet y^{-4}$
$= \dfrac{3x^3}{2y^4}$

71. $\left(-7x^2y\right)\left(-3x^5y^6\right)^4$
$= \left(-7x^2y\right)\left[(-3)^4 x^{5\cdot 4} y^{6\cdot 4}\right]$
$= -7 \bullet 81 \bullet x^{2+20} \bullet y^{1+24} = -567x^{22}y^{25}$

73. $\left(2x^2y^{-3}\right)\left(3x^{-4}y^{-2}\right) = 6 \bullet x^{2+(-4)} \bullet y^{-3+(-2)}$
$= 6 \bullet x^{-2} \bullet y^{-5} = \dfrac{6}{x^2y^5}$

75. $\dfrac{\left(x^{-3}\right)\left(y^2\right)}{y^{-3}} = x^{-3} \bullet y^{2-(-3)} = x^{-3} \bullet y^5 = \dfrac{y^5}{x^3}$

77. $\dfrac{15x^{-3}y^2z^{-4}}{20x^{-4}y^{-3}z^2} = \dfrac{15}{20} \bullet x^{-3-(-4)} \bullet y^{2-(-3)} \bullet z^{-4-2}$
$= \dfrac{3}{4} \bullet x^1 \bullet y^5 \bullet z^{-6} = \dfrac{3xy^5}{4z^6}$

79. $\dfrac{x^{-5}y^{-7}}{x^0y^{-4}} = x^{-5-0} \bullet y^{-7-(-4)} = x^{-5} \bullet y^{-3} = \dfrac{1}{x^5y^3}$

81. $\dfrac{x^{-2}y^2}{x^3y^{-2}} \bullet \dfrac{x^{-4}y^2}{x^{-2}y^{-2}} = \dfrac{x^{-2+(-4)} \bullet y^{2+2}}{x^{3+(-2)} \bullet y^{-2+(-2)}} = \dfrac{x^{-6} \bullet y^4}{x \bullet y^{-4}}$
$= x^{-6-1} \bullet y^{4-(-4)} = x^{-7} \bullet y^8$
$= \dfrac{y^8}{x^7}$

< Objective 3 >

83. $9\times 10^4 = 90{,}000$

85. $1.21\times 10^1 = 12.1$

87. $8\times 10^{-3} = 0.008$

89. $2.8\times 10^{-5} = 0.000028$

91. $3{,}420{,}000 = 3.42\times 10^6$

93. $8 = 8\times 10^0$

95. $0.0005 = 5\times 10^{-4}$

97. $0.00037 = 3.7\times 10^{-4}$

99. $93{,}000{,}000 = 9.3\times 10^7$

101. $130{,}000{,}000{,}000 = 1.3\times 10^{11}$

103. $1.98\times 10^{30} = 198\times 10^{28}$
28 zeros follow the digit 8.

105. $\left(2\times 10^5\right)\left(4\times 10^4\right) = (2\times 4)\left(10^{5+4}\right) = 8\times 10^9$

107. $\dfrac{6\times 10^9}{3\times 10^7} = 2\times 10^{9-7} = 2\times 10^2$

109. $\dfrac{\left(3.3\times 10^{15}\right)\left(6\times 10^{15}\right)}{\left(1.1\times 10^8\right)\left(3\times 10^6\right)} = \left(\dfrac{3\times 3\times 6}{1.1\times 3}\right)10^{15+15-8-6}$
$= 6\times 10^{16}$

111. $\left(4\times 10^{-3}\right)\left(2\times 10^{-5}\right) = (4\times 2)\left(10^{-3}\times 10^{-5}\right)$
$= 8\times 10^{-3+(-5)} = 8\times 10^{-8}$

113. $\dfrac{9\times 10^3}{3\times 10^{-2}} = 3\times 10^{3-(-2)} = 3\times 10^5$

< Objective 4 >

115. $\dfrac{6.6\times 10^{17}}{10^{16}} = 6.6\times 10^{17-16} = 6.6\times 10^1 = 66$
It takes light approximately 66 years to travel from Megrez to Earth.

117. $15{,}500\times 10^{19} = 1.55\times 10^4\times 10^{19} = 1.55\times 10^{23}$
The amount of water on Earth is approximately 1.55×10^{23} liters.

119. $(7.1\times10^9)(6.57\times10^5)\approx46.6\times10^{14}$
$=4.66\times10^{15}$
There is 4.66×10^{15} L of freshwater on the Earth.

121. False

123. True

125. $P=7\times2^{\frac{(2025-2010)}{45}}=7\times2^{\frac{1}{3}}\approx8.82$
(rounded to two decimal places)
Earth's population in 2025 will be approximately 8.82 billion.

127. $P=310\times2^{\frac{(2025-2010)}{60}}=310\times2^{\frac{1}{4}}\approx369$
The U.S. population in 2025 will be approximately 369 million, rounded to the nearest million.

129. **(a)** $12\times10^4\,\Omega=120{,}000\,\Omega$
(b) $1.2\times10^5\,\Omega$

131. $c=\frac{A}{\varepsilon d}=\frac{0.254}{(3.6\times10^3)(1.4)}=\frac{2.54\times10^{-1}}{5.04\times10^3}$
$\approx0.504\times10^{-1-3}=0.504\times10^{-4}$
$=5.04\times10^{-5}$
The concentration is approximately 5.04×10^{-5} moles per liter.

133. $x^{2n}\bullet x^{3n}=x^{2n+3n}=x^{5n}$

135. $\frac{x^{n+3}}{x^{n+1}}=x^{(n+3)-(n+1)}=x^2$

137. $(y^n)^{3n}=y^{n\bullet3n}=y^{3n^2}$

139. $\frac{x^{2n}\bullet x^{n+2}}{x^{3n}}=x^{2n+(n+2)-3n}=x^2$

141. **(a)** 30 cuts gives 2^{30} pieces of paper.
$(2^{30})(0.002)\approx2{,}147{,}484\text{ in.}=178{,}957\text{ ft}$
$\approx33.9\text{ mi}$
The stack would be approximately 33.9 miles high.
(b) $(2^{50})(0.002)\approx2.2517998\times10^{12}$ in
$\approx1.8764998\times10^{11}$ ft
$\approx35{,}539{,}770$ mi
The stack would be approximately 35.5 million mi high.

143. Above and Beyond

145. Above and Beyond

Exercises 5.3

< Objective 1 >

1. $7x^3$ is a polynomial.

3. $2x^5y^3-4x^2y^4$ is a polynomial.

5. –7 is a polynomial.

7. $\frac{3+x}{x^2}$ is not a polynomial.

9. Terms: $2x^2$, $-3x$
Coefficients: 2, –3

11. Terms: $4x^3$, $-3x$, 2
Coefficients: 4, –3, 2

13. $4x^3-2x^2$ is a binomial.

15. $7y^2+4y+5$ is a trinomial.

17. $2x^4-3x^2+5x-2$ is not classified.

19. $7x^{10}$ is a monomial.

21. $x^5-\frac{3}{x^2}$ is not a polynomial.

< Objectives 2 and 3 >

23. $4x^5 - 3x^2$ has degree 5. The leading coefficient is 4.

25. $-5x^9 + 7x^7 + 4x^3$ has degree 9. The leading coefficient is –5.

27. $-9x$ has degree 1. The leading coefficient is –9.

29. $x^6 - 3x^5 + 5x^2 - 7$ has degree 6. The leading coefficient is 1.

31.
$$x = 1$$
$$9x - 2 = 9(1) - 2 = 7$$
$$x = -1$$
$$9x - 2 = 9(-1) - 2 = -11$$

33.
$$x = 2$$
$$x^3 - 2x = (2)^3 - 2(2) = 8 - 4 = 4$$
$$x = -2$$
$$x^3 - 2x = (-2)^3 - 2(-2) = -8 + 4 = -4$$

35.
$$x = 4$$
$$3x^2 + 4x - 2 = 3(4)^2 + 4(4) - 2 = 3 \cdot 16 + 16 - 2 = 62$$
$$x = -4$$
$$3x^2 + 4x - 2 = 3(-4)^2 + 4(-4) - 2 = 3 \cdot 16 - 16 - 2 = 30$$

37.
$$x = 3$$
$$-x^2 - x + 12 = -(3)^2 - (3) + 12 = -9 - 3 + 12 = 0$$
$$x = -4$$
$$-x^2 - x + 12 = -(-4)^2 - (-4) + 12 = -16 + 4 + 12 = 0$$

39. The polynomial to describe the cost of typing x number of pages is $5x + 35$. For 50-page paper: $5(50) + 35 = 285$. The cost of typing a 50-page paper is $285.

41. $4(12)^2 - 95 = 481$

The revenue, when 12 pairs of slippers are sold, is $481.

43. always

45. sometimes

47. sometimes

49. sometimes

51. $\text{Wh} = 58t + 144 = 58(0) + 144 = 144$

In the 24-hour period of non-operation, the TV consumed 144 Wh.

53.
$$g = 0.472t^3 - 5.298t^2 + 11.802t + 93.143$$
$$= 0.472(5)^3 - 5.298(5)^2 + 11.802(5) + 93.143$$
$$= 78.703$$

55. $P(1) = (1)^3 - 2(1)^2 + 5 = 4$

57. $Q(2) = 2(2)^2 + 3 = 11$

59. $P(3) = (3)^3 - 2(3)^2 + 5 = 14$

61. $P(0) = (0)^3 - 2(0)^2 + 5 = 5$

63. $P(2) + Q(-1)$
$$= \left[(2)^3 - 2(2)^2 + 5\right] + \left[2(-1)^2 + 3\right] = 5 + 5$$
$$= 10$$

65. $P(3) - Q(-3) \div Q(0)$
$$= \left[(3)^3 - 2(3)^2 + 5\right] - \left[2(-3)^2 + 3\right] \div \left[2(0)^2 + 3\right]$$
$$= 14 - (21 \div 3) = 7$$

67. $|Q(4)|-|P(4)|$
$=\left|2(4)^2+3\right|-\left|(4)^3-2(4)^2+5\right|=|35|-|37|$
$=-2$

Exercises 5.4

< Objective 1 >

1. $(5a-7)+(4a+11)=5a-7+4a+11$
$=(5a+4a)+(-7+11)$
$=9a+4$

3. $(8b^2-11b)+(5b^2-7b)$
$=8b^2-11b+5b^2-7b=13b^2-18b$

5. $(3x^2-2x)+(-5x^2+2x)=-2x^2$

7. $(2x^2+5x-3)+(3x^2-7x+4)$
$=2x^2+5x-3+3x^2-7x+4=5x^2-2x+1$

9. $(3b^2-7)+(2b-7)=3b^2-7+2b-7$
$=3b^2+2b-14$

11. $(8y^3-5y^2)+(5y^2-2y)=8y^3-2y$

13. $(2a^2-4a^3)+(3a^3+2a^2)=-a^3+4a^2$

15. $(7x^2-5+4x)+(8-5x-9x^2)$
$=-2x^2-x+3$

17. $-(4a+5b)=-4a-5b$

19. $5a-(2b-3c)=5a-2b+3c$

21. $9r-(3r+5s)=9r-3r-5s=6r-5s$

23. $5p-(-3p+2q)=5p+3p-2q=8p-2q$

< Objective 2 >

25. $(2x-3)-(x+4)=2x-3-x-4=x-7$

27. $(4m^2-5m)-(3m^2-2m)$
$=4m^2-5m-3m^2+2m=m^2-3m$

29. $(4y^2+5y)-(6y^2+5y)$
$=4y^2+5y-6y^2-5y=-2y^2$

31. $(3x^2-5x-2)-(x^2-4x-3)$
$=3x^2-5x-2-x^2+4x+3=2x^2-x+1$

33. $(8a^2-9a)-(3a+7)=8a^2-9a-3a-7$
$=8a^2-12a-7$

35. $(5b-2b^2)-(4b^2-3b)$
$=5b-2b^2-4b^2+3b=-6b^2+8b$

37. $(7x^2-11x+31)-(5x^2+19-11x)$
$=7x^2-11x+31-5x^2-19+11x=2x^2+12$

39. $(2x^3-3x+1)-(4x^2+8x+6)$
$=2x^3-3x+1-4x^2-8x-6$
$=2x^3-4x^2-11x-5$

41. $[(5b-4)+(3b+8)]-(2b+5)$
$=(8b+4)-(2b+5)=8b+4-2b-5$
$=6b-1$

43. $\left[(x^2+5x-2)+(2x^2+7x-8)\right]$
$-(3x^2+2x-1)$
$=(3x^2+12x-10)-(3x^2+2x-1)$
$=3x^2+12x-10-3x^2-2x+1=10x-9$

45. $\left[(4x^2-5)+(2x-7)\right]-(2x^2-3x)$
$=(4x^2+2x-12)-(2x^2-3x)$
$=4x^2+2x-12-2x^2+3x=2x^2+5x-12$

47. $(2y^2-8y)-[(3y^2-3y)+(5y^2+3y)]$
$=(2y^2-8y)-(8y^2)=-6y^2-8y$

49. $[(9x^2-3x+5)-(3x^2+2x-1)]$
$-(x^2-2x-3)$
$=(9x^2-3x+5-3x^2-2x+1)$
$-(x^2-2x-3)$
$=(6x^2-5x+6)-(x^2-2x-3)$
$=6x^2-5x+6-x^2+2x+3=5x^2-3x+9$

51.
$$\begin{array}{r} 4w^2 \qquad\quad +11 \\ 7w \quad -9 \\ \underline{2w^2-9w \qquad\quad} \\ 6w^2-2w \quad +2 \end{array}$$

53.
$$\begin{array}{l} 3x^2+3x-4 \\ 4x^2-3x-3 \\ \underline{2x^2-\ x+7} \\ 9x^2-x \end{array}$$

55.
$$\begin{array}{r} 9a^2-4a \\ \underline{-(7a^2-9a)} \end{array} \qquad \begin{array}{r} 9a^2-4a \\ \underline{-7a^2+9a} \end{array}$$

57.
$$\begin{array}{r} 8x^2-5x+7 \\ \underline{-(5x^2-6x+7)} \end{array} \qquad \begin{array}{r} 8x^2-5x+7 \\ \underline{-5x^2+6x-7} \\ 3x^2+x \end{array}$$

59.
$$\begin{array}{r} 8x^2-9 \\ \underline{-(5x^2-3x)} \end{array} \qquad \begin{array}{r} 8x^2 \qquad -9 \\ \underline{-5x^2+3x \qquad} \\ 3x^2+3x-9 \end{array}$$

61. $P=2L+2W=2(8x+9)+2(6x-7)$
$=16x+18+12x-14=28x+4$
The perimeter is given by $28x+4$.

63. Profit $(P)=$ Revenue $(R)-$ Cost (C)
$=(90x-x^2)-(150+25x)$
$=90x-x^2-150-25x$
$=-x^2+65x-150$
The profit can be represented by $-x^2+65x-150$.

65. sometimes

67. The shear polynomial S_v for a polymer after vulcanization is
$S_v=(0.4x^2-144x+318)$
$+(0.2x^2-14x+144)$
$=0.6x^2-158x+452$

69. If $\alpha=3.9\times10^{-3}$ and $R_0=1.72\times10^{-8}$, then
$R_t=R_0(1+\alpha t)$
$=(1.72\times10^{-8})(1+3.9\times10^{-3}t)$
$=(1.72\times10^{-8})(1)+(1.72\times10^{-8})(1+3.9\times10^{-3}t)$
$=1.72\times10^{-8}+6.708\times10^{-11}t$
The resistance R_t of the copper piece is $R_t=1.72\times10^{-8}+6.708\times10^{-11}t$.

71. $3ax^4-5x^3+x^2-cx+2$
$=9x^4-bx^3+x^2-2d$
$3a=9 \quad -5=-b \quad -c=0 \quad 2=-2d$
$a=3 \quad b=5 \quad c=0 \quad d=-1$

Exercises 5.5

< Objectives 1 and 2 >

1. $(5x^2)(3x^3)=(5\cdot3)(x^2x^3)=15x^5$

3. $(-2b^2)(14b^8)=(-2\cdot14)(b^2b^8)=-28b^{10}$

5. $(-5p^7)(-8p^6)=[(-5)(-8)](p^7p^6)$
$=40p^{13}$

7. $(4m^5)(-3m) = [4 \cdot (-3)](m^5 m) = -12m^6$

9. $(4x^3y^2)(8x^2y) = (4 \cdot 8)(x^3x^2)(y^2y)$
$= 32x^5y^3$

11. $(-3m^5n^2)(2m^4n) = (-3 \cdot 2)(m^5m^4)(n^2n)$
$= -6m^9n^3$

13. $10(x+3) = 10 \cdot x + 10 \cdot 3 = 10x + 30$

15. $3a(4a+5) = 3a \cdot 4a + 3a \cdot 5 = 12a^2 + 15a$

17. $3s^2(4s^2 - 7s) = 3s^2 \cdot 4s^2 - 3s^2 \cdot 7s$
$= 12s^4 - 21s^3$

19. $3x(5x^2 - 3x - 1) = 3x \cdot 5x^2 - 3x \cdot 3x - 3x \cdot 1$
$= 15x^3 - 9x^2 - 3x$

21. $3xy(2x^2y + xy^2 + 5xy)$
$= 3xy \cdot 2x^2y + 3xy \cdot xy^2 + 3xy \cdot 5xy$
$= 6x^3y^2 + 3x^2y^3 + 15x^2y^2$

23. $6m^2n(3m^2n - 2mn + mn^2)$
$= 6m^2n \cdot 3m^2n - 6m^2n \cdot 2mn + 6m^2n \cdot mn^2$
$= 18m^4n^2 - 12m^3n^2 + 6m^3n^3$

25. $(x+3)(x+2) = x \cdot x + x \cdot 2 + 3 \cdot x + 3 \cdot 2$
$= x^2 + 5x + 6$

27. $(m-5)(m-9) = m^2 - 9m - 5m + 45$
$= m^2 - 14m + 45$

29. $(p-8)(p+7) = p^2 + 7p - 8p - 56$
$= p^2 - p - 56$

31. $(w-7)(w+6) = w^2 + 6w - 7w - 42$
$= w^2 - w - 42$

33. $(3x-5)(x-8) = 3x^2 - 24x - 5x + 40$
$= 3x^2 - 29x + 40$

35. $(2x-3)(3x+4) = 6x^2 + 8x - 9x - 12$
$= 6x^2 - x - 12$

37. $(3a-b)(4a-9b)$
$= 12a^2 - 27ab - 4ab + 9b^2$
$= 12a^2 - 31ab + 9b^2$

39. $(3p-4q)(7p+5q)$
$= 21p^2 + 15pq - 28pq - 20q^2$
$= 21p^2 - 13pq - 20q^2$

41.
$$\begin{array}{rrrrr}
 & & 2x^2 & -x & +8 \\
 & & x^2 & +4x & +6 \\
\hline
 & & 12x^2 & -6x & +48 \\
 & 8x^3 & -4x^2 & +32x & \\
2x^4 & -x^3 & +8x^2 & & \\
\hline
2x^4 & +7x^3 & +16x^2 & +26x & +48
\end{array}$$

< Objective 3 >

43. $(x+3)^2 = (x)^2 + 2(3)(x) + (3)^2$
$= x^2 + 6x + 9$

45. $(w-6)^2 = w^2 - 2 \cdot 6 \cdot w + (-6)^2$
$= w^2 - 12w + 36$

47. $(z+12)^2 = z^2 + 2 \cdot 12 \cdot z + 144$
$= z^2 + 24z + 144$

49. $(2a-1)^2 = (2a)^2 - 2 \cdot 1 \cdot 2a + 1$
$= 4a^2 - 4a + 1$

51. $(6m+1)^2 = (6m)^2 + 2 \cdot 1 \cdot 6m + 1$
$= 36m^2 + 12m + 1$

53. $(3x-y)^2 = (3x)^2 - 2(3x)(y) + (-y)^2$
$= 9x^2 - 6xy + y^2$

55. $(2r+5s)^2 = 4r^2 + 2(2r)(5s) + 25s^2$
$= 4r^2 + 20rs + 25s^2$

57. $(6a-5b)^2 = 36a^2 - 2(6a)(5b) + 25b^2$
$= 36a^2 - 60ab + 25b^2$

59. $\left(x+\frac{1}{2}\right)^2 = x^2 + 2\bullet\frac{1}{2}\bullet x + \left(\frac{1}{2}\right)^2$
$= x^2 + x + \frac{1}{4}$

< Objective 4 >

61. $(x-6)(x+6) = (x)^2 - (6)^2 = x^2 - 36$

63. $(m+7)(m-7) = m^2 - 49$

65. $\left(x-\frac{1}{2}\right)\left(x+\frac{1}{2}\right) = x^2 - \left(\frac{1}{2}\right)^2 = x^2 - \frac{1}{4}$

67. $(p-0.4)(p+0.4) = p^2 - (0.4)^2$
$= p^2 - 0.16$

69. $(a-3b)(a+3b) = a^2 - (3b)^2 = a^2 - 9b^2$

71. $(3x+2y)(3x-2y) = (3x)^2 - (2y)^2$
$= 9x^2 - 4y^2$

73. $(8w+5z)(8w-5z) = 64w^2 - 25z^2$

75. $(5x-9y)(5x+9y) = 25x^2 - 81y^2$

77. $2x(3x-2)(4x+1) = 2x(12x^2 - 5x - 2)$
$= 24x^3 - 10x^2 - 4x$

79. $5a(4a-3)(4a+3) = 5a(16a^2 - 9)$
$= 80a^3 - 45a$

81. $3s(5s-2)(4s-1) = 3s(20s^2 - 13s + 2)$
$= 60s^3 - 39s^2 + 6s$

83. $(x-2)(x+1)(x-3)$
$= (x-2)(x^2 - 2x - 3)$
$= x^3 - 2x^2 - 3x - 2x^2 + 4x + 6$
$= x^3 - 4x^2 + x + 6$

85. $(a-1)^3 = (a-1)(a-1)^2$
$= (a-1)(a^2 - 2a + 1)$
$= a^3 - 2a^2 + a - a^2 + 2a - 1$
$= a^3 - 3a^2 + 3a - 1$

87. $\left(\frac{x}{2}+\frac{2}{3}\right)\left(\frac{2x}{3}-\frac{2}{5}\right)$
$= \frac{x}{2}\bullet\frac{2x}{3} - \frac{x}{2}\bullet\frac{2}{5} + \frac{2}{3}\bullet\frac{2x}{3} - \frac{2}{3}\bullet\frac{2}{5}$
$= \frac{x^2}{3} - \frac{x}{5} + \frac{4x}{9} - \frac{4}{15} = \frac{x^2}{3} + \frac{11x}{45} - \frac{4}{15}$

89. $[x+(y-2)][x-(y-2)]$
$= x^2 - (y-2)^2 = x^2 - (y^2 - 4y + 4)$
$= x^2 - y^2 + 4y - 4$

91. $A = LW = (3x+5)(2x-7) = 6x^2 - 11x - 35$
The area of the rectangle is $(6x^2 - 11x - 35)$ cm.

93. *R* represents the revenue generated from the sales of *x* vases.
$R = xp = x(12 - 0.05x) = 12x - 0.05x^2$

95. $(5x-4)^2 = 25x^2 - 40x + 16$
There are $25x^2 - 40x + 16$ trees in the orchard.

97. Let *x* be the length of the rectangle. Then $x+2$ represents its width.
$A = LW = x(x+2) = x^2 + 2x.$

99. Let x be the number.
$(x+6)(x-6)=x^2-36$

101. Let x be the number.
$(x-4)^2=x^2-8x+16$

103. $(x+y)^2=x^2+2xy+y^2\neq x^2+y^2$
False

105. $(x+y)^2=x^2+2xy+y^2$
True

107. $M=\dfrac{wc^2}{8}\bullet\dfrac{c^2}{2}=\dfrac{wc^4}{16}$
The new bending moment with a load on the cantilever is $M=\dfrac{wc^4}{16}$.

109. The allowable stress of a material after allowing is
$(86.2x-0.6x^2+258)\left(\dfrac{p}{8.6}\right)$
$=(86.2x)\left(\dfrac{p}{8.6}\right)-0.6x^2\left(\dfrac{p}{8.6}\right)+258\left(\dfrac{p}{8.6}\right)$
$\approx 10.02xp-0.07x^2p+30p$
(rounded to two decimal places).

111. Above and Beyond

113. Above and Beyond

115. Above and Beyond

117. $(49)(51)=(50-1)(50+1)=2{,}500-1$
$=2{,}499$

119. $(34)(26)=(30+4)(30-4)=900-16$
$=884$

121. $(55)(65)=60^2-5^2=3{,}600-25=3{,}575$

Exercises 5.6

< Objective 1 >

1. $\dfrac{22x^9}{11x^5}=2x^{9-5}=2x^4$

3. $\dfrac{35m^3n^2}{7mn^2}=5m^2$

5. $\dfrac{3a+6}{3}=\dfrac{3a}{3}+\dfrac{6}{3}=a+2$

7. $\dfrac{9b^2-12}{3}=\dfrac{9b^2}{3}-\dfrac{12}{3}=3b^2-4$

9. $\dfrac{28a^3-42a^2}{7a}=\dfrac{28a^3}{7a}-\dfrac{42a^2}{7a}=4a^2-6a$

11. $\dfrac{12m^2+6m}{-3m}=\dfrac{12m^2}{-3m}+\dfrac{6m}{-3m}=-4m-2$

13. $\dfrac{18a^4-45a^3+63a^2}{9a^2}=\dfrac{18a^4}{9a^2}-\dfrac{45a^3}{9a^2}+\dfrac{63a^2}{9a^2}$
$=2a^2-5a+7$

15. $\dfrac{20x^5-15x^4+5x^3+10x^2-5x}{5x}$
$=\dfrac{20x^5}{5x}-\dfrac{15x^4}{5x}+\dfrac{5x^3}{5x}+\dfrac{10x^2}{5x}-\dfrac{5x}{5x}$
$=4x^4-3x^3+x^2+2x-1$

17. $\dfrac{16a^3b^2-4a^2b^2+8a^2b^3}{4a^2b^2}$
$=\dfrac{16a^3b^2}{4a^2b^2}-\dfrac{4a^2b^2}{4a^2b^2}+\dfrac{8a^2b^3}{4a^2b^2}=4a-1+2b$

19. $\dfrac{20x^4y^2-15x^2y^3+10x^3y}{5x^2y}$
$=\dfrac{20x^4y^2}{5x^2y}-\dfrac{15x^2y^3}{5x^2y}+\dfrac{10x^3y}{5x^2y}$
$=4x^2y-3y^2+2x$

< Objective 2 >

21.
$$\begin{array}{r} x+3 \\ x-4\overline{)x^2-x-12} \\ \underline{x^2-4x} \\ 3x-12 \\ \underline{3x-12} \\ 0 \end{array}$$
$$\frac{x^2-x-12}{x-4}=x+3$$

23.
$$\begin{array}{r} x-5 \\ x+4\overline{)x^2-x-20} \\ \underline{x^2+4x} \\ -5x-20 \\ \underline{-5x-20} \\ 0 \end{array}$$
$$\frac{x^2-x-20}{x+4}=x-5$$

25.
$$\begin{array}{r} x+6 \\ x-5\overline{)x^2+x-30} \\ \underline{x^2-5x} \\ 6x-30 \\ \underline{6x-30} \\ 0 \end{array}$$
$$\frac{x^2+x-30}{x-5}=x+6$$

27.
$$\begin{array}{r} 2x+3 \\ x-3\overline{)2x^2-3x-5} \\ \underline{2x^2-6x} \\ 3x-5 \\ \underline{3x-9} \\ 4 \end{array}$$
$$\frac{2x^2-3x-5}{x-3}=2x+3+\frac{4}{x-3}$$

29.
$$\begin{array}{r} 4x+2 \\ x-5\overline{)4x^2-18x-15} \\ \underline{4x^2-20x} \\ 2x-15 \\ \underline{2x-10} \\ -5 \end{array}$$
$$\frac{4x^2-18x-15}{x-5}=4x+2+\frac{-5}{x-5}$$

31.
$$\begin{array}{r} 2x+3 \\ 3x-5\overline{)6x^2-x-10} \\ \underline{6x^2-10x} \\ 9x-10 \\ \underline{9x-15} \\ 5 \end{array}$$
$$\frac{6x^2-x-10}{3x-5}=2x+3+\frac{5}{3x-5}$$

33.
$$\begin{array}{r} x^2-x-2 \\ x+2\overline{)x^3+x^2-4x-4} \\ \underline{x^3+2x^2} \\ -x^2-4x-4 \\ \underline{-x^2-2x} \\ -2x-4 \\ \underline{-2x-4} \\ 0 \end{array}$$
$$\frac{x^3+x^2-4x-4}{x+2}=x^2-x-2$$

35.
$$\begin{array}{r} x^2+2x+3 \\ 4x-1\overline{)4x^3+7x^2+10x+5} \\ \underline{4x^3-x^2} \\ 8x^2+10x+5 \\ \underline{8x^2-2x} \\ 12x+5 \\ \underline{12x-3} \\ 8 \end{array}$$
$$\frac{4x^3+7x^2+10x+5}{4x-1}=x^2+2x+3+\frac{8}{4x-1}$$

37.
$$\begin{array}{r}
x^2 + x + 2 \\
x-2 \overline{) x^3 - x^2 + 0x + 5} \\
\underline{x^3 - 2x^2} \quad\quad\quad \\
x^2 + 0x + 5 \\
\underline{x^2 - 2x} \quad \\
2x + 5 \\
\underline{2x - 4} \\
9
\end{array}$$

$$\frac{x^3 - x^2 + 5}{x-2} = x^2 + x + 2 + \frac{9}{x-2}$$

39.
$$\begin{array}{r}
7x^2 + 3x + 1 \\
7x-3 \overline{) 49x^3 + 0x^2 - 2x + 0} \\
\underline{7x^3 - 21x^2} \quad\quad\quad \\
21x^2 - 2x + 0 \\
\underline{21x^2 - 9x} \quad \\
7x + 0 \\
\underline{7x - 3} \\
3
\end{array}$$

$$\frac{49x^3 - 2x}{7x-3} = 7x^2 + 3x + 1 + \frac{3}{7x-3}$$

41. $2x^2 - 8 - 3x + x^3 = x^3 + 2x^2 - 3x - 8$
$$\begin{array}{r}
x^2 + 4x + 5 \\
x-2 \overline{) x^3 + 2x^2 - 3x - 8} \\
\underline{x^3 - 2x^2} \quad\quad\quad \\
4x^2 - 3x - 8 \\
\underline{4x^2 - 8x} \quad \\
5x - 8 \\
\underline{5x - 10} \\
2
\end{array}$$

$$\frac{2x^2 - 8 - 3x + x^3}{x-2} = x^2 + 4x + 5 + \frac{2}{x-2}$$

43.
$$\begin{array}{r}
x^3 + x^2 + x + 1 \\
x-1 \overline{) x^4 + 0x^3 + 0x^2 + 0x - 1} \\
\underline{x^4 - x^3} \quad\quad\quad\quad\quad \\
x^3 + 0x^2 + 0x - 1 \\
\underline{x^3 - x^2} \quad\quad\quad \\
x^2 + 0x - 1 \\
\underline{x^2 - x} \quad
\end{array}$$

$$\frac{x^4 - 1}{x-1} = x^3 + x^2 + x + 1$$

45.
$$\begin{array}{r}
x - 3 \\
x^2-1 \overline{) x^3 - 3x^2 - x + 3} \\
\underline{x^3 + 0x^2 - x} \quad \\
-3x^2 + 0x + 3 \\
\underline{-3x^2 + 0x + 3} \\
0
\end{array}$$

$$\frac{x^3 - 3x^2 - x + 3}{x^2 - 1} = x - 3$$

47.
$$\begin{array}{r}
x^2 \quad - 1 \\
x^2+3 \overline{) x^4 + 0x^3 + 2x^2 + 0x - 2} \\
\underline{x^4 + 0x^3 + 3x^2 \quad\quad\quad} \\
-x^2 + 0x - 2 \\
\underline{-x^2 + 0x - 3} \\
1
\end{array}$$

$$\frac{x^4 + 2x^2 - 2}{x^2 + 3} = x^2 - 1 + \frac{1}{x^2 + 3}$$

49.
$$\begin{array}{r}
y^2 + y + 1 \\
y-1 \overline{) y^3 + 0y^2 + 0y - 1} \\
\underline{y^3 - y^2} \quad\quad\quad \\
y^2 + 0y - 1 \\
\underline{y^2 - y} \quad \\
y - 1 \\
\underline{y - 1} \\
0
\end{array}$$

$$\frac{y^3 - 1}{y-1} = y^2 + y + 1$$

51.
$$\begin{array}{r} x^2 \qquad +1 \\ x^2-1 \overline{\smash{)}\, x^4+0x^3+0x^2+0x-1} \\ \underline{x^4+0x^3 \;\; -x^2 \qquad\qquad} \\ x^2+0x-1 \\ \underline{x^2+0x-1} \\ 0 \end{array}$$

$$\frac{x^4-1}{x^2-1}=x^2+1$$

53. sometimes

55. always

57. $\dfrac{y^2-y+c}{y+1}=y-2$

$$y^2-y+c=(y-2)(y+1)=y^2-y-2$$
$$c=-2$$

59. Above and Beyond

61. **(a)**
$$\begin{array}{r} x+1 \\ x-1 \overline{\smash{)}\, x^2+0x-1} \\ \underline{x^2 \;\; -x \quad\;} \\ x-1 \\ \underline{x-1} \\ 0 \end{array}$$

$$\frac{x^2-1}{x-1}=x+1$$

(b)
$$\begin{array}{r} x^2 \;\; +x+1 \\ x-1 \overline{\smash{)}\, x^3+0x^2+0x-1} \\ \underline{x^2-x^2 \qquad\qquad} \\ x^2+0x \quad\;\; \\ \underline{x^2 \;\; -x \quad\;\;} \\ x-1 \\ \underline{x-1} \\ 0 \end{array}$$

$$\frac{x^3-1}{x-1}=x^2+x+1$$

(c)
$$\begin{array}{r} x^3 \;\; +x^2 \;\; +x+1 \\ x-1 \overline{\smash{)}\, x^4+0x^3+0x^2+0x-1} \\ \underline{x^4 \;\; -x^3 \qquad\qquad\qquad} \\ x^3+0x^2 \qquad\quad \\ \underline{x^3 \;\; -x^2 \qquad\quad} \\ x^2+0x \quad\;\; \\ \underline{x^2 \;\; -x \quad\;\;} \\ x-1 \\ \underline{x-1} \\ 0 \end{array}$$

$$\frac{x^4-1}{x-1}=x^3+x^2+x+1$$

(d) In each problem (a), (b), and (c) the quotient is a polynomial of degree one less than the degree of the dividend. The quotient has no missing terms and all of its coefficients are 1. Therefore, it would seem like

$$\frac{x^{50}-1}{x-1}=x^{49}+x^{48}+\cdots+x+1.$$

Summary Exercises

1. $r^4r^9=r^{4+9}=r^{13}$

3. $(2w)^{-3} = \dfrac{1}{(2w)^3} = \dfrac{1}{2^3 w^3} = \dfrac{1}{8w^3}$

5. $y^{-5}y^2 = y^{-5+2} = y^{-3} = \dfrac{1}{y^3}$

7. $\dfrac{x^{12}}{x^{15}} = x^{12-15} = x^{-3} = \dfrac{1}{x^3}$

9. $(5a^2b^3)(2a^{-2}b^{-6}) = 5 \bullet 2a^{2+(-2)}b^{3+(-6)}$

$= 10a^0b^{-3} = \dfrac{10}{b^3}$

11. $\left(\dfrac{m^{-3}n^{-3}}{m^{-4}n^4}\right)^3 = \left(m^{-3-(-4)}n^{-3-4}\right)^3 = \left(mn^{-7}\right)^3$

$= m^{1\cdot 3}n^{-7\cdot 3} = m^3n^{-21} = \dfrac{m^3}{n^{21}}$

13. $\left(\dfrac{x^3y^4}{x^6y^2}\right)^3 = \left(\dfrac{y^2}{x^3}\right)^3 = \dfrac{y^6}{x^9}$

15. $(2a^3)^0(-3a^4)^2 = 1 \bullet (-3)^2(a^4)^2 = 9a^8$

17. $0.0000425 = 4.25\times 10^{-5}$

19. Let $P = \$2{,}000$, $r = 0.027$, $t = 5$ years.

$A = P(1+r)^t = 2{,}000(1+0.027)^5$

$= 2{,}000 \bullet 1.027^5 = 2{,}284.98$

Thus, the value of the investment is \$2,284.98.

21. Writing in scientific notation, 3,794,101 mi^2, rounding to one decimal place is 3.8×10^6 mi^2.

23. The percentage of U.S. surface area not covered by water $= (100 - 6.76) = 93.24\%$.

$\dfrac{93.24}{100}\times 3{,}794{,}101 = 3{,}537{,}619.77$

$\approx 3{,}500{,}000$

$= 3.5\times 10^6$ mi^2

25. $6x^4 - 3x$ is a binomial.

27. $4x^5 - 8x^3 + 5$ is a trinomial.

29. $-7a^4 - 9a^3$ is a binomial.

31. $5x^5 + 3x^2$ has degree 5.

33. $6x^2 + 4x^4 + 6 = 4x^4 + 6x^2 + 6$ has degree 4.

35. -8 has degree 0.

37. $(7a^2 + 3a) + (14a^2 - 5a)$

$= 7a^2 + 3a + 14a^2 - 5a = 21a^2 - 2a$

39. $(5y^3 - 3y^2) + (4y + 3y^2)$

$= 5y^3 - 3y^2 + 4y + 3y^2 = 5y^3 + 4y$

41. $(7x^2 - 2x + 3) - (2x^2 - 5x - 7)$

$= 7x^2 - 2x + 3 - 2x^2 + 5x + 7 = 5x^2 + 3x + 10$

43. $[(9x+2) + (-3x-7)] - (5x-3)$

$= 9x + 2 - 3x - 7 - 5x + 3 = x - 2$

45. $(7w^2 - 5w + 2) - [(16w^2 - 3w) + (8w + 2)]$

$= (7w^2 - 5w + 2) - (16w^2 + 5w + 2)$

$= 7w^2 - 5w + 2 - 16w^2 - 5w - 2$

$= -9w^2 - 10w$

47.

$$\begin{array}{r} 9b^2 \qquad -7 \\ \underline{8b + 5} \\ 9b^2 + 8b - 2 \end{array}$$

49.

$$\begin{array}{r} 7x^2 - 5x - 7 \\ \underline{-(5x^2 - 3x + 2)} \end{array} \qquad \begin{array}{r} 7x^2 - 5x - 7 \\ \underline{-5x^2 + 3x - 2} \\ 2x^2 - 2x - 9 \end{array}$$

51. Let x be the number of copies of the book sold by Castor Books. Therefore, the total cost of books is \$18.45x. Hence, the equation to represent the weekly cost (C) for Castor Books to sell this book is $C = 18.45x + 525$.

53. $P = 21.5(20) - 525 = -\$95$ (loss)
Thus, there is a loss of \$95 if they sell only 20 copies in one week.

55. $(3a^4)(2a^3) = 3 \cdot 2 \cdot a^4 \cdot a^3 = 6a^{4+3} = 6a^7$

57. $(-9p^3)(-6p^2) = 54p^5$

59. $5(3x-8) = 5 \cdot 3x - 5 \cdot 8 = 15x - 40$

61. $(-5rs)(2r^2s - 5rs)$
$= -5rs \cdot 2r^2s - (-5rs)(5rs)$
$= -10r^3s^2 + 25r^2s^2$

63. $(x+5)(x+4) = x \cdot x + x \cdot 4 + 5 \cdot x + 20$
$= x^2 + 9x + 20$

65. $(a-9b)(a+9b) = (a)^2 - (9b)^2 = a^2 - 81b^2$

67. $(a+4b)(a+3b) = a^2 + 3ab + 4ab + 12b^2$
$= a^2 + 7ab + 12b^2$

69. $(3x-5y)(2x-3y)$
$= 6x^2 - 9xy - 10xy + 15y^2$
$= 6x^2 - 19xy + 15y^2$

71. $(y+2)(y^2 - 2y + 3)$
$= y \cdot y^2 - y \cdot 2y + y \cdot 3 + 2 \cdot y^2 - 2 \cdot 2y$
$+2 \cdot 3$
$= y^3 - 2y^2 + 3y + 2y^2 - 4y + 6 = y^3 - y + 6$

73. $(x-2)(x^2 + 2x + 4)$
$= x^3 + 2x^2 + 4x - 2x^2 - 4x - 8 = x^3 - 8$

75. $2x(x+5)(x-6) = 2x(x^2 - x - 30)$
$= 2x^3 - 2x^2 - 60x$

77. $(x+7)^2 = x^2 + 2 \cdot 7 \cdot x + 7^2 = x^2 + 14x + 49$

79. $(2w-5)^2 = (2w)^2 - 2 \cdot 5 \cdot 2w + 5^2$
$= 4w^2 - 20w + 25$

81. $(a+7b)^2 = a^2 + 2 \cdot a \cdot 7 \cdot b + (7b)^2$
$= a^2 + 14ab + 49b^2$

83. $(x-5)(x+5) = x^2 - 5^2 = x^2 - 25$

85. $(2m+3)(2m-3) = (2m)^2 - 3^2 = 4m^2 - 9$

87. $(5r-2s)(5r+2s) = (5r)^2 - (2s)^2$
$= 25r^2 - 4s^2$

89. $3x(x-4)^2 = 3x(x^2 - 8x + 16)$
$= 3x^3 - 24x^2 + 48x$

91. $(y-4)(y+5)(y+4)$
$= (y+5)(y-4)(y+4) = (y+5)(y^2 - 16)$
$= y^3 + 5y^2 - 16y - 80$

93. The area of the rectangle as a polynomial is
$A = LW = [4(x+1)](2x+3)$
$= (4x+4)(2x+3) = 8x^2 + 12x + 8x + 12$
$= (8x^2 + 20x + 12)$ in^2.

95. $\dfrac{9a^5}{3a^2} = 3a^{5-2} = 3a^3$

97. $\dfrac{15a-10}{5} = \dfrac{15a}{5} - \dfrac{10}{5} = 3a - 2$

99. $\dfrac{9r^2s^3 - 18r^3s^2}{-3rs^2} = \dfrac{9r^2s^3}{-3rs^2} - \dfrac{18r^3s^2}{-3rs^2}$
$= -3rs + 6r^2$

101. $$\begin{array}{r} x-5 \\ x+3\overline{\big)x^2-2x-15} \\ \underline{x^2+3x} \\ -5x-15 \\ \underline{-5x-15} \\ 0 \end{array} \qquad \frac{x^2-2x-15}{x+3}=x-5$$

103. $$\begin{array}{r} x-3 \\ x-5\overline{\big)x^2-8x+17} \\ \underline{x^2-5x} \\ -3x+17 \\ \underline{-3x+15} \\ 2 \end{array}$$

$$\frac{x^2-8x+17}{x-5}=x-3+\frac{2}{x-5}$$

105. $$\begin{array}{r} x^2+2x-1 \\ 6x+2\overline{\big)6x^3+14x^2-2x-6} \\ \underline{6x^3+2x^2} \\ 12x^2-2x-6 \\ \underline{12x^2+4x} \\ -6x-6 \\ \underline{-6x-2} \\ -4 \end{array}$$

$$\frac{6x^3+14x^2-2x-6}{6x+2}=x^2+2x-1+\frac{-4}{6x+2}$$

107. $3x^2+x^3+5+4x=x^3+3x^2+4x+5$

$$\begin{array}{r} x^2+x+2 \\ x+2\overline{\big)x^3+3x^2+4x+5} \\ \underline{x^3+2x^2} \\ x^2+4x+5 \\ \underline{x^2+2x} \\ 2x+5 \\ \underline{2x+4} \\ 1 \end{array}$$

$$\frac{3x^2+x^3+5+4x}{x+2}=x^2+x+2+\frac{1}{x+2}$$

Chapter Test 5

1. $(3x^2y)(-2xy^3)=-6x^3y^4$

3. $(x^4y^5)^2=x^8y^{10}$

5. $(3x^2y)^3(-2xy^2)^2=(27x^6y^3)(4x^2y^4)$
$=108x^8y^7$

7. $\dfrac{3x^0}{(2y)^0}=\dfrac{3\cdot 1}{1}=3$

9. $5x^2+8x-8$ is a trinomial.

11. $(3x^2-7x+2)+(7x^2-5x-9)$
$=3x^2-7x+2+7x^2-5x-9$
$=10x^2-12x-7$

13. $(8x^2+9x-7)-(5x^2-2x+5)$
$=8x^2+9x-7-5x^2+2x-5$
$=3x^2+11x-12$

15. $$\begin{array}{r} x^2 \qquad +3 \\ 5x-9 \\ \underline{+3x^2 \qquad\qquad} \\ 4x^2+5x-6 \end{array}$$

17. $5ab(3a^2b-2ab+4ab^2)$
$=15a^3b^2-10a^2b^2+20a^2b^3$

19. $(3m+2n)=9m^2+12mn+4n^2$

21. Canada's surface area in water is $\dfrac{8.92}{100}\times 9{,}984{,}670\approx 890{,}000\text{ km}^2$; in scientific notation is $8.9\times10^5\text{ km}^2$.

23. Let x be the number of copies of the book sold by the Funghi Books. Therefore, the total cost of books is $6.27x$. The weekly cost for Funghi Books to sell this book is $6.27x+285$.

25. $P=8.68(75)-285=\$366$

Thus, there is profit of $366 if they sell 75 copies in one week.

Cumulative Review: Chapters 0–5

1. $-2x^2-3y^2+5z=-2(3)^2-3(-2)^2+5(4)$
$=-2\cdot 9-3\cdot 4+20=-10$

3. $11x-7=10x$
$x-7=0$
$x=7;\{7\}$

5. $7x-5=3x+11$
$4x-5=11$
$4x=16$
$x=4;\{4\}$

7. $\frac{3}{5}x-8=15-\frac{2}{5}x$
$x-8=15$
$x=23;\{23\}$

9. $4x-(2-x)=5x+3$
$4x-2+x=5x+3$
$5x-2=5x+3$
$-2=3$
False statement; $\varnothing$

11. $7x-5>8x+10$
$-5>x+10$
$-15>x$
$\{x|x<-15\}$

13. $f(-2)=2(-2)^3-(-2)^2+7$
$=2(-8)-(4)+7=-16-4+7$
$=-13$

15. $R(x)=39.95x$

17. $4x+y=9$
$y=-4x+9$
The slope is –4; the y-intercept is (0, 9).

19. $y=-2x+4$

21. $m=\frac{5-(-3)}{3-1}=4$
$y=4x-7$

23. $(x^2-3x+5)+(2x^2+5x-9)$
$=x^2-3x+5+2x^2+5x-9=3x^2+2x-4$

25. $4x(3x-5)=12x^2-20x$

27. $(x+2)(x^2-3x+5)$
$=x^3-3x^2+5x+2x^2-6x+10$
$=x^3-x^2-x+10$

29. $(3x-5)^2=9x^2-30x+25$

31. $\frac{32x^2y^3-16x^4y^2+8xy^2}{8xy^2}=4xy-2x^3+1$

33. $(3x^2)^2(2x^3)=(9x^4)(2x^3)=18x^7$

35. $(2x^0)^3(-3x^2y)^2=8(9x^4y^2)=72x^4y^2$

37. $\frac{(4.2\times 10^7)(6.0\times 10^{-3})}{1.2\times 10^{-5}}$
$=\left(\frac{4.2\cdot 6.0}{1.2}\right)\left(\frac{10^7\cdot 10^{-3}}{10^{-5}}\right)=21\times 10^9$
$=2.1\times 10^{10}$

39. Let n be an odd integer.
Then $n+2$ is the next odd integer.

$$n+(n+2)=132$$
$$2n+2=132$$
$$2n=130$$
$$n=65$$
$$n+2=67$$

The numbers are 65 and 67.

Chapter 6
Factoring

Exercises 6.1

< Objective 1 >

1. $20 = 2 \cdot 2 \cdot 5$
$22 = 2 \cdot 11$
The GCF is 2.

3. $16 = 2 \cdot 2 \cdot 2 \cdot 2$
$32 = 2 \cdot 2 \cdot 2 \cdot 2 \cdot 2$
$88 = 2 \cdot 2 \cdot 2 \cdot 11$
The GCF is 8.

5. $x^2 = x \cdot x$
$x^5 = x \cdot x \cdot x \cdot x \cdot x$
The GCF is x^2.

7. $a^3 = a^3$
$a^6 = a^3 \cdot a^3$
$a^9 = a^3 \cdot a^6$
The GCF is a^3.

9. $5x^4 = 5x^4$
$10x^5 = 2x \cdot 5x^4$
The GCF is $5x^4$.

11. $4a^4 = 2 \cdot 2a^4$
$10a^7 = 5a^3 \cdot 2a^4$
$12a^{14} = 6a^{10} \cdot 2a^4$
The GCF is $2a^4$.

13. $9x^2y = 3x \cdot 3xy$
$12xy^2 = 4y \cdot 3xy$
$15x^2y^2 = 5xy \cdot 3xy$
The GCF is $3xy$.

15. $15ab^3 = 3ab^2 \cdot 5b$
$10a^2bc = 2a^2c \cdot 5b$
$25b^2c^3 = 5bc^3 \cdot 5b$
The GCF is $5b$.

17. $15a^2bc^2 = 5a \cdot 3abc^2$
$9ab^2c^2 = 3b \cdot 3abc^2$
$6a^2b^2c^2 = 2ab \cdot 3abc^2$
The GCF is $3abc^2$.

19. $5 = 5$
$12xy = 4 \cdot 3xy$
$4ab = 4ab$
The GCF is 1.

21. $(x+y)^2 = (x+y)^2$
$(x+y)^5 = (x+y)^3 \cdot (x+y)^2$
The GCF is $(x+y)^2$.

< Objective 2 >

23. $10x + 5 = 5(2x+1)$

25. $24m - 32n = 8(3m - 4n)$

27. The expression $4x - 7$ is a prime polynomial since it cannot be factored.

29. $12m^2 + 8m = 4m(3m+2)$

31. $10s^2 + 5s = 5s(2s+1)$

33. $24x^2 - 60x = 12x(2x-5)$

35. The expression $m^2 + 1$ is a prime polynomial since it cannot be factored.

37. $15a^3 - 25a^2 = 5a^2(3a-5)$

39. $6pq + 18p^2q = 6pq(1+3p)$

41. $7m^3n - 21mn^3 = 7mn(m^2 - 3n^2)$

43. $6x^2 - 18x + 30 = 6(x^2 - 3x + 5)$

45. $5a^3 - 15a^2 + 25a = 5a(a^2 - 3a + 5)$

47. $12x + 8xy - 28xy^2 = 4x(3 + 2y - 7y^2)$

49. $10x^2y + 15xy - 5xy^2 = 5xy(2x + 3 - y)$

51. $10r^3s^2 + 25r^2s^2 - 15r^2s^3$
$= 5r^2s^2(2r + 5 - 3s)$

53. $9a^5 - 15a^4 + 21a^3 - 27a$
$= 3a(3a^4 - 5a^3 + 7a^2 - 9)$

55. $15m^3n^2 - 20m^2n + 35mn^3 - 10mn$
$= 5mn(3m^2n - 4m + 7n^2 - 2)$

57. $x(x-9) + 5(x-9) = (x+5)(x-9)$

59. $p(p-2q) - q(p-2q) = (p-q)(p-2q)$

61. $x(y-z) + 3(y-z) = (x+3)(y-z)$

63. $-t^2 + 6t + 10 = -(t^2 - 6t - 10)$

65. $-3x^2 + 4x - 5 = -(3x^2 - 4x + 5)$

67. $-4x^2 - 16x + 4 = -4(x^2 + 4x - 1)$

69. $-4m^2n^3 - 6mn^3 - 10n^2$
$= -2n^2(2m^2n + 3mn + 5)$

< Objective 3 >

71. $ab - ac + b^2 - bc = a(b-c) + b(b-c)$
$= (a+b)(b-c)$

73. $6r^2 + 12rs - r - 2s = 6r(r+2s) - (r+2s)$
$= (6r-1)(r+2s)$

75. $ab^2 - 2b^2 + 3a - 6 = b^2(a-2) + 3(a-2)$
$= (b^2+3)(a-2)$

77. $x^2 + 3x - 4xy - 12y = x(x+3) - 4y(x+3)$
$= (x-4y)(x+3)$

79. $m^2 - 6n^3 + 2mn^2 - 3mn$
$= m^2 - 3mn + 2mn^2 - 6n^3$
$= m(m-3n) + 2n^2(m-3n)$
$= (m+2n^2)(m-3n)$

81. $20x + 60 - 5xy - 15y$
$= 5(4x + 12 - xy - 3y)$
$= 5(4x - xy + 12 - 3y)$
$= 5(x(4-y) + 3(4-y)) = 5(4-y)(x+3)$

83. $-18ab - 6b^2 + 9ac + 3bc$
$= -3(6ab + 2b^2 - 3ac - bc)$
$= -3(2b(3a+b) - c(3a+b))$
$= -3(3a+b)(2b-c)$

85. $(x-3)(x+2) = x^2 + 2x - 3x - 6$
$= x^2 - x - 6$
The factoring is correct.

87. $(x+6)(x-2) = x^2 + 4x - 12$
The factoring is incorrect.

89. $(2x+1)(x-3) = 2x^2 - 5x - 3$
The factoring is correct.

91. False

93. always

95. $-t^3-6t^2+11t+66=-t^2(t+6)+11(t+6)$
$=(t+6)(-t^2+11)$

97. $v_0t-4.9t^2=t(v_0-4.9t)$

99. $(2x-6)(5x+10)=2(x-3)\cdot 5(x+2)$
$=10(x-3)(x+2)$
The GCF is $2\cdot 5=10$.

101. $(2x^3-4x)(3x+6)=2x(x^2-2)\cdot 3(x+2)$
$=6x(x^2-2)(x+2)$
The GCF is $2x\cdot 3=6x$.

103. Above and Beyond

105. $33t-t^2=t(33-t)$
The length of the rectangle is $33-t$.

107. Above and Beyond

109. Above and Beyond

Exercises 6.2

< Objective 1 >

1. $m^2-n^2=(m+n)(m-n)$

3. $x^2-169=(x+13)(x-13)$

5. $49-y^2=(7+y)(7-y)$

7. $9b^2-16=(3b+4)(3b-4)$

9. $16w^2-49=(4w+7)(4w-7)$

11. The binomial x^2-10 is not a difference of squares. Hence, it is not factorable.

13. The binomial x^2+16 is a sum of squares. Hence, it is not factorable.

15. $4s^2-9r^2=(2s+3r)(2s-3r)$

17. $9w^2-49z^2=(3w+7z)(3w-7z)$

19. $4x^2y^2-169y^2=(2xy-13y)(2xy+13y)$
$=y^2(2x-13)(2x+13)$

21. The binomial $121x^8+64$ is a sum of squares. Hence, it is not factorable

23. $49a^2-9b^2=(7a+3b)(7a-3b)$

25. $x^4-36=(x^2+6)(x^2-6)$

27. $x^2y^2-16=(xy+4)(xy-4)$

29. $25-a^2b^2=(5+ab)(5-ab)$

31. The binomial $49-xy$ is not a difference of squares. Hence, it is not factorable.

33. $r^4-4s^2=(r^2+2s)(r^2-2s)$

35. $81a^2-100b^6=(9a+10b^3)(9a-10b^3)$

37. $18x^3-2xy^2=2x(9x^2-y^2)$
$=2x(3x+y)(3x-y)$

39. $12m^3n-75mn^3=3mn(4m^2-25n^2)$
$=3mn(2m+5n)(2m-5n)$

41. $16a^4-81b^4$
$=(4a^2+9b^2)(4a^2-9b^2)$
$=(4a^2+9b^2)(2a+3b)(2a-3b)$

< Objective 2 >

43. $y^3+125=y^3+5^3=(y+5)(y^2-5y+25)$

45. $m^3-125=(m-5)(m^2+5m+25)$

47. $a^3b^3-27=(ab)^3-3^3$
$=(ab-3)(a^2b^2+3ab+9)$

49. $8w^3+z^3=(2w)^3+z^3$
$=(2w+z)(4w^2-2wz+z^2)$

51. $r^3-64s^3=(r-4s)(r^2+4rs+16s^2)$

53. $8x^3-27y^3=(2x-3y)(4x^2+6xy+9y^2)$

55. x^3+5 is not factorable.

57. $3a^3+81b^3=3(a^3+27b^3)$
$=3(a+3b)(a^2-3ab+9b^2)$

< Objective 3 >

59. $x^2-4x+4=(x-2)^2$

61. The trinomial y^2+4y+8 is not a perfect square trinomial. Hence, it is not factorable.

63. $x^2+2x+1=(x+1)^2$

65. $16a^2+24a+9=(4a+3)^2$

67. $25x^2+120x+144=(5x+12)^2$

69. $9x^3+12x^2y+4xy^2=x(9x^2+12xy+4y^2)$
$=x(3x+2y)^2$

71. $x^2(x+y)-y^2(x+y)$
$=(x^2-y^2)(x+y)=(x+y)(x-y)(x+y)$
$=(x+y)^2(x-y)$

73. $2m^2(m-2n)-18n^2(m-2n)$
$=(2m^2-18n^2)(m-2n)$
$=2(m^2-9n)(m-2n)$
$=2(m+3n)(m-3n)(m-2n)$

75. False

77. True

79. always

81. always

83. $d=\left(\frac{w}{8EI}\right)(l_1^2-l_2^2)(l_1^2+l_2^2)$
$=\left(\frac{w}{8EI}\right)(l_1+l_2)(l_1-l_2)(l_1^2+l_2^2)$

85. $338-2t^2=2(169-t^2)=2(13-t)(13+t)$

87. $(2x+5)(2x-5)=4x^2-25$
$k=4$

89. $2x(x-3y)(x+3y)=2x(x^2-9y^2)$
$=2x^3-18xy^2$
$k=18$

91. Above and Beyond

93. $(5x+3y)(25x^2-15xy+9y^2)$
$=125x^3+27y^3$

95. Above and Beyond

Exercises 6.3

1. $(x+3)(x-1)=x^2-x+3x-3=x^2+2x-3$
True

3. $(x-6)(x+4)=x^2-2x-24$
$\neq x^2-10x-24$
False

5. $(x-8)(x-8)=x^2-16x+64$
True

< Objective 1 >

7. $x^2-8x+15=(x-3)(x-5)$

9. $m^2+8m+12=(m+2)(m+6)$

11. $p^2-8p-20=(p+2)(p-10)$

13. $x^2-16x+64=(x-8)(x-8)$

15. $x^2-7xy+10y^2=(x-2y)(x-5y)$

17. $x^2+8x+15$
Consider only the positive factors of 15 because the last two terms are positive.
$15=(1)(15)$
$=(3)(5)$

Possible Factorizations	Middle Terms
$(x+1)(x+15)$	$16x$
$(x+3)(x+5)$	$8x$

$x^2+8x+15=(x+3)(x+5)$

19. $x^2-11x+28$
Consider only the negative factors of 28 because the middle term is negative.
$28=(-1)(-28)$
$=(-2)(-14)$
$=(-4)(-7)$

Possible Factorizations	Middle Terms
$(x-1)(x-28)$	$-29x$
$(x-2)(x-14)$	$-16x$
$(x-4)(x-7)$	$-11x$

$x^2-11x+28=(x-4)(x-7)$

21. $s^2+13s+30$
Consider only the positive factors of 30 because the last two terms are positive.
$30=(1)(30)$
$=(2)(15)$
$=(3)(10)$
$=(5)(6)$

Possible Factorizations	Middle Terms
$(s+1)(s+30)$	$31s$
$(s+2)(s+15)$	$17s$
$(s+3)(s+10)$	$13s$
$(s+5)(s+6)$	$11s$

$s^2+13s+30=(s+3)(s+10)$

23. $a^2 - 2a - 48 = (a+6)(a-8)$

The factors of –48 will have different signs.

$$\begin{aligned} -48 &= (1)(-48) \\ &= (-1)(48) \\ &= (2)(-24) \\ &= (-2)(24) \\ &= (3)(-16) \\ &= (-3)(16) \\ &= (4)(-12) \\ &= (-4)(12) \\ &= (6)(-8) \\ &= (-6)(8) \end{aligned}$$

Possible Factorizations	Middle Terms
$(a+1)(a-48)$	$-47a$
$(a-1)(a+48)$	$47a$
$(a+2)(a-24)$	$-22a$
$(a-2)(a+24)$	$22a$
$(a+3)(a-16)$	$-13a$
$(a-3)(a+16)$	$13a$
$(a+4)(a-12)$	$-8a$
$(a-4)(a+12)$	$8a$
$(a+6)(a-8)$	$-2a$
$(a-6)(a+8)$	$2a$

$a^2 - 2a - 48 = (a+6)(a-8)$

25. $x^2 - 8x + 7$

Consider only the negative factors of 7 because the middle term is negative.

$7 = (-1)(-7)$

Possible Factorizations	Middle Terms
$(x-1)(x-7)$	$-8x$

$x^2 - 8x + 7 = (x-1)(x-7)$

27. $m^2 + 3m - 28 = (m-4)(m+7)$

The factors of –28 will have different signs.

$$\begin{aligned} -28 &= (1)(-28) \\ &= (-1)(28) \\ &= (2)(-14) \\ &= (-2)(14) \\ &= (4)(-7) \\ &= (-4)(7) \end{aligned}$$

Possible Factorizations	Middle Terms
$(m+1)(m-28)$	$-27m$
$(m-1)(m+28)$	$27m$
$(m+2)(m-14)$	$-12m$
$(m-2)(m+14)$	$12m$
$(m+4)(m-7)$	$-3m$
$(m-4)(m+7)$	$3m$

$m^2 + 3m - 28 = (m-4)(m+7)$

29. $x^2 - 6x - 40$

The factors of –40 will have different signs.

$$\begin{aligned} -40 &= (1)(-40) \\ &= (-1)(40) \\ &= (2)(-20) \\ &= (-2)(20) \\ &= (4)(-10) \\ &= (-4)(10) \\ &= (5)(-8) \\ &= (-5)(8) \end{aligned}$$

Possible Factorizations	Middle Terms
$(x+1)(x-40)$	$-39x$
$(x-1)(x+40)$	$39x$
$(x+2)(x-20)$	$-18x$
$(x-2)(x+20)$	$18x$
$(x+4)(x-10)$	$-6x$
$(x-4)(x+10)$	$6x$
$(x+5)(x-8)$	$-3x$
$(x-5)(x+8)$	$3x$

$$x^2 - 6x - 40 = (x+4)(x-10)$$

31. $x^2 - 14x + 49$

Consider only the negative factors of 49 because the middle term is negative.

$$\begin{aligned} 49 &= (-1)(-49) \\ &= (-7)(-7) \end{aligned}$$

Possible Factorizations	Middle Terms
$(x-1)(x-49)$	$-50x$
$(x-7)(x-7)$	$-14x$

$$x^2 - 14x + 49 = (x-7)(x-7)$$

33. $p^2 - 10p - 24$

The factors of –24 will have different signs.

$$\begin{aligned} -24 &= (1)(-24) \\ &= (-1)(24) \\ &= (2)(-12) \\ &= (-2)(12) \\ &= (3)(-8) \\ &= (-3)(8) \\ &= (4)(-6) \\ &= (-4)(6) \end{aligned}$$

Possible Factorizations	Middle Terms
$(p+1)(p-24)$	$-23p$
$(p-1)(p+24)$	$23p$
$(p+2)(p-12)$	$-10p$
$(p-2)(p+12)$	$10p$
$(p+3)(p-8)$	$-5p$
$(p-3)(p+8)$	$5p$
$(p+4)(p-6)$	$-2p$
$(p-4)(p+6)$	$2p$

$$p^2 - 10p - 24 = (p+2)(p-12)$$

35. $x^2+5x-66$

The factors of –66 will have different signs.

$$\begin{aligned}-66&=(1)(-66)\\&=(-1)(66)\\&=(2)(-33)\\&=(-2)(33)\\&=(3)(-22)\\&=-(3)(22)\\&=(6)(-11)\\&=(-6)(11)\end{aligned}$$

Possible Factorizations	**Middle Terms**
$(x+1)(x-66)$	$-65x$
$(x-1)(x+66)$	$65x$
$(x+2)(x-33)$	$-31x$
$(x-2)(x+33)$	$31x$
$(x+3)(x-22)$	$-19x$
$(x-3)(x+22)$	$19x$
$(x+6)(x-11)$	$-5x$
$(x-6)(x+11)$	$5x$

$x^2+5x-66=(x-6)(x+11)$

37. $c^2+19c+60$

Consider only the positive factors of 60 because the last two terms are both positive.

$$\begin{aligned}60&=(1)(60)\\&=(2)(30)\\&=(3)(20)\\&=(4)(15)\\&=(5)(12)\\&=(6)(10)\end{aligned}$$

Possible Factorizations	**Middle Terms**
$(c+1)(c+60)$	$61c$
$(c+2)(c+30)$	$32c$
$(c+3)(c+20)$	$23c$
$(c+4)(c+15)$	$19c$

It is unnecessary to list all the pairs, since we found the right pair.

$c^2+19c+60=(c+4)(c+15)$

39. $x^2-7xy+10y^2$

Consider only the positive factors of $10y^2$ because the last two terms are positive.

$$\begin{aligned}10y^2&=(y)(10y)\\&=(2y)(5y)\end{aligned}$$

Possible Factorizations	**Middle Terms**
$(x+y)(x+10y)$	$11xy$
$(x+2y)(x+5y)$	$7xy$

$x^2+7xy+10y^2=(x+2y)(x+5y)$

41. $a^2 - ab - 42b^2$

The factors of –42 will have different signs.

$$\begin{aligned}-42 &= (1)(-42)\\ &= (-1)(42)\\ &= (2)(-21)\\ &= (-2)(21)\\ &= (3)(-14)\\ &= (-3)(14)\\ &= (6)(-7)\\ &= (-6)(7)\end{aligned}$$

Possible Factorizations	Middle Terms
$(a+b)(a-42b)$	$-41ab$
$(a-b)(a+42b)$	$41ab$
$(a+2b)(a-21b)$	$-19ab$
$(a-2b)(a+21b)$	$19ab$
$(a+3b)(a-14b)$	$-11ab$
$(a-3b)(a+14b)$	$11ab$
$(a+6b)(a-7b)$	$-ab$

It is unnecessary to list all the pairs, since we found the right pair.

$a^2 - ab - 42b^2 = (a+6b)(a-7b)$

43. $x^2 + x + 7$

Consider only the positive factors of 7 because the middle term is positive.

$7 = (7)(7)$

Possible Factorizations	Middle Terms
$(x+7)(x+1)$	$+8x$

The trinomial is not factorable.

45. $x^2 - 13xy + 40y^2$

Consider only the negative factors of 40 because the middle term is negative.

$$\begin{aligned}40 &= (-1)(-40)\\ &= (-2)(-20)\\ &= (-4)(-10)\\ &= (-5)(-8)\end{aligned}$$

Possible Factorizations	Middle Terms
$(x-y)(x-40y)$	$-41xy$
$(x-2y)(x-20y)$	$-22xy$
$(x-4y)(x-10y)$	$-14xy$
$(x-5y)(x-8y)$	$-13xy$

$x^2 - 13xy + 40y^2 = (x-5y)(x-8y)$

47. $x^2 - 2xy - 8y^2$

The factors of –8 will have different signs.

$$\begin{aligned}-8 &= (1)(-8)\\ &= (-1)(8)\\ &= (2)(-4)\\ &= (-2)(4)\end{aligned}$$

Possible Factorizations	Middle Terms
$(x+y)(x-8y)$	$-7xy$
$(x-y)(x+8y)$	$7xy$
$(x+2y)(x-4y)$	$-2xy$
$(x-2y)(x+4y)$	$2xy$

$x^2 - 2xy - 8y^2 = (x+2y)(x-4y)$

49. $s^2 - 2st - 2t^2$

The factors of – 2 will have different signs.

$$\begin{aligned}-2 &= (1)(-2)\\ &= (-1)(2)\end{aligned}$$

Possible Factorizations	Middle Terms
$(s+t)(s-2t)$	$-st$
$(s-t)(s+2t)$	st

The trinomial is not factorable.

51. $25m^2 + 10mn + n^2$

Consider only the positive factors of 25 because the last two terms are positive.

$$25 = (1)(25)$$
$$= (5)(5)$$

Possible Factorizations	Middle Terms
$(m+n)(n+25n)$	$26mn$
$(5m+n)(5m+n)$	$10mn$

$$25m^2 + 10mn + n^2 = (5m+n)(5m+n)$$

< Objective 2 >

53. $3a^2 - 3a - 126 = 3(a^2 - a - 42)$
$$= 3(a+6)(a-7)$$

55. $r^3 + 7r^2 - 18r = r(r^2 + 7r - 18)$
$$= r(r-2)(r-9)$$

57. $2x^3 - 20x^2 - 48x = 2x(x^2 - 10x - 24)$
$$= 2x(x-12)(x+2)$$

59. $x^2y - 9xy^2 - 36y^3 = y(x^2 - 9xy - 36y^2)$
$$= y(x+3y)(x-12y)$$

61. $m^3 - 29m^2n + 120mn^2$
$$= m(m^2 - 29mn + 120n^2)$$
$$= m(m-5n)(m-24n)$$

63. $-s^2 - 13s - 30 = -(s^2 + 13s + 30)$
$$= -(s+10)(s+3)$$

65. $-x^2 + 14x - 49 = -(x-7)(x-7)$
$$= -(x-7)^2$$

< Objective 3 >

67. $u^4 - 5u^2 + 4 = (u^2 - 4)(u^2 - 1)$
$$= (u+2)(u-2)(u+1)(u-1)$$

69. $w^4 - 5w^2 - 36 = (w^2 - 9)(w^2 + 4)$
$$= (w+3)(w-3)(w^2+4)$$

71. $2y^4 - 12y^2 - 54 = 2(y^4 - 6y - 27)$
$$= 2(y^2 - 9)(y^2 + 3)$$
$$= 2(y+3)(y-3)(y^2+3)$$

73. False

75. True

77. sometimes

79. always

81. $310(x^2 - 36x + 128) = 310(x-4)(x-32)$

83. $s = 325 + 60l - l^2 = (5+l)(65-l)$

or $s = -(l+5)(l-65)$

85. $x^2 - 64 = (x+8)(x-8)$

87. $x^2 + kx + 8$

Consider the positive factors of 8.

$$8 = (1)(8)$$
$$= (2)(4)$$

Possible Factorizations	Middle Terms
$(x+1)(x+8)$	$9x$
$(x+2)(x+4)$	$6x$

$k = 6$ or 9

89. $x^2 - kx + 16$
Consider the negative factors of 16.
$16 = (-1)(-16)$
$= (-2)(-8)$
$= (-4)(-4)$

Possible Factorizations	Middle Terms
$(x-1)(x-16)$	$-17x$
$(x-2)(x-8)$	$-10x$
$(x-4)(x-4)$	$-8x$

$k = 8$, 10, or 17

91. $x^2 - kx - 5$
The factors of –5 will have different signs.
$-5 = (1)(-5)$
$= (-1)(5)$

Possible Factorizations	Middle Terms
$(x+1)(x-5)$	$-4x$
$(x-1)(x+5)$	$4x$

$k = 4$

93. $x^2 + 3x + k$
Since the sign of the last two terms is positive, find all pairs of positive integers whose sum is 3.
$1 + 2 = 3$, and $2 = (1)(2)$
$(x+1)(x+2) = x^2 + 3x + 2$ then $k = 2$.

95. $x^2 + 2x - k$
Since the last term is negative, find pairs of integers of opposite sign whose sum is 2.
$-1 + 3 = 2$
$-2 + 4 = 2$
$-3 + 5 = 2$, etc.
So, since
$-3 = (-1)(3)$ and $(x-1)(x+3) = x^2 + 2x - 3$
$-8 = (-2)(4)$ and $(x-2)(x+4) = x^2 + 2x - 8$
$-15 = (-3)(5)$ and $(x-3)(x+5) = x^2 + 2x - 15$
$-24 = (-4)(6)$ and $(x-4)(x+6) = x^2 + 2x - 24$
etc., then $k = 3$, 8, 15, 24, …

Exercises 6.4

1. $a = 1$, $b = 7$, and $c = -5$

3. $a = 1$, $b = -3$, and $c = 8$

5. $a = 3$, $b = 5$, and $c = -8$

7. $a = 4$, $b = 8$, and $c = 11$

9. $a = -7$, $b = -5$, and $c = 2$

< Objective 1 >

11. $ac = -6$; $b = 1$
$m = 3$; $n = -2$
$mn = -6 = ac$
$m + n = 1 = b$
$x^2 + x - 6$ is factorable.

13. $ac = -1$; $b = 3$
$x^2 + 3x - 1$ is not factorable.

15. $ac = 6$; $b = -5$
$m = -3$; $n = -2$
$mn = 6 = ac$
$m + n = -5 = b$
$x^2 - 5x + 6$ is factorable.

17. $ac = -6$; $b = 5$
$m = 6$; $n = -1$
$mn = -6 = ac$
$m + n = 5 = b$
$2x^2 + 5x - 3$ is factorable.

19. $ac = 60$; $b = -19$
$m = -15$; $n = -4$
$mn = 60 = ac$
$m + n = -19 = b$
$6x^2 - 19x + 10$ is factorable.

< Objectives 2 and 3 >

21. $4x^2 - 4x - 3 = (2x+1)(2x-3)$

23. $6a^2 + 13a + 6 = (2a+3)(3a+2)$

25. $15x^2 - 16x + 4 = (3x-2)(5x-2)$

27. $16a^2 + 8ab + b^2 = (4a+b)(4a+b)$

29. $4m^2 + 5mn - 6n^2 = (m+2n)(4m-3n)$

31. $$\begin{aligned} x^2 + 8x + 15 &= x^2 + 5x + 3x + 15 \\ &= (x^2 + 5x) + (3x + 15) \\ &= x(x+5) + 3(x+5) \\ &= (x+5)(x+3) \end{aligned}$$

33. $$\begin{aligned} s^2 + 13s + 30 &= s^2 + 3s + 10s + 30 \\ &= (s^2 + 3s) + (10s + 30) \\ &= s(s+3) + 10(s+3) \\ &= (s+10)(s+3) \end{aligned}$$

35. $x^2 + 3x + 11$
There are no factors of 11 that add to +3; therefore, it is not factorable.

37. $$\begin{aligned} x^2 - 6x - 40 &= x^2 - 10x + 4x - 40 \\ &= (x^2 - 10x) + (4x - 40) \\ &= x(x-10) + 4(x-10) \\ &= (x-10)(x+4) \end{aligned}$$

39. $$\begin{aligned} p^2 - 10p - 24 &= p^2 - 12p + 2p - 24 \\ &= (p^2 - 12p) + (2p - 24) \\ &= p(p-12) + 2(p-12) \\ &= (p-12)(p+2) \end{aligned}$$

41. $$\begin{aligned} x^2 + 5x - 66 &= x^2 + 11x - 6x - 66 \\ &= (x^2 + 11x) + (-6x - 66) \\ &= x(x+11) - 6(x+11) \\ &= (x+11)(x-6) \end{aligned}$$

43. $$\begin{aligned} c^2 + 19c + 60 &= c^2 + 15c + 4c + 60 \\ &= (c^2 + 15c) + (4c + 60) \\ &= c(c+15) + 4(c+15) \\ &= (c+4)(c+15) \end{aligned}$$

45. $$\begin{aligned} n^2 + 5n - 50 &= n^2 + 10n - 5n - 50 \\ &= (n^2 + 10n) + (-5n - 50) \\ &= n(n+10) - 5(n+10) \\ &= (n+10)(n-5) \end{aligned}$$

47. $m^2 - 6m + 1$
There are no factors of +1 that add to – 6; therefore, it is not factorable.

49. $$\begin{aligned} x^2 + 7xy + 10y^2 &= x^2 + 5xy + 2xy + 10y^2 \\ &= (x^2 + 5xy) + (2xy + 10y^2) \\ &= x(x+5y) + 2y(x+5y) \\ &= (x+2y)(x+5y) \end{aligned}$$

51. $a^2 - ab - 42b^2 = a^2 - 7ab + 6ab - 42b^2$
$= (a^2 - 7ab) + (6ab - 42b^2)$
$= a(a - 7b) + 6b(a - 7b)$
$= (a - 7b)(a + 6b)$

53. $x^2 - 13xy + 40y^2$
$= x^2 - 8xy - 5xy + 40y^2$
$= (x^2 - 8xy) + (-5xy + 40y^2)$
$= x(x - 8y) - 5y(x - 8y)$
$= (x - 5y)(x - 8y)$

55. $6x^2 + 19x + 10 = 6x^2 + 4x + 15x + 10$
$= (6x^2 + 4x) + (15x + 10)$
$= 2x(3x + 2) + 5(3x + 2)$
$= (3x + 2)(2x + 5)$

57. $15x^2 + x - 6 = 15x^2 - 9x + 10x - 6$
$= (15x^2 - 9x) + (10x - 6)$
$= 3x(5x - 3) + 2(5x - 3)$
$= (5x - 3)(3x + 2)$

59. $6m^2 + 25m - 25 = 6m^2 - 5m + 30m - 25$
$= (6m^2 - 5m) + (30m - 25)$
$= m(6m - 5) + 5(6m - 5)$
$= (6m - 5)(m + 5)$

61. $9x^2 - 12x + 4 = 9x^2 - 6x - 6x + 4$
$= (9x^2 - 6x) + (-6x + 4)$
$= 3x(3x - 2) - 2(3x - 2)$
$= (3x - 2)(3x - 2)$

63. $12x^2 - 8x - 15 = 12x^2 + 10x - 18x - 15$
$= (12x^2 + 10x) + (-18x - 15)$
$= 2x(6x + 5) - 3(6x + 5)$
$= (6x + 5)(2x - 3)$

65. $3y^2 + 7y - 6 = 3y^2 - 2y + 9y - 6$
$= (3y^2 - 2y) + (9y - 6)$
$= y(3y - 2) + 3(3y - 2)$
$= (3y - 2)(y + 3)$

67. $8x^2 - 27x - 20 = 8x^2 + 5x - 32x - 20$
$= (8x^2 + 5x) + (-32x - 20)$
$= x(8x + 5) - 4(8x + 5)$
$= (8x + 5)(x - 4)$

69. $4x^2 + 3x + 11$
There are no factors of +44 that add to +3; therefore, it is not factorable.

71. $2x^2 + 3xy + y^2 = 2x^2 + xy + 2xy + y^2$
$= (2x^2 + xy) + (2xy + y^2)$
$= x(2x + y) + y(2x + y)$
$= (2x + y)(x + y)$

73. $5a^2 - 8ab - 4b^2$
$= 5a^2 + 2ab - 10ab - 4b^2$
$= (5a^2 + 2ab) + (-10ab - 4b^2)$
$= a(5a + 2b) - 2b(5a + 2b)$
$= (5a + 2b)(a - 2b)$

75. $9x^2 + 4xy - 5y^2 = 9x^2 - 5xy + 9xy - 5y^2$
$= (9x^2 - 5xy) + (9xy - 5y^2)$
$= x(9x - 5y) + y(9x - 5y)$
$= (9x - 5y)(x + y)$

77. $6m^2 - 17mn + 12n^2$
$= 6m^2 - 8mn - 9mn + 12n^2$
$= (6m^2 - 8mn) + (-9mn + 12n^2)$
$= 2m(3m - 4n) - 3n(3m - 4n)$
$= (3m - 4n)(2m - 3n)$

79. $36a^2-3ab-5b^2$
$=36a^2-15ab+12ab-5b^2$
$=(36a^2-15ab)+(12ab-5b^2)$
$=3a(12a-5b)+b(12a-5b)$
$=(12a-5b)(3a+b)$

81. $x^2+4xy+4y^2=x^2+2xy+2xy+4y^2$
$=(x^2+2xy)+(2xy+4y^2)$
$=x(x+2y)+2y(x+2y)$
$=(x+2y)(x+2y)$
$=(x+2y)^2$

83. $2x^2+18x-1$
There are no factors of – 2, that add to +18; therefore, it is not factorable.

85. $20x^2-20x-15$
$=5(4x^2-4x-3)=5(4x^2-6x+2x-3)$
$=5[2x(2x-3)+1(2x-3)]$
$=5(2x-3)(2x+1)$

87. $8m^2+12m+4=4(2m^2+3m+1)$
$=4(2m^2+m+2m+1)$
$=4[m(2m+1)+(2m+1)]$
$=4(2m+1)(m+1)$

89. $15r^2-21rs+6s^2$
$=3(5r^2-7rs+2s^2)$
$=3(5r^2-2rs-5rs+2s^2)$
$=3[r(5r-2s)-s(5r-2s)]$
$=3(5r-2s)(r-s)$

91. $-p^2+10p+24$
$=-(p^2-10p-24)$
$=-(p^2-12p+2p-24)$
$=-[p(p-12)+2(p-12)]$
$=-(p-12)(p+2)$

93. $-n^2-5n+50=-(n^2+5n-50)$
$=-(n^2-5n+10n-50)$
$=-[n(n-5)+10(n-5)]$
$=-(n-5)(n+10)$

95. $x^2-2x+12$
There are no factors of 12 that add to –2; therefore, it is not factorable.

97. $-3x^2-x+9=-(3x^2+x-9)$

99. $-6x^2-19x-10$
$=-(6x^2+19x+10)$
$=-(6x^2+15x+4x+10)$
$=-[3x(2x+5)+2(2x+5)]$
$=-(2x+5)(3x+2)$

101. $2x^3-2x^2-4x=2x(x^2-x-2)$
$=2x(x^2-2x+x-2)$
$=2x[x(x-2)+(x-2)]$
$=2x(x-2)(x+1)$

103. $2y^4+5y^3+3y^2$
$=y^2(2y^2+5y+3)=y^2(2y^2+3y+2y+3)$
$=y^2[y(2y+3)+(2y+3)]$
$=y^2(2y+3)(y+1)$

105. $10(x+y)^2-11(x+y)-6$
$=[5(x+y)+2][2(x+y)-3]$
$=(5x+5y+2)(2x+2y-3)$

107. $5(x-1)^2-15(x-1)-350$

$=5\left[(x-1)^2-3(x-1)-70\right]$

$=5\left[(x-1)-10\right]\left[(x-1)+7\right]$

$=5(x-1-10)(x-1+7)=5(z-11)(x+6)$

109. $15+29x-48x^2$

$=-\left(48x^2-29x-15\right)$

$=-\left(48x^2+16x-45x-15\right)$

$=-\left[16x(3x+1)-15(3x+1)\right]$

$=-(3x+1)(16x-15)$ or $(1+3x)(15-16x)$

111. $-6x^2+19x-15$

$=-\left(6x^2-19x+15\right)$

$=-\left(6x^2-10x-9x+15\right)$

$=-\left[2x(3x-5)-3(3x-5)\right]$

$=-(2x-3)(3x-5)$

113. False

115. sometimes

117. sometimes

119. $Y=-0.05x^2+1.5x+140$

$=-0.05\left(x^2-30x-2800\right)$

$=-0.05\left(x^2-70x+40x-2800\right)$

$=-0.05(x+40)(x-70)$

121. $N=50+25t-3t^2=-3t^2+25t+50$

$=-\left(3t^2-25t-50\right)$

$=-\left(3t^2-30t+5t-50\right)$

$=-(3t+5)(t-10)$

Exercises 6.5

< Objectives 1–3 >

1. Factor out the GCF (x);
$x^2-3x=x(x-3)$

3. $x^2-5x-24$
Trial and error; $m=-8$ and $n=3$.
$=x^2-8x+3x-24=x(x-8)+3(x-8)$
$=(x-8)(x+3)$

5. Factor out the GCF $(x-y)$;
$x(x-y)+2(x-y)=(x-y)(x+2)$

7. Factor out the GCF ($2y$);
$2x^2y-6xy+8y^2=2y\left(x^2-3x+4y\right)$

9. $y^2-13y+40$
Trial and error; $m=-8$, $n=-5$.
$=y^2-5y-8y+40=y(y-5)-8(y-5)$
$=(y-5)(y-8)$

11. $3b^2+17b-28$
Factor by trial and error; $m=21$ and $n=-4$.
$=3b^2+21b-4b-28=3b(b+7)-4(b+7)$
$=(b+7)(3b-4)$

13. $3x^2-14xy-24y^2$
Factor by trial and error; $m=-18$, $n=4$.
$=3x^2-18xy+4xy-24y^2$
$=3x(x-6y)+4y(x-6y)$
$=(x-6y)(3x+4y)$

15. $2a^2+11a+12$
Factor by trial and error; $m=8$, $n=3$.
$=2a^2+8a+3a+12=2a(a+4)+3(a+4)$
$=(a+4)(2a+3)$

17. Factor out GCF (r^2);
$125r^3 + r^2 = r^2(125r + 1)$

19. $3x^2 - 30x + 63$
Factor our GCF (3), then trial and error method; $m = -7$, $n = -3$
$= 3(x^2 - 10x + 21) = 3(x^2 - 7x - 3x + 21)$
$= 3[x(x-7) - 3(x-7)] = 3(x-7)(x-3)$

21. Factor out GCF (5)
$40a^2 + 5 = 5(8a^2 + 1)$

23. $2w^2 - 14w - 36$
Factor out GCF (2), then trial and error method; $m = -9$ and $n = 2$.
$= 2(w^2 - 7w - 18) = 2(w^2 - 9w + 2w - 18)$
$= 2[w(w-9) + 2(w-9)]$
$= 2(w-9)(w+2)$

25. Factor out GCF ($3b$), then difference of squares;
$3a^2b - 48b^3 = 3b(a^2 - 16b^2)$
$= 3b(a+4b)(a-4b)$

27. $x^4 - 3x^2 - 10$
Trial and error; $m = -5$, $n = 2$.
$= x^4 - 5x^2 + 2x^2 - 10$
$= x^2(x^2 - 5) + 2(x^2 - 5) = (x^2 - 5)(x^2 + 2)$

29. Difference of two cubes;
$8p^3 - q^3r^3 = (2p - qr)(4p^2 + 2pqr + q^2r^2)$

31. $x^2 - 5x - 14 = x^2 + 2x - 7x - 14$
$= x(x+2) - 7(x+2)$
$= (x-7)(x+2)$

33. $a^2 + 10a + 24 = a^2 + 4a + 6a + 24$
$= a(a+4) + 6(a+4)$
$= (a+6)(a+4)$

35. $y^2 - 10y + 21 = y^2 - 3y - 7y - 21$
$= y(y-3) - 7(y-3)$
$= (y-7)(y-3)$

37. $w^3 - 125 = (w-5)(w^2 + 5w + 25)$

39. $2t^2 + 9t - 5 = 2t^2 + 10t - t - 5$
$= 2t(t+5) - 1(t+5)$
$= (2t-1)(t+5)$

41. $4y^2 - 81 = (2y+9)(2y-9)$

43. $3a^2 + 6a - 21 = 3(a^2 + 2a - 7)$

45. $8a^3 + 27 = (2a+3)(4a^2 - 6a + 9)$

47. $4t^3 - 4 = 4(t^3 - 1) = 4(t-1)(t^2 + t + 1)$

49. $3x^2 + 30x + 75 = 3(x^2 + 10x + 25)$
$= 3(x+5)(x+5) = 3(x+5)^2$

51. $xy - 2x + 6y - 12 = x(y-2) + 6(y-2)$
$= (x+6)(y-2)$

53. The polynomial $2x^2 + 5x + 4$ is not factorable.

55. $a^2 - 10a + 25 = (a-5)(a-5) = (a-5)^2$

57. $2abx - 14ax - 8bx + 56x$
$= 2x(ab - 7a - 4b + 28)$
$= 2x[a(b-7) - 4(b-7)]$
$= 2x(a-4)(b-7)$

59. $3x^2 + x - 10 = 3x^2 + 6x - 5x - 10$
$= 3x(x+2) - 5(x+2)$
$= (3x-5)(x+2)$

61. $5n^2 + 20n + 30 = 5(n^2 + 4n + 6)$

63. $3y^3-81=3(y^3-27)$
$=3(y-3)(y^2+3y+9)$

65. $3x^2y+9xy+9y=3y(x^2+3x+3)$

67. $9x^3-4x=x(9x^2-4)=x(3x-2)(3x+2)$

69. $2n^4-16n=2n(n^3-8)$
$=2n(n-2)(n^2+2n+4)$

71. $-3x^2+12x+15=-3(x^2-4x-5)$
$=-3(x^2-5x+x-5)$
$=-3[x(x-5)+1(x-5)]$
$=-3(x+1)(x-5)$

73. $x^2y^2-4x^2-9y^2+36$
$=x^2(y^2-4)-9(y^2-4)=(x^2-9)(y^2-4)$
$=(x-3)(x+3)(y-2)(y+2)$

75. $8x^2+8x+2=2(4x^2+4x+1)$
$=2(2x+1)(2x+1)$
$=2(2x+1)^2$

77. $x^3+5x^2-4x-20=x^2(x+5)-4(x+5)$
$=(x+5)(x^2-4)$
$=(x+5)(x-2)(x+2)$

79. $x^4+3x^2-10=x^4+5x^2-2x^2-10$
$=x^2(x^2+5)-2(x^2+5)$
$=(x^2-2)(x^2+5)$

81. $t^4+t^2-20=t^4+5t^2-4t^2-20$
$=t^2(t^2+5)-4(t^2+5)$
$=(t^2-4)(t^2+5)$
$=(t+2)(t-2)(t^2+5)$

83. $(x-5)^2-169=(x-5+13)(x-5-13)$
$=(x+8)(x-18)$

85. $x^2+4xy+4y^2-16$
$=(x^2+4xy+4y^2)-16$
$=(x+2y)(x+2y)-16=(x+2y)^2-16$
$=(x+2y+4)(x+2y-4)$

87. $6(x-2)^2+7(x-2)-5$
$=\left[6(x-2)^2+10(x-2)\right]-\left[3(x-2)-5\right]$
$=\left[2(x-2)\right]\left[3(x-2)+5\right]-\left[3(x-2)+5\right]$
$=\left[3(x-2)+5\right]\left[2(x-2)-1\right]$
$=(3x-6+5)(2x-4-1)=(3x-1)(2x-5)$

89. True

91. True

Exercises 6.6

< Objective 1 >

1. $(x-3)(x-4)=0$
$x-3=0$ or $x-4$
$x=3$ $\quad x=4$

3. $(3x+1)(x-6)=0$
$3x+1=0$ or $x-6=0$
$x=-\frac{1}{3}$ $\quad x=6$

5. $x^2-3x-10=0$
$(x-5)(x+2)=0$
$x-5=0$ or $x+2=0$
$x=5$ $\quad x=-2$

7. $x^2-2x-15=0$
$(x-5)(x+3)=0$
$x-5=0$ or $x+3=0$
$x=5$ $\quad x=-3$

9. $x^2-11x+30=0$
$(x-5)(x-6)=0$
$x-5=0$ or $x-6=0$
$x=5$ $\quad x=6$

11. $x^2-4x-21=0$
$(x-7)(x+3)=0$
$x-7=0$ or $x+3=0$
$x=7$ $\quad x=-3$

13. $x^2-5x=50$
$x^2-5x-50=0$
$(x-10)(x+5)=0$
$x-10=0$ or $x+5=0$
$x=10$ $\quad x=-5$

15. $x^2=5x+84$
$x^2-5x-84=0$
$(x-12)(x+7)=0$
$x-12=0$ or $x+7=0$
$x=12$ $\quad x=-7$

17. $x^2-8x=0$
$x(x-8)=0$
$x=0$ or $x-8=0$
$x=8$

19. $x^2+10x=0$
$x(x+10)=0$
$x=0$ or $x+10=0$
$x=-10$

21. $x^2=5x$
$x^2-5x=0$
$x(x-5)=0$
$x=0$ or $x-5=0$
$x=5$

23. $x^2-25=0$
$(x+5)(x-5)=0$
$x+5=0$ or $x-5=0$
$x=-5$ $\quad x=5$

25. $9x^2=25$
$9x^2-25=0$
$(3x+5)(3x-5)=0$
$3x+5=0$ or $3x-5=0$
$x=-\frac{5}{3}$ $\quad x=\frac{5}{3}$

27. $4x^2+12x+9=0$
$(2x+3)(2x+3)=0$
$2x+3=0$
$x=-\frac{3}{2}$

29. $2x^2-17x+36=0$
$(x-4)(2x-9)=0$
$x-4=0$ or $2x-9=0$
$x=4$ $\quad x=\frac{9}{2}$

31. $5x^2+9x=18$
$5x^2+9x-18=0$
$(x+3)(5x-6)=0$
$x+3=0$ or $5x-6=0$
$x=-3$ $\quad x=\frac{6}{5}$

33.
$$\begin{aligned} 6x^2 &= 7x-2 \\ 6x^2-7x+2 &= 0 \\ (2x-1)(3x-2) &= 0 \end{aligned}$$
$2x-1=0$ or $3x-2=0$
$x=\frac{1}{2}$ $\quad x=\frac{2}{3}$

35.
$$\begin{aligned} 2m^2 &= 12m+54 \\ 2m^2-12m-54 &= 0 \\ 2(m^2-6m-27) &= 0 \\ 2(m-9)(m+3) &= 0 \end{aligned}$$
$m-9=0$ or $m+3=0$
$m=9$ $\quad m=-3$

37. $7x^2-63x=0$
$7x(x-9)=0$
$x=0$ or $x-9=0$
$x=9$

39.
$$\begin{aligned} 5x^2 &= 15x \\ x^2 &= 3x \\ x^2-3x &= 0 \\ x(x-3) &= 0 \end{aligned}$$
$x=0$ or $x-3=0$
$x=3$

41.
$$\frac{x^2}{8}-\frac{x}{4}-1=0$$
$$8\left(\frac{x^2}{8}\right)-8\left(\frac{x}{4}\right)-8(1)=8(0)$$
$$x^2-2x-8=0$$
$$(x+2)(x-4)=0$$
$x+2=0$ or $x-4=0$
$x=-2$ $\quad x=4$

43.
$$\frac{x^2}{6}+\frac{3x}{2}+3=0$$
$$6\left(\frac{x^2}{6}\right)+6\left(\frac{3x}{2}\right)+6(3)=6(0)$$
$$x^2+9x+18=0$$
$$(x+3)(x+6)=0$$
$x+3=0$ or $x+6=0$
$x=-3$ $\quad x=-6$

45.
$$\frac{2x^2}{3}+\frac{x}{15}=\frac{1}{5}$$
$$15\left(\frac{2x^2}{3}\right)+15\left(\frac{x}{15}\right)=15\left(\frac{1}{5}\right)$$
$$10x^2+x=3$$
$$10x^2+x-3=0$$
$$(5x+3)(2x-1)=0$$
$5x+3=0$ or $2x-1=0$
$5x=-3$ $\quad 2x=1$
$x=-\frac{3}{5}$ $\quad x=\frac{1}{2}$

47.
$$\begin{aligned} x(x+2) &= 15 \\ x^2+2x &= 15 \\ x^2+2x-15 &= 0 \\ (x+5)(x-3) &= 0 \end{aligned}$$
$x+5=0$ or $x-3=0$
$x=-5$ $\quad x=3$

49.
$$\begin{aligned} x(2x-3) &= 9 \\ 2x^2-3x-9 &= 0 \\ (2x+3)(x-3) &= 0 \end{aligned}$$
$2x+3=0$ or $x-3=0$
$x=-\frac{3}{2}$ $\quad x=3$

51.
$$2x(3x+1)=28$$
$$x(3x+1)=14$$
$$3x^2+x-14=0$$
$$(3x+7)(x-2)=0$$
$$3x+7=0 \quad \text{or} \quad x-2=0$$
$$x=-\frac{7}{3} \qquad x=2$$

53.
$$(x-3)(x-1)=15$$
$$x^2-4x+3=15$$
$$x^2-4x-12=0$$
$$(x-6)(x+2)=0$$
$$x-6=0 \quad \text{or} \quad x+2=0$$
$$x=6 \qquad x=-2$$

55.
$$(x-5)(x+2)=18$$
$$x^2-3x-10=18$$
$$x^2-3x-28=0$$
$$(x-7)(x+4)=0$$
$$x-7=0 \quad \text{or} \quad x+4=0$$
$$x=7 \qquad x=-4$$

< Objective 2 >

57.
$$x^2-6x+5=0$$
$$(x-1)(x-5)=0$$
The zeros are 1 and 5.

59.
$$x^2-9x=0$$
$$x(x-9)=0$$
The zeros are 0 and 9.

< Objective 3 >

61. Let x be the (first) number. Then $2x+3$ represents the other integer.
$$x(2x+3)=65$$
$$2x^2+3x=65$$
$$2x^2+3x-65=0$$
$$(2x+13)(x-5)=0$$
$$x=5$$
$$2x+3=13$$
The integers are 5 and 13.

63. Let x be the (first) number. Then $10-x$ represents the other integer.
$$x(10-x)=24$$
$$10x-x^2=24$$
$$0=x^2-10x+24=(x-6)(x-4)$$
The integers are 4 and 6.

65. Let x be the (first) integer.
Then $x+1$ represents the next integer.
$$x(x+1)=72$$
$$x^2+x-72=0$$
$$(x-8)(x+9)=0$$
$$x=8 \quad \text{or} \quad x=-9$$
$$x+1=9 \qquad x+1=-8$$
The integers are –9 and –8, or 8 and 9.

67. Let x be the (first) whole number. Then $x+1$ is the next whole number.
$$x^2+(x+1)^2=61$$
$$x^2+x^2+2x+1=61$$
$$2x^2+2x-60=0$$
$$x^2+x-30=0$$
$$(x+6)(x-5)=0$$
$$x=5$$
$$x+1=6$$
The whole numbers are 5 and 6.

69. Let x be the (first) integer.
Then $x+1$ and $x+2$ are the next integers.

$$x^2+(x+1)^2+(x+2)^2=50$$
$$x^2+(x^2+2x+1)+(x^2+4x+4)=50$$
$$3x^2+6x-45=0$$
$$x^2+2x-15=0$$
$$(x+5)(x-3)=0$$
$$x=-5 \quad \text{or} \quad x=3$$
$$x+1=-4 \qquad x+1=4$$
$$x+2=-3 \qquad x+2=5$$

The integers are –5, –4, and –3, or 3, 4, and 5.

71. Let L be the length of the rectangle. Then $(L-3)$ is the width of the rectangle.

$$A=LW$$
$$70=L(L-3)=L^2-3L$$
$$0=L^2-3L-70=(L-10)(L+7)$$

Since lengths cannot be negative, $L=10$. The rectangle is 7 ft by 10 ft.

73. Let W be the width of the rectangle. Then $(3W+2)$ is the length of the rectangle.

$$W(3W+2)=85$$
$$3W^2+2W-85=0$$
$$(3W+17)(W-5)=0$$

Since lengths cannot be negative, $W=5$. The dimensions are 5 cm by 17 cm.

75. Let W be the width of the rectangle. Then $W+1$ is the length of the rectangle. And $2(W+1)=2W+2$ is the length doubled.

$$W(2W+2)=W(W+1)+30$$
$$2W^2+2W=W^2+W+30$$
$$W^2+W-30=0$$
$$(W+6)(W-5)=0$$

Since lengths cannot be negative,
$W=6$ cm
Length $=W+1=(5+1)$ cm $=6$ cm
The original rectangle is 5 cm by 6 cm.

77. Express the dimensions as shown in the figure in the text.

$$V=LWH$$
$$4{,}000=10(2x-20)(x-20)$$
$$400=(2x-20)(x-20)$$
$$200=(x-10)(x-20)$$
$$0=x^2-30x=x(x-30)$$

The dimensions of original piece of tin were 30 cm by 60 cm.

79. Let W and $(W+4)$ be the dimensions of the cardboard. Then the dimensions of the box are $(W-2)$, $(W+2)$, and 1.

$$V=LWH$$
$$32=(W+2)(W-2)\bullet 1=W^2-4$$
$$0=W^2-36=(W+6)(W-6)$$

The dimensions were 6 ft by 10 ft.

81. Let s be the length of a side of the square. Then $(s-10)$ is the length of a side (of the base) of the box.

$$5(s-10)^2=4{,}500$$
$$(s-10)^2=900$$
$$s^2-20s-800=0$$
$$(s-40)(s+20)=0$$

The length of a side of the original piece of cardboard is 40 cm.

83. $h=-16t^2+64t$

$$0=-16t^2+64t=-16t(t-4)$$

The ball lands after 4 s.

85. $h=-16t^2+96t$

$$0=-16t^2+96t=-16t(t-6)$$

The ball lands after 6 s.

87. $h=-16t^2+64t+192$

$$0=-16t^2+64t+192=t^2-4t-12$$
$$=(t-6)(t+2)$$

The ball takes 6 s to fall back to the ground.

89. $h = -16t^2 + 96t + 192$

$192 = -16t^2 + 96t + 192$

$0 = -16t^2 + 96t = -16t(t-6)$

The ball takes 6 s to return to the thrower.

91. $h = -5t^2 + 20t + 105$

$0 = -5t^2 + 20t + 105 = t^2 - 4t - 21$

$= (t-7)(t+3)$

The ball will take 7 s to fall to the ground.

93. $\frac{x^2}{20} + x = 120$

$x^2 + 20x = 2,400$

$x^2 + 20x - 2,400 = 0$

$(x-40)(x+60) = 0$

The car was travelling at a speed of 40 mi/hr.

95. $\frac{x^2}{20} + x = 315$

$x^2 + 20x = 6,300$

$x^2 + 20x - 6,300 = 0$

$(x-70)(x+90) = 0$

The maximum speed is 70 mi/hr.

97. False

99. always

101. always

103. $h = v_0 t - 16t^2$

$75 = 80t - 16t^2$

$16t^2 - 80t + 75 = 0$

$(4t-5)(4t-15)$

$4t - 5 = 0$ or $4t - 15 = 0$

$\frac{4t}{4} = \frac{5}{4}$ $\quad$ $\frac{4t}{4} = \frac{15}{4}$

$t = \frac{5}{4} = 1.25$ $\quad$ $t = \frac{15}{4} = 3.75$

1.25s need to pass before the drop reaches a height of 75 ft.

105. $2Q^2 + Q - 21 = 0$

$(2Q+7)(Q-3) = 0$

The flow rate is 3.

Note: $-\frac{7}{2}$ is not a solution to the application.

107. $y = -4x^2 + 500x$

$0 = -4x(x-125)$

The productivity is zero if there is 0 cm^3 or 125 cm^3 of water present.

109. $y = -7x^2 + 1,022x$

$0 = -7x(x-146)$

The productivity is zero if there is 0 cm^3 or 146 cm^3 of water present.

111. $T = 0.4t^2 - 2.6t + 103$

$100 = 0.4t^2 - 2.6t + 103$

$0 = 0.4t^2 - 2.6t + 3 = 4t^2 - 26t + 30$

$= (4t-6)(t-5)$

The patient's temperature rises back up to 100°F after 5 hr.

113. $R = C$

$x^2 - 8x = 20x + 60$

$x^2 - 28x - 60 = 0$

$(x-30)(x+2) = 0$

The shop breaks even if they sell 30 bicycles.
Note: –2 is not a solution to the application.

115. $(x+4)(x-5) = 0$

$x^2 - x - 20 = 0$

117. $(x-2)(x-6) = 0$

$x^2 - 8x - 12 = 0$

119. $x^3 - 3x^2 - 10x = 0$

$x(x^2 - 3x - 10) = 0$

$x(x-5)(x+2) = 0$

$x = 0,\ x = -5,\ x = -2$

$\{-2, 0, 5\}$

121. $x^3 - 9x = 0$

$x(x^2 - 9) = 0$

$x(x+3)(x-3) = 0$

$x = -3,\ x = 0,\ x = 3$

$\{-3, 0, 3\}$

123. $x^3 + x^2 - 4x - 4 = 0$

$x^2(x+1) - 4(x+1) = 0$

$(x^2 - 4)(x+1) = 0$

$(x+2)(x-2)(x+1) = 0$

$x = -2,\ x = -1,\ x = 2$

$\{-2, -1, 2\}$

125. $x^4 - 10x^2 + 9 = 0$

$(x^2 - 9)(x^2 - 1) = 0$

$(x+3)(x-3)(x+1)(x-1) = 0$

$x = -3,\ x = -1,\ x = 1,\ x = 3$

$\{-3, -1, 1, 3\}$

127. Above and Beyond

Summary Exercises

1. $14a - 35 = 7(2a - 5)$

3. $24s^2t - 16s^2 = 8s^2(3t - 2)$

5. $27s^4 + 18s^3 = 9s^3(3s + 2)$

7. $18m^2n^2 - 27m^2n + 36m^2n^3$

$= 9m^2n(2n - 3 + 4n^2)$

9. $8a^2b + 24ab - 16ab^2 = 8ab(a + 3 - 2b)$

11. $2x(3x+4y) - y(3x+4y)$

$= (2x - y)(3x + 4y)$

13. $x^2 - 4x + 5x - 20 = x(x-4) + 5(x-4)$

$= (x+5)(x-4)$

15. $8x^2 + 6x - 20x - 15$

$= 2x(4x+3) - 5(4x+3) = (2x-5)(4x+3)$

17. $6x^3 + 9x^2 - 4x^2 - 6x$

$= x(6x^2 + 9x - 4x - 6)$

$= x[3x(2x+3) - 2(2x+3)]$

$= x(3x-2)(2x+3)$

19. $p^2 - 49 = (p+7)(p-7)$

21. $9n^2 - 25m^2 = (3n+5m)(3n-5m)$

23. $25 - z^2 = (5+z)(5-z)$

25. $25a^2 - 36b^2 = (5a+6b)(5a-6b)$

27. $3w^3 - 12wz^2 = 3w(w^2 - 4z^2)$

$= 3w(w+2z)(w-2z)$

29. $2m^2 - 72n^4 = 2(m^2 - 36n^4)$

$= 2(m+6n^2)(m-6n^2)$

31. $x^2 - 18x + 81 = (x-9)(x-9) = (x-9)^2$

33. $x^2 + 14x + 49 = (x+7)(x+7) = (x+7)^2$

35. $4x^2 + 12x + 9 = (2x+3)(2x+3) = (2x+3)^2$

37. The binomial $x^2 - 20$ is not a difference of squares. Hence, it is not factorable.

39. $a^4 - 16b^4 = (a^2 + 4b^2)(a^2 - 4b^2)$

$= (a^2 + 4b^2)(a+2b)(a-2b)$

41. $8x^3+1=(2x+1)(4x^2-2x+1)$

43. $125m^3+64n^3$
$=(5m+4n)(25m^2-20mn+16n^2)$

45. $x^2+12x+20=(x+10)(x+2)$

47. $w^2-15w+54=(w-6)(w-9)$

49. $x^2-8xy-48y^2=(x-12y)(x+4y)$

51. $14x^2+43x-21=(7x-3)(2x+7)$

53. $-x^2-9x-20=-1(x+4)(x+5)$

55. $a^2-7a+12=(a-4)(a-3)$

57. $x^2+16x+64=(x+8)^2$

59. $b^2-4bc-21c^2=(b-7c)(b+3c)$

61. $m^3+2m^2-35m=m(m^2+2m-35)$
$=m(m+7)(m-5)$

63. $3y^3-48y^2+189y=3y(y^2-16y+63)$
$=3y(y-7)(y-9)$

65. $3x^2+8x+5=(3x+5)(x+1)$

67. $2b^2-9b+9=(2b-3)(b-3)$

69. $10x^2-11x+3=(5x-3)(2x-1)$

71. $16y^2-8xy-15x^2=(4y-5x)(4y+3x)$

73. $-8x^3+36x^2+20x=-4x(2x^2-9x-5)$
$=-4x(2x+1)(x-5)$

75. $6x^3-3x^2-9x=3x(2x^2-x-3)$
$=3x(2x-3)(x+1)$

77. $ac=-30$; $b=-1$
$m=-6$; $n=5$
x^2-x-30 is factorable.

79. $ac=24$; $b=-11$
$m=-3$; $n=-8$
$2x^2-11x+12$ is factorable.

81. $5x^4-5x^3-210x^2=5x^2(x^2-x-42)$
$=5x^2(x-7)(x+6)$

83. $72y-2y^3=2y(36-y^2)$
$=2y(6+y)(6-y)$

85. $10x^5+80x^4+160x^3=10x^3(x^2+8x+16)$
$=10x^3(x+4)^2$

87. $x^3+2x^2-25x-50=x^2(x+2)-25(x+2)$
$=(x^2-25)(x+2)$
$=(x+5)(x-5)(x+2)$

89. $x^2+5x-6=0$
$(x+6)(x-1)=0$
$x+6=0$ or $x-1=0$
$x=-6$ or $x=1$
$\{-6, 1\}$

91. $x^2+7x=30$
$x^2+7x-30=0$
$(x+10)(x-3)=0$
$x+10=0$ or $x-3=0$
$x=-10$ or $x=3$
$\{-10, 3\}$

93.
$$x^2+x=20$$
$$x^2+x-20=0$$
$$(x+5)(x-4)=0$$
$\{-5, 4\}$

95.
$$x^2-10x=0$$
$$x(x-10)=0$$
$\{0, 10\}$

97.
$$x^2-25=0$$
$$(x+5)(x-5)=0$$
$\{-5, 5\}$

99.
$$2x^2-x-3=0$$
$$(2x-3)(x+1)=0$$
$\left\{-1, \frac{3}{2}\right\}$

101.
$$3x^2+9x-30=0$$
$$x^2+3x-10=0$$
$$(x+5)(x-2)=0$$
$\{-5, 2\}$

103.
$$x(x-5)=36$$
$$x^2-5x-36=0$$
$$(x-9)(x+4)=0$$
$\{-4, 9\}$

105.
$$x^3-2x^2-15x=0$$
$$x(x^2-2x-15)=0$$
$$x(x-5)(x+3)=0$$
$\{-3, 0, 5\}$

107. Let x be the smaller integer.
Then $(x+15)$ is the other integer.
$$x(x+15)=-54$$
$$x^2+15x+54=0$$
$$(x+6)(x+9)=0$$
$$x+6=0 \quad \text{or} \quad x+9=0$$
$$x=-6 \qquad x=-9$$
$$x+15=9 \qquad x+15=6$$
The integers are –6 and 9 or –9 and 6.

109. Let W be the width of the rectangle. Then $(W+6)$ is its length.
$$W(W+6)=216$$
$$W^2+6W-216=0$$
$$(W-12)(W+18)=0$$
The rectangle is 12 ft by 18 ft.

111.
$$-16t^2+80t+20=84$$
$$-16t^2+80t-64=0$$
$$t^2-5t+4=0$$
$$(t-1)(t-4)=0$$
On the way down, the ball reaches a height of 84 ft in 4 s after being thrown.

113.
$$3x^2-60x+3000=7,500$$
$$3x^2-60x-4,500=0$$
$$x^2-20x-1,500=0$$
$$(x-50)(x+30)=0$$
50 systems can be produced for \$7,500.

Chapter Test 6

1.
$$32a^2b-50b^3=2b(16a^2-25b^2)$$
$$=2b(4a+5b)(4a-5b)$$

3. $7b+42=7(b+6)$

5. $9x^2-12xy+4y^2=(3x-2y)^2$

7. $16y^2 - 49x^2 = (4y+7x)(4y-7x)$

9. $27y^3 - 8x^3 = (3y-2x)(9y^2+6xy+4x^2)$

11. $6x^2 + 4x - 15x - 10$
$= 2x(3x+2) - 5(3x+2) = (2x-5)(3x+2)$

13. $-6x^3 - 3x^2 + 30x = -3x(2x^2 + x - 10)$
$= -3x(2x+5)(x-2)$

15.
$$x^2 - 11x = -30$$
$$x^2 - 11x + 30 = 0$$
$$(x-6)(x-5) = 0$$
$\{5, 6\}$

17.
$$x^2 - 2x - 3 = 0$$
$$(x-3)(x+1) = 0$$
$$x - 3 = 0 \quad \text{or} \quad x + 1 = 0$$
$$x = 3 \qquad x = -1$$
$\{-1, 3\}$

19. Let W be the width of the rectangle. Then $2W - 4$ is its length.
$$W(2W-4) = 240$$
$$2W^2 - 4W - 240 = 0$$
$$W^2 - 2W - 120 = 0$$
$$(W-12)(W+10) = 0$$
$$W = 12$$
$$L = 2W - 4 = 20$$
The length of the rectangle is 20 cm.

20.
$$-5t^2 + 20t + 18 = 38$$
$$-5t^2 + 20t - 20 = 0$$
$$t^2 - 4t + 4 = 0$$
$$(t-2)^2 = 0$$
2 s after being thrown, the ball reaches a maximum height of 38 ft.
$$\frac{4x^3 - 5x^2 + 7x - 9}{x-2} = 4x^2 + 3x + 13 + \frac{17}{x-2}$$

Cumulative Review: Chapters 0–6

1. $\dfrac{3x^2 - 2z + 1}{5y + 2x} = \dfrac{3(-3)^2 - 2(-4) + 1}{5(2) + 2(-3)} = \dfrac{36}{4} = 9$

3.
$$\frac{2}{3}x = -22$$
$$\frac{3}{2}\left(\frac{2}{3}x\right) = \frac{3}{2}(-22)$$
$$x = -33; \{-33\}$$

5.
$$P = 2L + 2W$$
$$P - 2L = 2W$$
$$\frac{P - 2L}{2} = \frac{2W}{2} = W$$

7.
$$2x - 9 \le 7x - 3(x-1)$$
$$2x - 9 \le 7x - 3x + 3$$
$$-12 \le 2x$$
$$-6 \le x$$
$$\{x \mid x \ge -6\}$$

9.
$$2x - 3y = 13$$
$$x = 3y + 9$$
Substitute x into the first equation.
$$2(3y+9) - 3y = 13 \qquad x = 3\left(-\frac{5}{3}\right) + 9$$
$$6y + 18 - 3y = 13 \qquad = -5 + 9 = 4$$
$$3y = -5$$
$$y = -\frac{5}{3}$$
$$\left\{\left(4, -\frac{5}{3}\right)\right\}$$

11.
$$m = \frac{7-1}{1-(-2)} = 2$$
$$y - 1 = 2(x+2)$$
$$y = 2x + 5$$

13. $7x^2y + 3xy - 5x^2y + 2xy = 2x^2y + 5xy$

15. $(5x^2+4x-3)-(4x^2-5x-1)$
$=5x^2+4x-3-4x^2+5x+1=x^2+9x-2$

17. $(2x-5)(x+1)=2x^2-3x-5$

19. $(4x-3)(2x+5)=8x^2+14x-15$

21. $(x-2y)^2=x^2-4xy+4y^2$

23. $(-2x^{-2}y^3)^3=-8x^{-6}y^9=-\dfrac{8y^9}{x^6}$

25. $(2x^3y^{-2})^{-2}(x^{-3}y^{-1})=\dfrac{1}{4}x^{-6}y^4x^{-3}y^{-1}$
$=\dfrac{1}{4}x^{-9}y^3=\dfrac{y^3}{4x^9}$

27. $12x+20=4(3x+5)$

29. $12x^2-15x+8x-10$
$=3x(4x-5)+2(4x-5)=(3x+2)(4x-5)$

31.
$$x^2+2x=15$$
$$x^2+2x-15=0$$
$$(x+5)(x-3)=0$$
$x+5=0$ or $x-3=0$
$x=-5$ $\quad x=3$
$\{-5, 3\}$

33. $\dfrac{24x^2y^5-15x^4y^2+9xy}{3xy}=8xy^4-5x^3y+3$

35. Let x be the number.
$3x-5=46$
$3x=51$
$x=17$
The number is 17.

37. $240=\dfrac{x^2}{20}+x$
$0=x^2+20x-4800=(x-60)(x+80)$
The speed of the car is 60 mi/hr.

Chapter 7
Radicals and Exponents

Exercises 7.1

< Objective 1 >

1. $\sqrt{49} = 7$

3. $-\sqrt{36} = -6$

5. $\pm\sqrt{81} = \pm 9$

7. $\sqrt{-49}$ is not a real number.

9. $\sqrt[3]{27} = 3$

11. $\sqrt[3]{-64} = -4$

13. $-\sqrt[3]{216} = -6$

15. $\sqrt[4]{81} = 3$

17. $\sqrt[5]{-32} = -2$

19. $-\sqrt[4]{16} = -2$

21. $\sqrt[4]{-16}$ is not a real number.

23. $-\sqrt[5]{243} = -3$

25. $\sqrt{\frac{4}{9}} = \frac{2}{3}$

27. $\sqrt[3]{\frac{8}{27}} = \frac{2}{3}$

< Objective 2 >

29. $\sqrt{6^2} = 6$

31. $\sqrt{(-3)^2} = 3$

33. $\sqrt[3]{4^3} = 4$

35. $\sqrt[4]{3^4} = 3$

37. $\sqrt{3}$

$1^2 = 1$ and $2^2 = 4$

So, $1 < \sqrt{3} < 4$

Consecutive integers that trap the root are 1 and 4.

39. $\sqrt{30}$

$5^2 = 25$ and $6^2 = 36$

So, $5 < \sqrt{30} < 6$

Consecutive integers that trap the root are 5 and 6.

41. $\sqrt{150}$

$12^2 = 144$ and $13^2 = 169$

So, $12 < \sqrt{30} < 13$

Consecutive integers that trap the root are 12 and 13.

43. $\sqrt{500}$

$22^2 = 484$ and $23^2 = 529$

So, $22 < \sqrt{500} < 23$

Consecutive integers that trap the root are 22 and 23.

45. $\sqrt[3]{100}$

$4^3 = 64$ and $5^3 = 125$

So, $4 < \sqrt[3]{100} < 5$

Consecutive integers that trap the root are 4 and 5.

47. $\sqrt[3]{-20}$

$(-3)^3 = -27$ and $(-2)^3 = -8$

So, $-3 < \sqrt[3]{-20} < -2$

Consecutive integers that trap the root are −3 and −2.

< Objective 3 >

49. $\sqrt{x^2} = |x|$

51. $\sqrt[5]{y^5} = y$

53. $\sqrt{9x^2} = |3x| = 3|x|$

55. $\sqrt{a^4b^6} = |a^2b^3| = a^2|b^3|$

57. $\sqrt{16x^4} = 4|x^2| = 4x^2$

< Objective 4 >

59. $c^2 = a^2 + b^2 = 12^2 + 9^2 = 225$

$c = \sqrt{225} = 15$

The missing side has length 15 units.

61. $c^2 = a^2 + b^2$

$17^2 = a^2 + 8^2$

$289 = a^2 + 64$

$225 = a^2$

$a = \sqrt{225} = 15$

The missing side has length 15 units.

63. $c^2 = a^2 + b^2$

$19^2 = 14^2 + b^2$

$361 = 196 + b^2$

$165 = b^2$

$\sqrt{165} = b$

The missing side has length $\sqrt{165}$ units.

< Objective 5 >

65. $d = \sqrt{(2-2)^2 + [(-9)-(-6)]^2}$

$= \sqrt{0^2 + (-3)^2} = 3$

67. $d = \sqrt{[-4-(-7)]^2 + (0-1)^2} = \sqrt{3^2 + 1^2}$

$= \sqrt{10}$

< Objective 6 >

69. $x^2 + y^2 = 4$

Center: (0, 0)

Radius: $\sqrt{4} = 2$

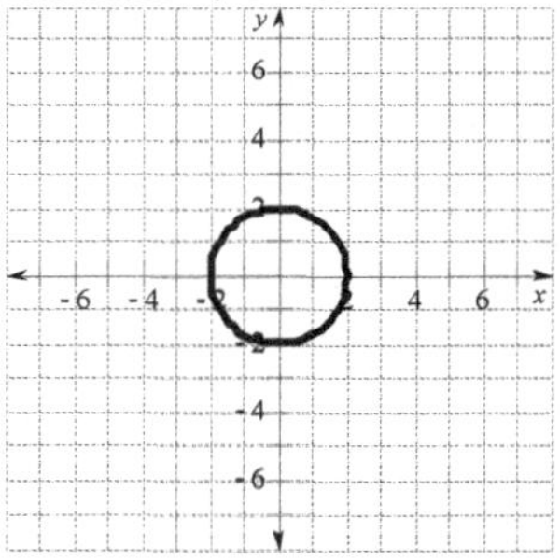

71. $(x-1)^2 + y^2 = 9$

Center: (1, 0)

Radius: $\sqrt{9} = 3$

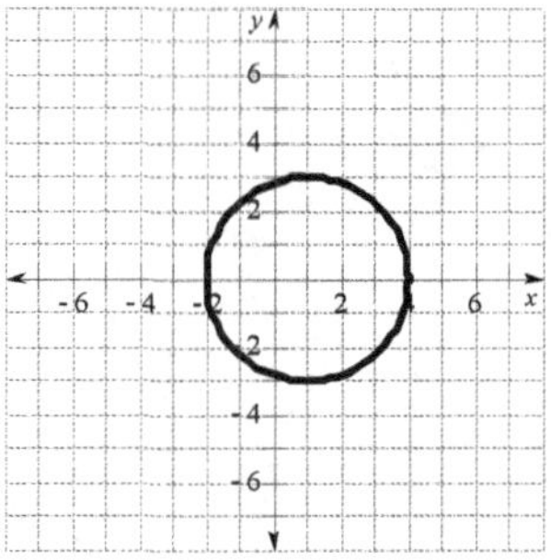

73. $(x-4)^2+(y+1)^2=16$

Center: (4, –1)

Radius: $\sqrt{16}=4$

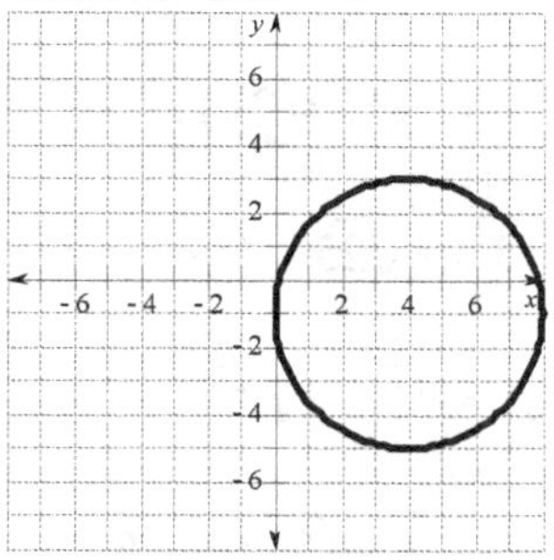

75. $(h,k)=(0,0)$

$r=5$

$x^2+y^2=5^2=25$

77. $(h,k)=(3,-2)$

$r=5$

$(x-h)^2+(y-k)^2=r^2$

$(x-3)^2+(y+2)^2=25$

79. $\sqrt{-16}=$ is not a real number

$-\sqrt{16}=-4$

Equivalent expressions: $-\sqrt{16}$, -4

81. $\sqrt[3]{-125}=-5$

$-\sqrt[3]{125}=-5$

$|-5|=5$

Equivalent expressions: $\sqrt[3]{-125}, -\sqrt[3]{125}$

83. $\sqrt[4]{10,000}=10$

$\sqrt[3]{1,000}=10$

Equivalent expressions: $\sqrt[4]{10,000}, \sqrt[3]{1,000}$

85. $\sqrt{16x^{16}}=4x^4$

$\sqrt{16x^{16}}=4x^8\neq 4x^4$

False

87. $\sqrt{16x^{-4}y^{-4}}$ is a real number.

$\sqrt{16x^{-4}y^{-4}}=4x^{-2}y^{-2}$

True

89. $\dfrac{\sqrt{x^2-25}}{x-5}=\dfrac{\sqrt{(x-5)(x+5)}}{x-5}\neq\sqrt{x+5}$

False

91. When $x=0$, $y=3$

$3+-3=0$

When $y=0$, $x=3$

$3+-3=0$

True

93. never

95. $\sqrt{15}=3.873$

97. $\sqrt{156}=12.49$

99. $\sqrt{-15}=$ not a real number

101. $\sqrt[3]{83}=4.362$

103. $\sqrt[5]{123}=2.618$

105. $\sqrt[3]{-15}=-2.466$

107. $\sqrt{10}=3.16$ ft

109. $\sqrt{17}=4.12$ ft

111. $\sqrt{32}=5.66$ ft

113. $\pi r^2=147$

$r^2=\dfrac{147}{\pi}$

$r=\sqrt{\dfrac{147}{\pi}}=6.84$ ft

115. $t=\dfrac{1}{4}\sqrt{d}=\dfrac{1}{4}\sqrt{800}\approx\dfrac{1}{4}(28.284)\approx 7.07$

It takes 7.07 s for the object to fall 800 ft.

117. $(x+3)^2+(y-2)^2=16$

$$r=\sqrt{16}=4$$
$$|x+3|\le 4$$
$$|y-2|\le 4$$
$$-4\le x+3\le 4 \quad\text{and}\quad -4\le y-2\le 4$$
$$-7\le x\le 1 \qquad -2\le y\le 6$$
$$D=\{x|-7\le x\le 1\}$$
$$R=\{y|-2\le y\le 6\}$$

119. $x^2+(y-3)^2=25$

$$r=\sqrt{25}=5$$
$$|x|\le 5$$
$$|y-3|\le 5$$
$$-5\le y-3\le 5$$
$$-2\le y\le 8$$
$$D=\{x|-5\le x\le 5\}$$
$$R=\{y|-2\le y\le 8\}$$

121. No

123. Above and Beyond

125. Above and Beyond

127. **(a)** $\sqrt{4\bullet 9}=\sqrt{36}=6$

(b) $\sqrt{4}\bullet\sqrt{9}=2\bullet 3=6$

(c) $\sqrt{9\bullet 16}=\sqrt{144}=12$

(d) $\sqrt{9}\bullet\sqrt{16}=3\bullet 4=12$

(e) $\sqrt{(-4)(-25)}=\sqrt{100}=10$

(f) $\sqrt{-4}\bullet\sqrt{-25}$ is not a real number.

(g) Above and Beyond

Exercises 7.2

< Objective 1 >

1. $\sqrt{12}=\sqrt{4\bullet 3}=\sqrt{4}\bullet\sqrt{3}=2\sqrt{3}$

3. $\sqrt{50}=\sqrt{25\bullet 2}=\sqrt{25}\bullet\sqrt{2}=5\sqrt{2}$

5. $-\sqrt{108}=-\sqrt{36\bullet 3}=-\sqrt{36}\bullet\sqrt{3}=-6\sqrt{3}$

7. $\sqrt{52}=\sqrt{4\bullet 13}=\sqrt{4}\bullet\sqrt{13}=2\sqrt{13}$

9. $\sqrt{60}=\sqrt{4\bullet 15}=\sqrt{4}\bullet\sqrt{15}=2\sqrt{15}$

11. $-\sqrt{125}=-\sqrt{25\bullet 5}=-\sqrt{25}\bullet\sqrt{5}=-5\sqrt{5}$

13. $\sqrt[3]{16}=\sqrt[3]{8\bullet 2}=\sqrt[3]{8}\bullet\sqrt[3]{2}=2\sqrt[3]{2}$

15. $\sqrt[3]{-48}=\sqrt[3]{-8\bullet 6}=\sqrt[3]{-8}\bullet\sqrt[3]{6}=-2\sqrt[3]{6}$

17. $\sqrt[3]{135}=\sqrt[3]{27\bullet 5}=\sqrt[3]{27}\bullet\sqrt[3]{5}=3\sqrt[3]{5}$

19. $\sqrt[4]{32}=\sqrt[4]{16\bullet 2}=\sqrt[4]{16}\bullet\sqrt[4]{2}=2\sqrt[4]{2}$

21. $\sqrt{18z^2}=\sqrt{9\bullet 2\bullet z^2}=\sqrt{9}\bullet\sqrt{2}\bullet\sqrt{z^2}=3z\sqrt{2}$

23. $\sqrt{63x^4}=\sqrt{9\bullet 7\bullet (x^2)^2}=\sqrt{9}\bullet\sqrt{7}\bullet\sqrt{(x^2)^2}$
$=3x^2\sqrt{7}$

25. $\sqrt{98m^3}=\sqrt{49\bullet 2\bullet m^2\bullet m}$
$=\sqrt{49}\bullet\sqrt{2m}\bullet\sqrt{m^2}=7m\sqrt{2m}$

27. $\sqrt{80x^2y^3}=\sqrt{16\bullet 5\bullet x^2\bullet y^2\bullet y}$
$=\sqrt{16x^2y^2}\bullet\sqrt{5y}=4xy\sqrt{5y}$

29. $\sqrt[3]{40b^3}=\sqrt[3]{8\bullet 5\bullet b^3}=\sqrt[3]{8b^3}\bullet\sqrt[3]{5}=2b\sqrt[3]{5}$

31. $\sqrt[3]{48p^9}=\sqrt[3]{8\bullet 6\bullet (p^3)^3}=\sqrt[3]{8(p^3)^3}\bullet\sqrt[3]{6}$
$=2p^3\sqrt[3]{6}$

33. $\sqrt[3]{54m^7}=\sqrt[3]{27\bullet 2\bullet (m^2)^3\bullet m}$
$=\sqrt[3]{27(m^2)^3}\bullet\sqrt[3]{2m}=3m^2\sqrt[3]{2m}$

35. $\sqrt[3]{56x^6y^5z^4} = \sqrt[3]{8 \bullet 7 \bullet (x^2)^3 \bullet y^3 \bullet y^2 \bullet z^3 \bullet z}$
$= \sqrt[3]{8(x^2)^3 y^3z^3} \bullet \sqrt[3]{7y^2z}$
$= 2x^2yz\sqrt[3]{7y^2z}$

37. $\sqrt[4]{32x^8} = \sqrt[4]{16 \bullet 2 \bullet (x^2)^4} = \sqrt[4]{16(x^2)^4} \bullet \sqrt[4]{2}$
$= 2x^2\sqrt[4]{2}$

39. $\sqrt[4]{128a^{12}b^{17}} = \sqrt[4]{16 \bullet 8 \bullet (a^3)^4 \bullet (b^4)^4 \bullet b}$
$= \sqrt[4]{16(a^3)^4 \bullet (b^4)^4} \bullet \sqrt[4]{8b}$
$= 2a^3b^4\sqrt[4]{8b}$

< Objective 2 >

41. $\sqrt{\dfrac{5}{16}} = \dfrac{\sqrt{5}}{\sqrt{16}} = \dfrac{\sqrt{5}}{4}$

43. $\sqrt{\dfrac{5}{9y^4}} = \dfrac{\sqrt{5}}{\sqrt{9(y^2)^2}} = \dfrac{\sqrt{5}}{3y^2}$

45. $\sqrt[3]{\dfrac{5}{8}} = \dfrac{\sqrt[3]{5}}{\sqrt[3]{8}} = \dfrac{\sqrt[3]{5}}{2}$

< Objective 3 >

47. $\dfrac{5}{\sqrt{7}} = \dfrac{5}{\sqrt{7}} \bullet \dfrac{\sqrt{7}}{\sqrt{7}} = \dfrac{5\sqrt{7}}{7}$

49. $\dfrac{7}{\sqrt{12}} = \dfrac{7}{\sqrt{12}} \bullet \dfrac{\sqrt{3}}{\sqrt{3}} = \dfrac{7\sqrt{3}}{\sqrt{36}} = \dfrac{7\sqrt{3}}{6}$

51. $\dfrac{2\sqrt{3}}{\sqrt{10}} = \dfrac{2\sqrt{3}}{\sqrt{10}} \bullet \dfrac{\sqrt{10}}{\sqrt{10}} = \dfrac{2\sqrt{30}}{10} = \dfrac{\sqrt{30}}{5}$

53. $\sqrt[3]{\dfrac{7}{4}} = \dfrac{\sqrt[3]{7}}{\sqrt[3]{4}} \bullet \dfrac{\sqrt[3]{2}}{\sqrt[3]{2}} = \dfrac{\sqrt[3]{14}}{2}$

55. $\dfrac{5}{\sqrt[3]{16}} = \dfrac{5}{\sqrt[3]{16}} \bullet \dfrac{\sqrt[3]{4}}{\sqrt[3]{4}} = \dfrac{5\sqrt[3]{4}}{4}$

57. $\sqrt{\dfrac{12}{w}} = \dfrac{\sqrt{4 \bullet 3}}{\sqrt{w}} \bullet \dfrac{\sqrt{w}}{\sqrt{w}} = \dfrac{2\sqrt{3w}}{w}$

59. $\dfrac{\sqrt{8m^3}}{\sqrt{5n}} = \dfrac{\sqrt{4m^2 \bullet 2m}}{\sqrt{5n}} \bullet \dfrac{\sqrt{5n}}{\sqrt{5n}} = \dfrac{2m\sqrt{2m \bullet 5n}}{5n}$
$= \dfrac{2m\sqrt{10mn}}{5n}$

61. $\sqrt[3]{\dfrac{5}{y}} = \dfrac{\sqrt[3]{5}}{\sqrt[3]{y}} \bullet \dfrac{\sqrt[3]{y^2}}{\sqrt[3]{y^2}} = \dfrac{\sqrt[3]{5y^2}}{y}$

63. $\dfrac{3}{\sqrt[3]{2x}} = \dfrac{3}{\sqrt[3]{2x}} \bullet \dfrac{\sqrt[3]{4x^2}}{\sqrt[3]{4x^2}} = \dfrac{3\sqrt[3]{4x^2}}{2x}$

65. $\sqrt[3]{\dfrac{2}{5x^2}} = \dfrac{\sqrt[3]{2}}{\sqrt[3]{5x^2}} \bullet \dfrac{\sqrt[3]{25x}}{\sqrt[3]{25x}} = \dfrac{\sqrt[3]{50x}}{5x}$

67. $\dfrac{\sqrt[3]{5}}{\sqrt[3]{4a^2}} = \dfrac{\sqrt[3]{5}}{\sqrt[3]{4a^2}} \bullet \dfrac{\sqrt[3]{2a}}{\sqrt[3]{2a}} = \dfrac{\sqrt[3]{10a}}{2a}$

69. $\sqrt[3]{\dfrac{a^5}{b^7}} = \dfrac{\sqrt[3]{a^3 \bullet a^2}}{\sqrt[3]{b^7}} \bullet \dfrac{\sqrt[3]{b^2}}{\sqrt[3]{b^2}} = \dfrac{a\sqrt[3]{a^2b^2}}{b^3}$

71. $d = \sqrt{[0-(-7)]^2 + (0-1)^2} = \sqrt{7^2+1^2}$
$= \sqrt{50} = \sqrt{25 \bullet 2} = 5\sqrt{2}$

73. $d = \sqrt{[-9-(-13)]^2 + (18-22)^2}$
$= \sqrt{4^2+4^2} = \sqrt{32} = \sqrt{16}\sqrt{2} = 4\sqrt{2}$

75. True;
$\sqrt{16x^{16}} = 4x^8$

77. True;

$$\frac{\sqrt{x^2-25}}{\sqrt{x-5}}=\frac{\sqrt{(x+5)(x-5)}}{\sqrt{x-5}}$$
$$=\frac{\sqrt{x+5}\bullet\sqrt{x-5}}{\sqrt{x-5}}=\sqrt{x+5}$$

79. True;

$$\sqrt[3]{\left(8b^6\right)^2}=\sqrt[3]{64b^{12}}=4b^4$$
$$\left(\sqrt[3]{8b^6}\right)^2=\left(2b^2\right)^2=4b^4$$

Thus, $\sqrt[3]{\left(8b^6\right)^2}=\left(\sqrt[3]{8b^6}\right)^2$

81. True

83. True

85. Using $T=k\sqrt{L}$,

L	10	20	30	40	50	60	70
T	0.633	0.9	1.1	1.267	1.4	1.533	1.633
k	.02	0.201	0.201	0.2	0.198	0.198	1.633

87. Answers will vary.

89. $$\frac{7\sqrt{x^2y^4}\bullet\sqrt{36xy}}{6\sqrt{x^{-6}y^{-2}}\bullet\sqrt{49x^{-1}y^{-3}}}$$
$$=\frac{7xy^2\bullet 6\sqrt{xy}}{6x^{-3}y^{-1}\bullet 7y^{-1}\sqrt{x^{-1}y^{-1}}}\bullet\frac{\sqrt{x^{-1}y^{-1}}}{\sqrt{x^{-1}y^{-1}}}$$
$$=\frac{42xy^2}{42x^{-3}y^{-2}\bullet x^{-1}y^{-1}}=\frac{xy^2}{x^{-4}y^{-3}}=x^5y^5$$

91. $\sqrt{10mn}$ is in simplest form.

93. The expression is not in simplest form. Remove the perfect square factors from the radical and simplify.

$$\frac{\sqrt{98x^2y}}{7x}=\frac{\sqrt{49x^2\bullet 2y}}{7x}=\frac{7x\sqrt{2y}}{7x}=\sqrt{2y}$$

95. Above and Beyond

97. Above and Beyond

99. **(a)** 3 [×] [2nd] [√] 5 [+]

4 [×] [2nd] [√] 5 [ENTER]

The result is 15.65.

(b) 7 [×] [2nd] [√] 5 [ENTER]

The result is 15.65.

(c) 2 [×] [2nd] [√] 6 [+]

3 [×] [2nd] [√] 6 [ENTER]

The result is 12.25.

(d) 5 [×] [2nd] [√] 6 [ENTER]

The result is 12.25.

(e) Above and Beyond

Exercises 7.3

< Objective 1 >

1. $3\sqrt{5}+4\sqrt{5}=(3+4)\sqrt{5}=7\sqrt{5}$

3. $11\sqrt{3a}-8\sqrt{3a}=(11-8)\sqrt{3a}=3\sqrt{3a}$

5. $7\sqrt{m}+6\sqrt{n}$; cannot be simplified.

7. $2\sqrt[3]{2}+7\sqrt[3]{2}=(2+7)\sqrt[3]{2}=9\sqrt[3]{2}$

9. $8\sqrt{6}-2\sqrt{6}+3\sqrt{6}=(8-2+3)\sqrt{6}=9\sqrt{6}$

11. $\sqrt{20}+\sqrt{5}=\sqrt{4\bullet 5}+\sqrt{5}=2\sqrt{5}+\sqrt{5}=3\sqrt{5}$

13. $4\sqrt{28}-\sqrt{63}=4\sqrt{4\bullet 7}-\sqrt{9\bullet 7}$
$=4\bullet 2\sqrt{7}-3\sqrt{7}=8\sqrt{7}-3\sqrt{7}$
$=5\sqrt{7}$

15. $\sqrt{98}-\sqrt{18}+\sqrt{8}=7\sqrt{2}-3\sqrt{2}+2\sqrt{2}=6\sqrt{2}$

17. $\sqrt[3]{81}+\sqrt[3]{3}=3\sqrt[3]{3}+\sqrt[3]{3}=4\sqrt[3]{3}$

19. $\sqrt{54w}-\sqrt{24w}=3\sqrt{6w}-2\sqrt{6w}=\sqrt{6w}$

21. $\sqrt{18x^3}+\sqrt{8x^3}=3x\sqrt{2x}+2x\sqrt{2x}=5x\sqrt{2x}$

23. $\sqrt[3]{16w^5}+2w\sqrt[3]{2w^2}-\sqrt[3]{2w^5}$
$=2w\sqrt[3]{2w^2}+2w\sqrt[3]{2w^2}-w\sqrt[3]{2w^2}=3w\sqrt[3]{2w^3}$

25. $\sqrt{3}+\frac{1}{\sqrt{3}}=\sqrt{3}+\frac{1}{\sqrt{3}}\cdot\frac{\sqrt{3}}{\sqrt{3}}=\sqrt{3}+\frac{\sqrt{3}}{3}$
$=\left(1+\frac{1}{3}\right)\sqrt{3}=\frac{4}{3}\sqrt{3}$

27. $\sqrt[3]{48}-\sqrt[3]{\frac{3}{4}}=2\sqrt[3]{6}-\frac{\sqrt[3]{3}}{\sqrt[3]{4}}\cdot\frac{\sqrt[3]{2}}{\sqrt[3]{2}}=2\sqrt[3]{6}-\frac{\sqrt[3]{6}}{2}$
$=\left(2-\frac{1}{2}\right)\sqrt[3]{6}=\frac{3}{2}\sqrt[3]{6}$

29. $\frac{\sqrt{6}}{2}+\frac{1}{\sqrt{6}}=\frac{\sqrt{6}}{2}+\frac{1}{\sqrt{6}}\cdot\frac{\sqrt{6}}{\sqrt{6}}=\frac{\sqrt{6}}{2}+\frac{\sqrt{6}}{6}$
$=\left(\frac{1}{2}+\frac{1}{6}\right)\sqrt{6}=\frac{4}{6}\sqrt{6}=\frac{2}{3}\sqrt{6}$

31. $\frac{\sqrt{12}}{3}-\frac{1}{\sqrt{3}}=\frac{2\sqrt{3}}{3}-\frac{1}{\sqrt{3}}\cdot\frac{\sqrt{3}}{\sqrt{3}}=\frac{2\sqrt{3}}{3}-\frac{\sqrt{3}}{3}$
$=\left(\frac{2}{3}-\frac{1}{3}\right)\sqrt{3}=\frac{1}{3}\sqrt{3}$

< Objective 2 >

33. $\sqrt{a}\cdot\sqrt{11}=\sqrt{11a}$

35. $\sqrt{3}\cdot\sqrt{7}\cdot\sqrt{2}=\sqrt{3\cdot7\cdot2}=\sqrt{42}$

37. $\sqrt[3]{4}\cdot\sqrt[3]{9}=\sqrt[3]{4\cdot9}=\sqrt[3]{36}$

39. $\sqrt{3}\cdot\sqrt{12}=\sqrt{3\cdot12}=\sqrt{36}=6$

41. $\sqrt[3]{9p^2}\cdot\sqrt[3]{6p}=\sqrt[3]{9p^2\cdot6p}=\sqrt[3]{54p^3}=3p\sqrt[3]{2}$

43. $\sqrt[3]{4x^2y}\cdot\sqrt[3]{10xy^3}=\sqrt[3]{4x^2y\cdot10xy^3}$
$=\sqrt[3]{40x^3y^4}=2xy\sqrt[3]{5y}$

45. $\sqrt{2}\left(\sqrt{3}+5\right)=\sqrt{2}\cdot\sqrt{3}+\sqrt{2}\cdot5$
$=\sqrt{2\cdot3}+5\sqrt{2}=\sqrt{6}+5\sqrt{2}$

47. $\sqrt{3}\left(5\sqrt{2}-\sqrt{18}\right)=\sqrt{3}\left(5\sqrt{2}-3\sqrt{2}\right)$
$=\sqrt{3}\left(2\sqrt{2}\right)=2\sqrt{3\cdot2}$
$=2\sqrt{6}$

49. $\sqrt{x}\left(\sqrt{3x}+\sqrt{27x}\right)=\sqrt{x}\left(\sqrt{3x}+3\sqrt{3x}\right)$
$=\sqrt{x}\left(4\sqrt{3x}\right)=4\sqrt{3x^2}$
$=4x\sqrt{3}$

51. $\sqrt[3]{4}\left(\sqrt[3]{4}+\sqrt[3]{32}\right)=\sqrt[3]{4}\left(\sqrt[3]{4}+2\sqrt[3]{4}\right)$
$=\sqrt[3]{4}\left(3\sqrt[3]{4}\right)=3\sqrt[3]{16}=6\sqrt[3]{2}$

53. $\left(\sqrt{2}+3\right)\left(\sqrt{2}-4\right)$
$=\sqrt{2}\cdot\sqrt{2}-4\cdot\sqrt{2}+3\cdot\sqrt{2}-3\cdot4$
$=2-\sqrt{2}-15=-10-\sqrt{2}$

55. $\left(\sqrt{2}+3\sqrt{5}\right)\left(\sqrt{2}-2\sqrt{5}\right)$
$=\sqrt{2}\cdot\sqrt{2}-2\cdot\sqrt{5}\cdot\sqrt{2}+3\cdot\sqrt{5}\cdot\sqrt{2}$
$-6\sqrt{5}\cdot\sqrt{5}$
$=2-2\sqrt{10}+3\sqrt{10}-30=-28+\sqrt{10}$

57. $\left(\sqrt{5}+3\right)\left(\sqrt{5}-3\right)$
$=\sqrt{5}\cdot\sqrt{5}-3\cdot\sqrt{5}+3\cdot\sqrt{5}-3\cdot3=5+0-9$
$=-4$

59. $\left(\sqrt{a}+\sqrt{3}\right)\left(\sqrt{a}-\sqrt{3}\right)$
$=\sqrt{a}\cdot\sqrt{a}-\sqrt{a}\cdot\sqrt{3}+\sqrt{a}\cdot\sqrt{3}-\sqrt{3}\cdot\sqrt{3}$
$=a+0-3=a-3$

61. $\left(\sqrt{3}-5\right)^2=\left(\sqrt{3}\right)^2-2\cdot\sqrt{3}\cdot(5)+5^2$
$=3-10\sqrt{3}+25=28-10\sqrt{3}$

63. $\left(\sqrt{a}+3\right)^2=\left(\sqrt{a}\right)^2+2\cdot\sqrt{a}\cdot3+(3)^2$
$=a+6\sqrt{a}+9$

65. $$\begin{aligned}\left(\sqrt{x}+\sqrt{y}\right)^2&=\left(\sqrt{x}\right)^2+2\bullet\sqrt{x}\bullet\sqrt{y}+\left(\sqrt{y}\right)^2\\&=x+2\sqrt{xy}+y\end{aligned}$$

< Objective 3 >

67. $$\frac{\sqrt{3}}{\sqrt{7}}=\frac{\sqrt{3}}{\sqrt{7}}\bullet\frac{\sqrt{7}}{\sqrt{7}}=\frac{\sqrt{21}}{7}$$

69. $$\frac{\sqrt{2a}}{\sqrt{3b}}=\frac{\sqrt{2a}}{\sqrt{3b}}\bullet\frac{\sqrt{3b}}{\sqrt{3b}}=\frac{\sqrt{6ab}}{3b}$$

71. $$\frac{3}{\sqrt[3]{4}}=\frac{3}{\sqrt[3]{4}}\bullet\frac{\sqrt[3]{4^2}}{\sqrt[3]{4^2}}=\frac{6\sqrt[3]{2}}{4}=\frac{3\sqrt[3]{2}}{2}$$

73. $$\begin{aligned}\frac{1}{2+\sqrt{3}}&=\frac{1}{2+\sqrt{3}}\bullet\frac{2-\sqrt{3}}{2-\sqrt{3}}=\frac{2-\sqrt{3}}{4-3}\\&=2-\sqrt{3}\end{aligned}$$

75. $$\begin{aligned}\frac{8}{3-\sqrt{5}}&=\frac{8}{3-\sqrt{5}}\bullet\frac{3+\sqrt{5}}{3+\sqrt{5}}=\frac{24+8\sqrt{5}}{9-5}\\&=\frac{24+8\sqrt{5}}{4}=6+2\sqrt{5}\end{aligned}$$

77. $$\begin{aligned}\frac{\sqrt{6}+\sqrt{3}}{\sqrt{6}-\sqrt{3}}&=\frac{\sqrt{6}+\sqrt{3}}{\sqrt{6}-\sqrt{3}}\bullet\frac{\sqrt{6}+\sqrt{3}}{\sqrt{6}+\sqrt{3}}\\&=\frac{\left(\sqrt{6}\right)^2+2\bullet\sqrt{3}\bullet\sqrt{6}+\left(\sqrt{3}\right)^2}{6-3}\\&=\frac{6+6\sqrt{2}+3}{3}=\frac{9+6\sqrt{2}}{3}\\&=3+2\sqrt{2}\end{aligned}$$

79. $$\begin{aligned}\frac{\sqrt{w}+3}{\sqrt{w}-3}&=\frac{\sqrt{w}+3}{\sqrt{w}-3}\bullet\frac{\sqrt{w}+3}{\sqrt{w}+3}\\&=\frac{\left(\sqrt{w}\right)^2+2\bullet\sqrt{w}\bullet3+9}{w-9}\\&=\frac{w+6\sqrt{w}+9}{w-9}\end{aligned}$$

81. $$\begin{aligned}\frac{\sqrt{x}-\sqrt{y}}{\sqrt{x}+\sqrt{y}}&=\frac{\sqrt{x}-\sqrt{y}}{\sqrt{x}+\sqrt{y}}\bullet\frac{\sqrt{x}-\sqrt{y}}{\sqrt{x}-\sqrt{y}}\\&=\frac{\left(\sqrt{x}\right)^2-2\bullet\sqrt{x}\bullet\sqrt{y}+\left(\sqrt{y}\right)^2}{x-y}\\&=\frac{x-2\sqrt{xy}+y}{x-y}\end{aligned}$$

83. $$\begin{aligned}x\sqrt[3]{8x^4}+4\sqrt[3]{27x^7}&=2x^2\sqrt[3]{x}+12x^2\sqrt[3]{x}\\&=\left(2x^2+12x^2\right)\sqrt[3]{x}\\&=14x^2\sqrt[3]{x}\end{aligned}$$

85. $$\begin{aligned}\frac{\sqrt{2x^2+3x}}{\sqrt{x}}&=\frac{\sqrt{x(2x+3)}}{\sqrt{x}}=\frac{\sqrt{x}\sqrt{2x+3}}{\sqrt{x}}\\&=\sqrt{2x+3}\end{aligned}$$

87. $$\begin{aligned}p=a+b+c&=\left(\sqrt{3}-\sqrt{2}\right)+\left(\sqrt{3}+\sqrt{2}\right)+3\\&=\sqrt{3}+\sqrt{3}+3=2\sqrt{3}+3\end{aligned}$$

89. **(a)** $$\begin{aligned}P&=2(l+w)=2\left(\sqrt{36}+\sqrt{49}\right)\\&=2(6+7)=2\bullet13=26\end{aligned}$$

The perimeter is 26 ft.

(b) $A=lw=\sqrt{36}\bullet\sqrt{49}=6\bullet7=42$

The area is 42 ft^2.

91. **(a)** $$\begin{aligned}P&=2(l+w)=2\left(\sqrt{3}+\sqrt{5}+\sqrt{3}+\sqrt{5}\right)\\&=2\left(2\sqrt{3}+2\sqrt{5}\right)=4\sqrt{3}+4\sqrt{5}\end{aligned}$$

The perimeter is $\left(4\sqrt{3}+4\sqrt{5}\right)$ ft.

(b) $$\begin{aligned}A&=lw=\left(\sqrt{3}+\sqrt{5}\right)\left(\sqrt{3}+\sqrt{5}\right)\\&=\left(\sqrt{3}\right)^2+2\sqrt{3}\sqrt{5}+\left(\sqrt{5}\right)^2\\&=3+2\sqrt{15}+5=8+2\sqrt{15}\end{aligned}$$

The area is $\left(8+2\sqrt{15}\right)$ ft^2.

93. True

95. True

97. True

99. Conjugate square root expressions will always have a product that contains no radical sign.

101. $\sqrt{3}-\sqrt{2}\approx 0.32$

103. $\sqrt{5}+\sqrt{3}\approx 3.97$

105. $4\sqrt{3}-7\sqrt{5}\approx -8.72$

107. $5\sqrt{7}+8\sqrt{13}\approx 42.07$

109. $T=2\pi\sqrt{\frac{L}{g}}=2\pi\sqrt{\frac{L}{32}}=2\pi\frac{\sqrt{L}}{4\sqrt{2}}$

$$=2\pi\frac{\sqrt{L}}{4\sqrt{2}}\bullet\frac{\sqrt{2}}{\sqrt{2}}=\frac{2\pi\sqrt{2L}}{8}=\frac{\pi}{4}\sqrt{2L}$$

111. Above and Beyond

113. Above and Beyond

Exercises 7.4

< Objectives 1–3 >

1. $\sqrt{x}=2$ Check: $\sqrt{4}=\sqrt{2^2}=2$

$$\left(\sqrt{x}\right)^2=2^2$$
$$x=4$$

The solution set is $\{4\}$.

3. $2\sqrt{y}-1=0$

$$2\sqrt{y}=1$$
$$\sqrt{y}=\frac{1}{2}$$
$$y=\frac{1}{4}$$

Check: $2\sqrt{\frac{1}{4}}-1=2\left(\frac{1}{2}\right)-1=0$

The solution set is $\left\{\frac{1}{4}\right\}$.

5. $\sqrt{m+5}=3$ Check: $\sqrt{4+5}=\sqrt{9}=3$

$$\left(\sqrt{m+5}\right)^2=9$$
$$m+5=9$$
$$m=4$$

The solution set is $\{4\}$.

7. $\sqrt{2x+4}-4=0$

$$\sqrt{2x+4}=4$$
$$\left(\sqrt{2x+4}\right)^2=4^2$$
$$2x+4=16$$
$$x=\frac{12}{2}=6$$

Check: $\sqrt{2\bullet 6+4}-4=\sqrt{16}-4=4-4=0$

The solution set is $\{6\}$.

9. $\sqrt{3x-2}+2=0$

$$\sqrt{3x-2}=-2$$

No solution; $\varnothing$

11. $\sqrt{x-1}=\sqrt{1-x}$

$$\left(\sqrt{x-1}\right)^2=\left(\sqrt{1-x}\right)^2$$
$$x-1=1-x$$
$$2x=2$$
$$x=1$$

Check: $\sqrt{(1)-1}=0=\sqrt{1-(1)}$

13. $\sqrt{w+3}=\sqrt{3+w}$

$$w+3=3+w$$
$$0=0$$

The equation is true everywhere in which $\sqrt{w+3}$ is defined. That is for all w, radicand $w+3$ should be non negative.

$$w+3\geq 0$$
$$w\geq -3$$
$$\{w|w\geq -3\}$$

15. $\sqrt{2x-3}+1=3$

$$\sqrt{2x-3}=2$$
$$2x-3=4$$
$$2x=7$$
$$x=\frac{7}{2}$$

Check: $\sqrt{2\left(\frac{7}{2}\right)-3}+1=\sqrt{4}+2=2+1=3$

The solution set is $\left\{\frac{7}{2}\right\}$.

17. $2\sqrt{3z+2}-1=5$

$$2\sqrt{3z+2}=6$$
$$\sqrt{3z+2}=3$$
$$\left(\sqrt{3z+2}\right)^2=3^2$$
$$3z+2=9$$
$$z=\frac{7}{3}$$

Check:

$2\sqrt{3\left(\frac{7}{3}\right)+2}-1=2\sqrt{9}-2=2(3)-1=5$

The solution set is $\left\{\frac{7}{3}\right\}$.

19. $\sqrt{15-2x}=x$

$$\left(\sqrt{15-2x}\right)^2=x^2$$
$$15-2x=x^2$$
$$x^2+2x-15=0$$
$$(x+5)(x-3)=0$$

$x=-5$ or $x=3$

Check:

For $x=-5$: $\sqrt{15-2(-5)}=\sqrt{25}=5\neq -5$

For $x=3$: $\sqrt{15-2(3)}=\sqrt{9}=3$

-5 is an extraneous solution.
The solution set is $\{3\}$.

21. $\left(\sqrt{x+5}\right)^2=(x-1)^2$

$$x+5=x^2-2x+1$$
$$x^2-3x-4=0$$
$$(x+1)(x-4)=0$$

$x=-1$ or $x=4$

Check:

For $x=-1$: $\sqrt{(-1)+5}=\sqrt{2}=2$

$(-1)-1=-2$

For $x=4$: $\sqrt{4+5}=\sqrt{9}=3$

$4-1=3$

-1 is an extraneous solution.
The solution set is $\{4\}$.

23. $\sqrt{3m-2}=-m+10$

$$\left(\sqrt{3m-2}\right)^2=(-m+10)^2$$
$$3m-2=m^2-20m+100$$
$$m^2-23m+102=0$$
$$(m-6)(m-17)=0$$

$m=6$ or $m=17$

Check:

For $m=17$: $\sqrt{3(17)-2}=\sqrt{49}=7$

$-(17)+10=-7$

For $m=6$: $\sqrt{3(6)-2}=\sqrt{16}=4$

$-(6)+10=4$

17 is an extraneous solution.
The solution set is $\{6\}$.

25. $\sqrt{t+9}+3=t$

$\sqrt{t+9}=t-3$

$\left(\sqrt{t+9}\right)^2=(t-3)^2$

$t+9=t^2-6t+9$

$t^2-7t=0$

$t(t-7)=0$

$t=0$ or $t=7$

Check:

For $t=0$: $\sqrt{0+9}+3=\sqrt{9}+3=6\neq 0$

For $t=7$: $\sqrt{7+9}+3=\sqrt{16}+3=7$

0 is an extraneous solution.

The solution set is $\{7\}$.

27. $\sqrt{6x+1}-1=2x$

$\sqrt{6x+1}=2x+1$

$\left(\sqrt{6x+1}\right)^2=(2x+1)^2$

$6x+1=4x^2+4x+1$

$4x^2-2x=0$

$(2x)(2x-1)=0$

$x=0$ or $x=\frac{1}{2}$

Check:

For $x=0$: $\sqrt{6(0)+1}-1=\sqrt{1}-1=0=2(0)$

For $x=\frac{1}{2}$: $\sqrt{6\left(\frac{1}{2}\right)+1}-1=\sqrt{4}-1=2$

$=2\left(\frac{1}{2}\right)$

The solution set is $\left\{0,\frac{1}{2}\right\}$.

29. $\sqrt[3]{x-5}=3$ Check: $\sqrt[3]{32-5}=\sqrt[3]{27}=3$

$\left(\sqrt[3]{x-5}\right)^3=3^3$

$x-5=27$

$x=32$

The solution set is $\{32\}$.

31. $\sqrt[3]{x^2-1}=2$

$\left(\sqrt[3]{x^2-1}\right)^3=2^3$

$x^2-1=8$

$x^2-9=0$

$(x-3)(x+3)=0$

$x=3$ or $x=-3$

Check:

For $x=3$: $\sqrt[3]{(3)^2-1}=\sqrt[3]{8}=2$

For $x=-3$: $\sqrt[3]{(-3)^2-1}=\sqrt[3]{8}=2$

The solutions set is $\{\pm 3\}$.

33. $\sqrt{2x}=\sqrt{x+1}$

$\left(\sqrt{2x}\right)^2=\left(\sqrt{x+1}\right)^2$

$2x=x+1$

$x=1$

Check: $\sqrt{2(1)}=\sqrt{2}=\sqrt{(1)+1}$

The solution set is $\{1\}$.

35. $2\sqrt{3r}=\sqrt{r+11}$

$\left(2\sqrt{3r}\right)^2=\left(\sqrt{r+11}\right)^2$

$12r=r+11$

$11r=11$

$x=1$

Check: $2\sqrt{3(1)}=2\sqrt{3}=\sqrt{(1)+11}$

The solution set is $\{1\}$.

37.
$$\sqrt{x+2}+1=\sqrt{x+4}$$
$$\left(\sqrt{x+2}+1\right)^2=\left(\sqrt{x+4}\right)^2$$
$$x+2+2\sqrt{x+2}+1=x+4$$
$$2\sqrt{x+2}=1$$
$$\sqrt{x+2}=\frac{1}{2}$$
$$\left(\sqrt{x+2}\right)^2=\left(\frac{1}{2}\right)^2$$
$$x+2=\frac{1}{4}$$
$$x=-\frac{7}{4}$$

Check: $\sqrt{\left(-\frac{7}{4}\right)+2}+1=\sqrt{\frac{1}{4}}+1=\frac{1}{2}+1=\frac{3}{2}$

$\sqrt{\left(-\frac{7}{4}\right)+4}=\sqrt{\frac{9}{4}}=\frac{3}{2}$

The solution set is $\left\{-\frac{7}{4}\right\}$.

39.
$$\sqrt{4m-3}-2=\sqrt{2m-5}$$
$$\left(\sqrt{4m-3}-2\right)^2=\left(\sqrt{2m-5}\right)^2$$
$$4m-3-2\sqrt{4m-3}+4=2m-5$$
$$2m+6=4\sqrt{4m-3}$$
$$m+3=2\sqrt{4m-3}$$
$$(m+3)^2=\left(2\sqrt{4m-3}\right)^2$$
$$m^2+6x+9=16m-12$$
$$m^2-10m+21=0$$
$$(m-7)(m-3)=0$$

$m=3$ or $m=7$

Check:

For $m=3$: $\sqrt{4(3)-3}-2=\sqrt{9}-2=1$
$=\sqrt{2(3)-5}$

For $m=7$: $\sqrt{4(7)-3}-2=\sqrt{25}-2=3$
$=\sqrt{2(7)-5}$

The solutions set is $\{3, 7\}$.

41. $\sqrt{x+1}+\sqrt{x}=1$
$$\sqrt{x+1}=1-\sqrt{x}$$
$$\left(\sqrt{x+1}\right)^2=\left(1-\sqrt{x}\right)^2$$
$$x+1=x+1-2\sqrt{x}$$
$$2\sqrt{x}=0$$
$$x=0$$

Check: $\sqrt{0+1}+\sqrt{0}=1$

The solution set is $\{0\}$.

43. $\sqrt{5x+6}-\sqrt{x+3}=3$

$$\sqrt{5x+6}=3+\sqrt{x+3}$$
$$\left(\sqrt{5x+6}\right)^2=\left(3+\sqrt{x+3}\right)^2$$
$$5x+6=x+3+6\sqrt{x+3}+9$$
$$4x-6=6\sqrt{x+3}$$
$$2x-3=3\sqrt{x+3}$$
$$(2x-3)^2=\left(3\sqrt{x+3}\right)^2$$
$$4x^2-12x+9=9x+27$$
$$4x^2-21x-18=0$$
$$(x-6)(4x+3)=0$$
$$x=6 \text{ or } x=-\frac{3}{4}$$

Check:

For $z=-\frac{3}{4}$: $\sqrt{5\left(-\frac{3}{4}\right)+6}-\sqrt{\left(-\frac{3}{4}\right)+3}$

$$=\sqrt{\left(\frac{9}{4}\right)}-\sqrt{\frac{9}{4}}=0\neq 3$$

For $z=6$: $\sqrt{5(6)+6}-\sqrt{6+3}=\sqrt{36}-\sqrt{9}$

$$=3$$

$-\frac{3}{4}$ is an extraneous solution.

The solution set is $\{6\}$.

45. $\sqrt{y^2+12y}-3\sqrt{5}=0$

$$\sqrt{y^2+12y}=3\sqrt{5}$$
$$\left(\sqrt{y^2+12y}\right)^2=\left(3\sqrt{5}\right)^2$$
$$y^2+12y=45$$
$$y^2+12y-45=0$$
$$(y+15)(y-3)=0$$
$$y=-15 \text{ or } y=3$$

Check:

For $y=-15$: $\sqrt{(-15)^2+12(-15)}-3\sqrt{5}$

$$=45-3\sqrt{5}=3\sqrt{5}-3\sqrt{5}=0$$

For $y=3$: $\sqrt{(3)^2+12(3)}-3\sqrt{5}$

$$=45-3\sqrt{5}=3\sqrt{5}-3\sqrt{5}=0$$

The solutions set is $\{-15, 3\}$.

47. $\sqrt{\frac{x-3}{x+2}}=\frac{2}{3}$

$$\left(\sqrt{\frac{x-3}{x+2}}\right)^2=\left(\frac{2}{3}\right)^2$$
$$\frac{x-3}{x+2}=\frac{4}{9}$$
$$9(x-3)=4(x+2)$$
$$9x-27=4x+8$$
$$5x=35$$
$$x=7$$

Check: $\sqrt{\frac{7-3}{7+2}}=\sqrt{\frac{4}{9}}=\frac{2}{3}$

The solution set is $\{7\}$.

49. $\sqrt{\sqrt{t}+5}=3$

$$\left(\sqrt{\sqrt{t}+5}\right)^2=3^2$$
$$\sqrt{t}+5=9$$
$$\sqrt{t}=4$$
$$t=16$$

Check: $\sqrt{\sqrt{16}+5}=\sqrt{4+5}=\sqrt{9}=3$

The solution set is $\{16\}$.

51.
$$x+\sqrt{x}=12$$
$$\sqrt{x}=12-x$$
$$\left(\sqrt{x}\right)^2=(12-x)^2$$
$$x=144-24x+x^2$$
$$x^2-25x+144=0$$
$$(x-9)(x-16)=0$$
$$x=9 \text{ or } x=16$$
Check: $16+\sqrt{16}=16+4=20\neq 12$

$9+\sqrt{9}=9+3=12$

16 is an extraneous solution.
The integer is 9.

53.
$$x+2\sqrt{x}=24$$
$$2\sqrt{x}=24-x$$
$$\left(2\sqrt{x}\right)^2=(24-x)^2$$
$$4x=x^2-48x+576$$
$$x^2-52x+576=0$$
$$(x-16)(x-36)=0$$
$$x=16 \text{ or } x=36$$
Check: $36+2\sqrt{36}=36+12=48\neq 24$

$16+2\sqrt{16}=16+8=24$

36 is an extraneous solution.
The integer is 16.

55. $d=\sqrt{2(30,000)}=\sqrt{60,000}\approx 245$

The horizon is approximately 245 mi away.

57. $d=\sqrt{2h}$
$$d^2=2h$$
$$h=\frac{d^2}{2}$$

59. $s=2\sqrt{5x}=2\sqrt{5(55)}=2\sqrt{275}\approx 33$

The skidding distance is approximately 33 ft.

61. $s=2\sqrt{5x}=2\sqrt{5(75)}=2\sqrt{375}\approx 39$

The skidding distance is approximately 39 ft.

63. $s=2\sqrt{5x}$
$$\sqrt{5x}=\frac{s}{2}$$
$$5x=\frac{s^2}{4}$$
$$x=\frac{s^2}{20}$$

65. When we raise both sides of an equation to an even power, we sometimes obtain extraneous solutions.

67. If we raise both sides of an equation to an odd power, we never obtain extraneous solutions.

69. The square of a binomial is always a trinomial.

71. $\sqrt{x+4}=x-3$
$$f(x)=\sqrt{x+4}$$
$$g(x)=x-3$$

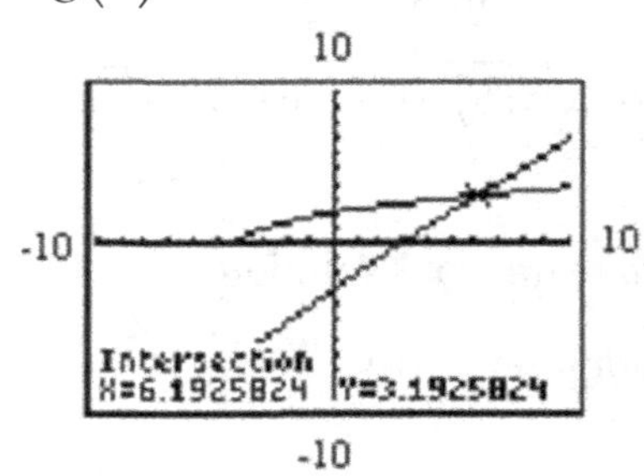

$\{6.19\}$

73. $3-2\sqrt{x+4}=2x-5$
$$f(x)=3-2\sqrt{x+4}$$
$$g(x)=2x-5$$

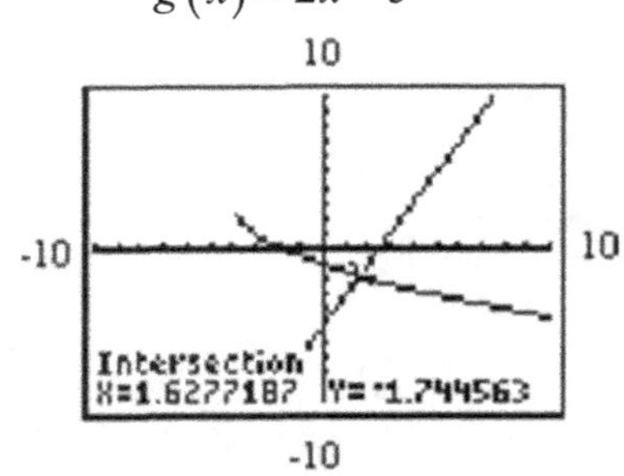

$\{1.63\}$

75. $T = 2\pi\sqrt{\frac{L}{g}}$

$\frac{T}{2\pi} = \sqrt{\frac{L}{g}}$

$\left(\frac{T}{2\pi}\right)^2 = \left(\sqrt{\frac{L}{g}}\right)^2$

$\frac{T^2}{4\pi^2} = \frac{L}{g}$

$\frac{gT^2}{4\pi^2} = L$

$L = \frac{g}{4\pi^2}T^2$

77. $c^2 = a^2 + b^2 = \left(\sqrt{147}\right)^2 + \left(\sqrt{108}\right)^2$

$= 147 + 108 = 255$

$c = \sqrt{255}$

The length of the diagonal is $\sqrt{255}$ ft.

79. $\sqrt{(x-1)^2} = |x-1|$

$|x-1| = x-1 \text{ if } x-1 \geq 0 \rightarrow x \geq 1$

$\{x | x \geq 1\}$

81. $h = \sqrt{pq}$

$h^2 = pq$

$q = \frac{h^2}{p}$

83. $v = \sqrt{2gR}$

$v^2 = 2gR$

$g = \frac{v^2}{2R}$

85. $r = \sqrt{\frac{S}{2\pi}}$

$r^2 = \frac{S}{2\pi}$

$S = 2\pi r^2$

87. $r = \sqrt{\frac{2V}{\pi h}}$

$r^2 = \frac{2V}{\pi h}$

$V = \frac{\pi r^2 h}{2}$

89. $d = \sqrt{(x-1)^2 + (y-2)^2}$

$d^2 = (x-1)^2 + (y-2)^2$

$(x-1)^2 = d^2 - (y-2)^2$

$x - 1 = \pm\sqrt{d^2 - (y-2)^2}$

$x = 1 \pm \sqrt{d^2 - (y-2)^2}$

91. $T = 2\pi\sqrt{\frac{L}{g}} = 2\pi\sqrt{\frac{0.9}{10}} \approx 1.88$

The period is 1.88 s.

93. $T = 2\pi\sqrt{\frac{L}{g}}$

$\frac{T}{2\pi} = \sqrt{\frac{L}{g}}$

$\frac{L}{g} = \frac{T^2}{4\pi^2}$

$L = \frac{gT^2}{4\pi^2}$

Exercises 7.5

< Objective 1 >

1. $49^{\frac{1}{2}} = \sqrt{49} = 7$

3. $-25^{\frac{1}{2}} = -\sqrt{25} = -5$

5. $(-49)^{\frac{1}{2}} = \sqrt{-49}$ is not a real number.

7. $27^{\frac{1}{3}} = \sqrt[3]{27} = 3$

9. $81^{\frac{1}{4}} = \sqrt[4]{81} = 3$

11. $\left(\frac{4}{9}\right)^{\frac{1}{2}} = \sqrt{\frac{4}{9}} = \frac{\sqrt{4}}{\sqrt{9}} = \frac{2}{3}$

13. $27^{\frac{2}{3}} = \left(\sqrt[3]{27}\right)^2 = 3^2 = 9$

15. $(-8)^{\frac{4}{3}} = \left(\sqrt[3]{-8}\right)^4 = (-2)^4 = 16$

17. $32^{\frac{2}{5}} = \left(\sqrt[5]{32}\right)^2 = 2^2 = 4$

19. $81^{\frac{3}{2}} = \left(\sqrt{81}\right)^3 = 9^3 = 729$

21. $\left(\frac{8}{27}\right)^{\frac{2}{3}} = \left(\sqrt[3]{\frac{8}{27}}\right)^2 = \left(\frac{2}{3}\right)^2 = \frac{4}{9}$

< Objective 2 >

23. $49^{-\frac{1}{2}} = \frac{1}{49^{\frac{1}{2}}} = \frac{1}{7}$

25. $81^{-1/4} = \frac{1}{81^{1/4}} = \frac{1}{3}$

27. $9^{-3/2} = \frac{1}{9^{3/2}} = \frac{1}{\left(\sqrt{9}\right)^3} = \frac{1}{3^3} = \frac{1}{27}$

29. $64^{-5/6} = \frac{1}{64^{5/6}} = \frac{1}{\left(\sqrt[6]{64}\right)^5} = \frac{1}{2^5} = \frac{1}{32}$

31. $\left(\frac{4}{25}\right)^{-1/2} = \left(\frac{25}{4}\right)^{1/2} = \sqrt{\frac{25}{4}} = \frac{5}{2}$

33. $x^{\frac{1}{2}} \bullet x^{\frac{1}{2}} = x^{\frac{1}{2}+\frac{1}{2}} = x^1 = x$

35. $y^{\frac{5}{7}} \bullet y^{\frac{1}{7}} = y^{\frac{5}{7}+\frac{1}{7}} = y^{\frac{6}{7}}$

37. $b^{\frac{2}{3}} \bullet b^{\frac{3}{2}} = b^{\frac{2}{3}+\frac{3}{2}} = b^{\frac{13}{6}}$

39. $\frac{x^{\frac{2}{3}}}{x^{\frac{1}{3}}} = x^{\frac{2}{3}-\frac{1}{3}} = x^{\frac{1}{3}}$

41. $\frac{s^{\frac{7}{5}}}{s^{\frac{2}{5}}} = s^{\frac{7}{5}-\frac{2}{5}} = s^1 = s$

43. $\frac{w^{\frac{7}{6}}}{w^{\frac{1}{2}}} = w^{\frac{7}{6}-\frac{1}{2}} = w^{\frac{2}{3}}$

45. $\left(x^{\frac{3}{4}}\right)^{\frac{4}{3}} = x^{\left(\frac{3}{4}\right)\left(\frac{4}{3}\right)} = x^1 = x$

47. $\left(a^{\frac{2}{5}}\right)^{\frac{3}{2}} = a^{\left(\frac{2}{5}\right)\left(\frac{3}{2}\right)} = a^{\frac{3}{5}}$

49. $\left(y^{-\frac{2}{3}}\right)^6 = y^{\left(-\frac{2}{3}\right)\bullet 6} = y^{-4} = \frac{1}{y^4}$

51. $\left(a^{\frac{2}{3}} \bullet b^{\frac{3}{2}}\right)^6 = \left(a^{\frac{2}{3}}\right)^6\left(b^{\frac{3}{2}}\right)^6 = a^4b^9$

53. $\left(x^{\frac{2}{7}}y^{\frac{3}{7}}\right)^7 = x^2y^3$

55. $\left(s^{\frac{3}{4}} \bullet t^{\frac{1}{4}}\right)^{\frac{4}{3}} = \left(s^{\frac{3}{4}}\right)^{\frac{4}{3}}\left(t^{\frac{1}{4}}\right)^{\frac{4}{3}} = st^{\frac{1}{3}}$

57. $\left(8p^{\frac{3}{2}} \bullet q^{\frac{5}{2}}\right)^{\frac{2}{3}} = 8^{\frac{2}{3}}\left(p^{\frac{3}{2}}\right)^{\frac{2}{3}}\left(q^{\frac{5}{2}}\right)^{\frac{2}{3}} = 4pq^{\frac{5}{3}}$

59. $\left(x^{\frac{3}{5}} \bullet y^{\frac{3}{4}} \bullet z^{\frac{3}{2}}\right)^{\frac{2}{3}} = \left(x^{\frac{3}{5}}\right)^{\frac{2}{3}}\left(y^{\frac{3}{4}}\right)^{\frac{2}{3}}\left(z^{\frac{3}{2}}\right)^{\frac{2}{3}}$
$= x^{\frac{2}{5}} y^{\frac{1}{2}} z$

61. $\dfrac{a^{\frac{5}{6}} \bullet b^{\frac{3}{4}}}{a^{\frac{1}{3}} \bullet b^{\frac{1}{2}}} = a^{\frac{5}{6}-\frac{1}{3}} b^{\frac{3}{4}-\frac{1}{2}} = a^{\frac{1}{2}} b^{\frac{1}{4}}$

63. $\dfrac{\left(r^{-1} \bullet s^{\frac{1}{2}}\right)^3}{r \bullet s^{-\frac{1}{2}}} = \dfrac{r^{-3} s^{\frac{3}{2}}}{r \bullet s^{-\frac{1}{2}}} = r^{-3-1} s^{\frac{3}{2}-\left(-\frac{1}{2}\right)} = r^{-4} s^2$
$= \dfrac{s^2}{r^4}$

65. $\left(\dfrac{x^{12}}{y^8}\right)^{\frac{1}{4}} = \dfrac{\left(x^{12}\right)^{\frac{1}{4}}}{\left(y^8\right)^{\frac{1}{4}}} = \dfrac{x^3}{y^2}$

67. $\left(\dfrac{m^{-\frac{1}{4}}}{n^{\frac{1}{2}}}\right)^4 = \dfrac{\left(m^{-\frac{1}{4}}\right)^4}{\left(n^{\frac{1}{2}}\right)^4} = \dfrac{m^{-1}}{n^2} = \dfrac{1}{mn^2}$

69. $\left(\dfrac{r^{-\frac{1}{2}} \bullet s^{3/4}}{t^{1/4}}\right)^4 = \dfrac{\left(r^{-\frac{1}{2}}\right)^4\left(s^{\frac{3}{4}}\right)^4}{\left(t^{\frac{1}{4}}\right)^4} = \dfrac{r^{-2}s^3}{t} = \dfrac{s^3}{r^2 t}$

71. $\left(\dfrac{8x^3 \bullet y^{-6}}{z^{-9}}\right)^{\frac{1}{3}} = \dfrac{8^{\frac{1}{3}}\left(x^3\right)^{\frac{1}{3}}\left(y^{-6}\right)^{\frac{1}{3}}}{\left(z^{-9}\right)^{1/3}} = \dfrac{2xy^{-2}}{z^{-3}}$
$= \dfrac{2xz^3}{y^2}$

73. $\left(\dfrac{16m^{-\frac{3}{5}} \bullet n^2}{m^{\frac{1}{5}} \bullet n^{-2}}\right)^{\frac{1}{4}} = \dfrac{16^{\frac{1}{4}}\left(m^{-\frac{3}{5}}\right)^{\frac{1}{4}}\left(n^2\right)^{\frac{1}{4}}}{\left(m^{1/5}\right)^{\frac{1}{4}}\left(n^{-2}\right)^{\frac{1}{4}}}$
$= \dfrac{2m^{-\frac{3}{20}} n^{\frac{1}{2}}}{m^{\frac{1}{20}} n^{-\frac{1}{2}}}$
$= 2m^{-\frac{3}{20}-\frac{1}{20}} n^{\frac{1}{2}-\left(-\frac{1}{2}\right)} = 2m^{-\frac{1}{5}} n^1$
$= \dfrac{2n}{m^{\frac{1}{5}}}$

75. $\left(\dfrac{x^{\frac{3}{2}} \bullet y^{\frac{1}{2}}}{z^2}\right)^{\frac{1}{2}}\left(\dfrac{x^{\frac{3}{4}} \bullet y^{\frac{3}{2}}}{z^{-3}}\right)^{\frac{1}{3}} = \dfrac{x^{\frac{3}{4}} y^{\frac{1}{4}}}{z} \bullet \dfrac{x^{\frac{1}{4}} y^{\frac{1}{2}}}{z^{-1}}$
$= xy^{\frac{3}{4}}$

< Objective 3 >

77. $a^{\frac{3}{4}} = \sqrt[4]{a^3}$

79. $2x^{\frac{2}{3}} = 2\sqrt[3]{x^2}$

81. $3x^{\frac{2}{5}} = 3\sqrt[5]{x^2}$

83. $(3x)^{\frac{2}{5}} = \sqrt[5]{3^2 x^2} = \sqrt[5]{9x^2}$

85. $\sqrt{7a} = (7a)^{\frac{1}{2}}$

87. $\sqrt[3]{8m^6n^9} = \left(8m^6n^9\right)^{\frac{1}{3}} = 2m^2n^3$

89. $\left(4\times10^8\right)^{\frac{1}{2}} = 4^{\frac{1}{2}}\times\left(10^8\right)^{\frac{1}{2}} = 2\times10^4$

91. $\left(16\times10^{-12}\right)^{\frac{1}{4}} = 16^{\frac{1}{4}}\times\left(10^{-12}\right)^{\frac{1}{4}} = 2\times10^{-3}$

93. $\left(16\times10^{-8}\right)^{\frac{1}{2}}=4\times10^{-4}$

95. $\left(27\times10^{6}\right)^{\frac{1}{3}}=3\times10^{2}$

97. $a^{\frac{1}{2}}\left(a^{\frac{3}{2}}+a^{\frac{3}{4}}\right)=a^{\frac{1}{2}}\bullet a^{\frac{3}{2}}+a^{\frac{1}{2}}\bullet a^{\frac{2}{4}}=a^{2}+a^{\frac{5}{4}}$

99. $\left(a^{\frac{1}{2}}+2\right)\left(a^{\frac{1}{2}}-2\right)=\left(a^{\frac{1}{2}}\right)^{2}-2^{2}=a-4$

101. $\left(m^{\frac{1}{2}}+n^{\frac{1}{2}}\right)\left(m^{\frac{1}{2}}-n^{\frac{1}{2}}\right)=\left(m^{\frac{1}{2}}\right)^{2}-\left(n^{\frac{1}{2}}\right)^{2}$
$=m-n$

103. $\left(x^{\frac{1}{2}}+2\right)^{2}=\left(x^{\frac{1}{2}}\right)^{2}+2\bullet2x^{\frac{1}{2}}+2^{2}$
$=x+4x^{\frac{1}{2}}+4$

105. $\left(r^{\frac{1}{2}}+s^{\frac{1}{2}}\right)^{2}=r+2r^{\frac{1}{2}}s^{\frac{1}{2}}+s$

107. False

109. A negative number raised to a rational number power is sometimes a real number.

111. $46^{\frac{3}{5}}\approx9.946$

113. $12^{-\frac{2}{5}}\approx0.370$

115. $t=C\left(\frac{L}{xy^{2}}\right)^{\frac{1}{3}}=20\left(\frac{600}{3\bullet5^{2}}\right)^{\frac{1}{3}}=20(8)^{\frac{1}{3}}$
$=20\bullet2=40$

117. $V=\frac{1.5x^{\frac{2}{3}}y^{\frac{1}{2}}}{z}=\frac{1.5(27)^{\frac{2}{3}}(16)^{\frac{1}{2}}}{12}$
$=\frac{(1.5)(9)(4)}{12}=\frac{54}{12}=4.5$
The average velocity is 4.5.

119. $x^{3n}\bullet x^{2n}=x^{3n+2n}=x^{5n}$

121. $\left(y^{2}\right)^{2n}=y^{2\bullet2n}=y^{4n}$

123. $\frac{r^{n+2}}{r^{n}}=r^{(n+2)-n}=r^{2}$

125. $\left(a^{3}\bullet b^{2}\right)^{2n}=a^{6n}b^{4n}$

127. $\left(\frac{x^{n+2}}{x^{n}}\right)^{\frac{1}{2}}=\left(x^{(n+2)-n}\right)^{\frac{1}{2}}=\left(x^{2}\right)^{\frac{1}{2}}=x$

129. $\sqrt{\sqrt{x}}=\left(x^{\frac{1}{2}}\right)^{\frac{1}{2}}=x^{\frac{1}{4}}=\sqrt[4]{x}$

131. $\sqrt[4]{\sqrt{y}}=\left(y^{\frac{1}{2}}\right)^{\frac{1}{4}}=y^{\frac{1}{8}}=\sqrt[8]{y}$

133. Let $u=x^{\frac{1}{3}}$. Then $u^{2}=x^{\frac{2}{3}}$.
$x^{\frac{2}{3}}+4x^{\frac{1}{3}}+3=u^{2}+4u+3=(u+3)(u+1)$
$=\left(x^{\frac{1}{3}}+3\right)\left(x^{\frac{1}{3}}+1\right)$

135. Let $u=a^{\frac{2}{5}}$. Then $u^{2}=a^{\frac{4}{5}}$.
$a^{\frac{4}{5}}-7a^{\frac{2}{5}}+12=u^{2}-7u+12$
$=(u-3)(u-4)$
$=\left(a^{\frac{2}{5}}-3\right)\left(a^{\frac{2}{5}}-4\right)$

137. Let $u = x^{\frac{2}{3}}$.

$$x^{\frac{4}{3}} - 4 = u^2 - 4 = (u+2)(u-2)$$
$$= \left(x^{\frac{2}{3}} + 2\right)\left(x^{\frac{2}{3}} - 2\right)$$

139. The prices increased by 34.8 % over 12 years. This inflation rate is equivalent to an annual growth rate of 2.47%

$$(1.348)^{\frac{1}{x}} = 2.47\%$$
$$(1.348)^{\frac{1}{x}} = 1.0247$$
$$(1.348)^{\frac{4}{49}} = 1.0247$$

1.348 is raised to the an exponent of $\frac{4}{49}$ to obtain the average annual growth rate.

141. Above and Beyond

143. (a) $\left(a^{\frac{2}{3}}\right)^x = a$

$$x = \frac{3}{2}$$

Answers will vary.

(b) $\left(a^{\frac{5}{6}}\right)^x = \frac{1}{a}$

$$x = -\frac{6}{5}$$

Answers will vary.

(c) $a^{2x} \bullet a^{\frac{3}{2}} = 1$

$$x = -\frac{3}{4}$$

Answers will vary.

(d) $\left(\sqrt{a^{\frac{2}{3}}}\right)^x = a$

$$x = 3$$

Answers will vary.

Exercises 7.6

< Objective 1 >

1. $\sqrt{-16} = \sqrt{16}i = 4i$

3. $-\sqrt{-121} = -\sqrt{121}i = -11i$

5. $\sqrt{-21} = \sqrt{21}i = i\sqrt{21}$

7. $\sqrt{-12} = \sqrt{12}i = 2\sqrt{3}i = 2i\sqrt{3}$

9. $-\sqrt{-108} = -\sqrt{108}i = -6i\sqrt{3}$

< Objective 2 >

11. $(5+4i)-(6+5i) = 5+4i-6-5i = -1-i$

13. $(-3-2i)+(2+3i) = -1+i$

15. $(5+4i)-(3+2i) = 5+4i-3-2i = 2+2i$

17. $(8-5i)-(3+2i) = 8-5i-3-2i = 5-7i$

19. $(3+i)+(4+5i)-7i = 7-i$

21. $(2+3i)-(3-5i)+(4+3i) = 3+11i$

23. $(7+3i)-[(3+i)-(2-5i)]$
$= (7+3i)-(1+6i) = 6-3i$

25. $(5+3i)+(-5-3i) = 0$

< Objective 3 >

27. $3i(3+5i) = 3i\bullet 3 + 3i\bullet 5i = 9i + 15(-1)$
$= -15+9i$

29. $6i(2-5i) = 6i\bullet 2 - 6i\bullet 5i = 12i - 30(-1)$
$= 30+12i$

31. $-2i(4-3i) = -8i + 6(-1) = -6-8i$

33. $6i\left(\frac{2}{3}+\frac{5}{6}i\right)=-5+4i$

35. $(4+3i)(4-3i)=4^2+3^2=16+9=25$

37. $(4-3i)(2+5i)=8+20i-6i+15=23+14i$

39. $(-2-3i)(-3+4i)=6+i+12=18+i$

41. $(7-3i)^2=49-42i-9=40-42i$

43. $3-2i \rightarrow 3+2i$

$(3-2i)(3+2i)=3^2+2^2=13$

45. $3+2i \rightarrow 3-2i$

$(3+2i)(3-2i)=13$

47. $-3-2i \rightarrow -3+2i$

$(-3-2i)(-3+2i)=13$

49. $5i \rightarrow -5i$

$(5i)(-5i)=25$

51. $\frac{3+2i}{i}=\frac{3+2i}{i}\bullet\frac{i}{i}=\frac{3i-2}{-1}=2-3i$

53. $\frac{5-2i}{3i}\bullet\frac{i}{i}=\frac{5i+2}{-3}=-\frac{2}{3}-\frac{5}{3}i$

55. $\frac{3}{2+5i}=\frac{3}{2+5i}\bullet\frac{2-5i}{2-5i}=\frac{6-15i}{29}=\frac{6}{29}-\frac{15}{29}i$

57. $\frac{13}{2+3i}=\frac{13}{2+3i}\bullet\frac{2-3i}{2-3i}=\frac{13(2-3i)}{13}=2-3i$

59. $\frac{2+3i}{4+3i}=\frac{2+3i}{4+3i}\bullet\frac{4-3i}{4-3i}=\frac{17+6i}{25}=\frac{17}{25}+\frac{6}{25}i$

61. $\frac{3-4i}{3+4i}=\frac{3-4i}{3+4i}\bullet\frac{3-4i}{3-4i}=\frac{-7-24i}{25}$

$=-\frac{7}{25}-\frac{24}{25}i$

63. The product of two complex numbers is sometimes a real number.

65. The product of a complex number and its conjugate is always a real number.

67. Above and Beyond

69. $\sqrt{-5}\bullet\sqrt{-7}=\sqrt{5}i\bullet\sqrt{7}i=-\sqrt{35}$

71. $\sqrt{-2}\bullet\sqrt{-18}=-\sqrt{36}=-6$

73. $\sqrt{-6}\bullet\sqrt{-15}=-\sqrt{90}=-3\sqrt{10}$

75. $\sqrt{-10}\bullet\sqrt{-10}=-10$

77. $i^{10}=i^5\bullet i^5=i\bullet i=-1$

79. $i^{20}=i^{10}\bullet i^{10}=-1\bullet -1=1$

81. $i^{38}=i^{10}\bullet i^{10}\bullet i^{10}\bullet i^8=-1\bullet -1\bullet -1\bullet 1=-1$

83. $i^{51}=i^{10}\bullet i^{10}\bullet i^{10}\bullet i^{10}\bullet i^{10}\bullet i$

$=-1\bullet -1\bullet -1\bullet -1\bullet i=-i$

85. Above and Beyond

Summary Exercises

1. $\sqrt{121}=11$

3. $\sqrt{-81}$ is not a real number.

5. $\sqrt[3]{-64}=\sqrt[3]{(-4)^3}=-4$

7. $\sqrt{\frac{9}{16}}=\frac{3}{4}$

9. $\sqrt{8^2}=8$

11. $\sqrt{a^4}=a^2$

13. $\sqrt{49w^4z^6}=7w^2z^3$

15. $\sqrt[3]{-27b^6} = -3b^2$

17. $\sqrt[4]{16x^4y^8} = 2xy^2$

19. The distance between (4, 8) and (6, 2) is

$$d = \sqrt{(6-4)^2 + (2-8)^2} = \sqrt{(2)^2 + (-6)^2}$$
$$= \sqrt{4+36} = \sqrt{40} = 2\sqrt{10}$$

21. $d = \sqrt{[1-(-1)]^2 + [3-(-2)]^2} = \sqrt{2^2 + 5^2}$
$= \sqrt{29}$

23. $x^2 + y^2 = 81$
Center: (0, 0)
Radius: $r = \sqrt{81} = 9$

25. $(x+2)^2 + (y-1)^2 = 25$
Center: (–2, 1)
Radius: $r = \sqrt{25} = 5$

27. $(x-2)^2 + y^2 = 9$

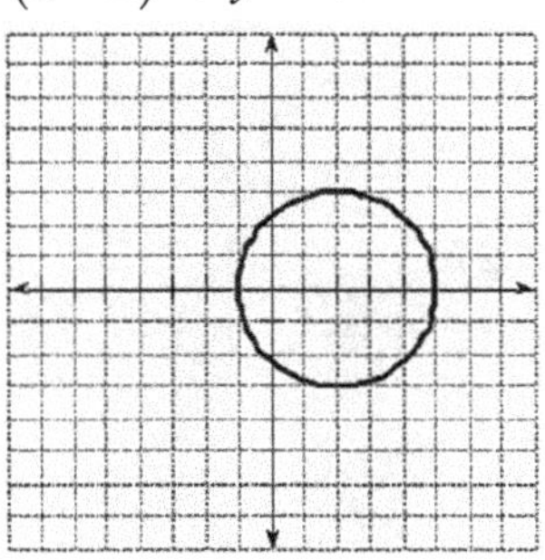

29. $\sqrt{45} = \sqrt{9 \bullet 5} = \sqrt{9} \bullet \sqrt{5} = 3\sqrt{5}$

31. $\sqrt{60x^2} = \sqrt{4x^2 \bullet 15} = 2x\sqrt{15}$

33. $\sqrt[3]{32} = \sqrt[3]{8 \bullet 4} = 2\sqrt[3]{4}$

35. $\sqrt{\dfrac{9}{16}} = \dfrac{\sqrt{9}}{\sqrt{16}} = \dfrac{3}{4}$

37. $\sqrt{\dfrac{y^4}{49}} = \dfrac{\sqrt{y^4}}{\sqrt{49}} = \dfrac{y^2}{7}$

39. $\sqrt{\dfrac{5}{16x^2}} = \dfrac{\sqrt{5}}{4x}$

41. $7\sqrt{10} + 4\sqrt{10} = (7+4)\sqrt{10} = 11\sqrt{10}$

43. $7\sqrt[3]{2x} + 3\sqrt[3]{2x} = (7+3)\sqrt[3]{2x} = 10\sqrt[3]{2x}$

45. $\sqrt{72} + \sqrt{50} = 6\sqrt{2} + 5\sqrt{2} = 11\sqrt{2}$

47. $\sqrt{20} - \sqrt{45} + 2\sqrt{125} = 2\sqrt{5} - 3\sqrt{5} + 10\sqrt{5}$
$= 9\sqrt{5}$

49. $\sqrt{27w^3} - w\sqrt{12w} = 3w\sqrt{3w} - 2w\sqrt{3w}$
$= w\sqrt{3w}$

51. $\sqrt{20} + \dfrac{3}{\sqrt{5}} = 2\sqrt{5} + \dfrac{3}{\sqrt{5}} \bullet \dfrac{\sqrt{5}}{\sqrt{5}} = 2\sqrt{5} + \dfrac{3}{5}\sqrt{5}$
$= \dfrac{13}{5}\sqrt{5}$

53. $\sqrt[3]{81a^4} - a\sqrt[3]{\dfrac{a}{9}} = 3a\sqrt[3]{3a} - \dfrac{a\sqrt[3]{a}}{\sqrt[3]{9}} \bullet \dfrac{\sqrt[3]{3}}{\sqrt[3]{3}}$
$= 3a\sqrt[3]{3a} - \dfrac{a\sqrt[3]{3a}}{3} = \dfrac{8a}{3}\sqrt[3]{3a}$

55. $\sqrt{3x} \bullet \sqrt{7y} = \sqrt{3x \bullet 7y} = \sqrt{21xy}$

57. $\sqrt[3]{4a^2b} \bullet \sqrt[3]{ab^2} = \sqrt[3]{4a^2b \bullet ab^2} = \sqrt[3]{4a^3b^3}$
$= ab\sqrt[3]{4}$

59. $\sqrt{6}\left(\sqrt{8} - \sqrt{2}\right) = \sqrt{6}\left(2\sqrt{2} - \sqrt{2}\right) = \sqrt{6}\sqrt{2}$
$= 2\sqrt{3}$

61. $\left(\sqrt{3}+5\right)\left(\sqrt{3}-7\right) = \left(\sqrt{3}\right)^2 - 2\sqrt{3} - 35$
$= 3 - 2\sqrt{3} - 35$
$= -32 - 2\sqrt{3}$

63. $\left(\sqrt{5}-2\right)\left(\sqrt{5}+2\right) = \left(\sqrt{5}\right)^2 - 2^2 = 5 - 4 = 1$

65. $\left(2+\sqrt{3}\right)^2 = 4+2\cdot 2\sqrt{3}+3 = 7+4\sqrt{3}$

67. $\sqrt{\dfrac{3}{7}} = \dfrac{\sqrt{3}}{\sqrt{7}} = \dfrac{\sqrt{3}}{\sqrt{7}}\cdot\dfrac{\sqrt{7}}{\sqrt{7}} = \dfrac{\sqrt{21}}{7}$

69. $\dfrac{\sqrt{10a}}{\sqrt{5b}} = \dfrac{\cancel{\sqrt{5}}\sqrt{2a}}{\cancel{\sqrt{5}}\sqrt{b}}\cdot\dfrac{\sqrt{b}}{\sqrt{b}} = \dfrac{\sqrt{2ab}}{b}$

71. $\dfrac{2}{\sqrt[3]{3x}} = \dfrac{2}{\sqrt[3]{3x}}\cdot\dfrac{\sqrt[3]{9x^2}}{\sqrt[3]{9x^2}} = \dfrac{2\sqrt[3]{9x^2}}{3x}$

73. $\dfrac{1}{3+\sqrt{2}} = \dfrac{1}{3+\sqrt{2}}\cdot\dfrac{3-\sqrt{2}}{3-\sqrt{2}} = \dfrac{3-\sqrt{2}}{7}$

75. $\dfrac{\sqrt{5}-2}{\sqrt{5}+2} = \dfrac{\sqrt{5}-2}{\sqrt{5}+2}\cdot\dfrac{\sqrt{5}-2}{\sqrt{5}-2} = 9-4\sqrt{5}$

77. $\sqrt{x-5} = 4$ Check: $\sqrt{21-5} = \sqrt{16} = 4$

$$\left(\sqrt{x-5}\right)^2 = 4^2$$
$$x-5 = 16$$
$$x = 21$$

The solution set is $\{21\}$.

79. $\sqrt{y+7} = y-5$

$$y+7 = y^2-10y+25$$
$$y^2-11y+18 = 0$$
$$(y-2)(y-9) = 0$$
$$y = 2 \quad \text{or} \quad y = 9$$

Check:

For $y=2$: $\sqrt{(2)+7} = 3 \neq -3 = (2)-5$

For $y=9$: $\sqrt{(9)+7} = 4 = (9)-5$

2 is an extraneous solution.
The solution set is $\{9\}$.

81. $\sqrt[3]{5x+2} = 3$

$$\left(\sqrt[3]{5x+2}\right)^3 = 3^3$$
$$5x+2 = 27$$
$$x = 5$$

Check: $\sqrt[3]{5(5)+2} = 3$
The solution set is $\{5\}$.

83. $\sqrt{z+7} = 1+\sqrt{z}$

$$z+7 = 1+2\sqrt{z}+z$$
$$3 = \sqrt{z}$$
$$9 = z$$

Check: $\sqrt{(9)+7} = 4 = 1+\sqrt{(9)}$
The solution set is $\{9\}$.

85. $49^{\frac{1}{2}} = \sqrt{49} = 7$

87. $(-27)^{\frac{1}{3}} = \sqrt[3]{-27} = -3$

89. $64^{\frac{2}{3}} = \left(\sqrt[3]{64}\right)^2 = 4^2 = 16$

91. $\left(\dfrac{4}{9}\right)^{\frac{3}{2}} = \left(\dfrac{2}{3}\right)^3 = \dfrac{8}{27}$

93. $49^{-\frac{1}{2}} = \dfrac{1}{49^{\frac{1}{2}}} = \dfrac{1}{7}$

95. $x^{\frac{3}{2}}\cdot x^{\frac{5}{2}} = x^{\frac{3}{2}+\frac{5}{2}} = x^4$

97. $\dfrac{r^{\frac{8}{5}}}{r^{\frac{3}{5}}} = r^{\frac{8}{5}-\frac{3}{5}} = r$

99. $\left(x^{\frac{3}{5}}\right)^{\frac{2}{3}} = x^{\frac{2}{5}}$

101. $\left(x^{\frac{4}{5}}y^{\frac{3}{2}}\right)^{10} = x^8y^{15}$

103. $\left(\dfrac{x^{-2}y^{-\frac{1}{6}}}{x^{-4}y}\right)^3 = \left(x^2y^{-\frac{7}{6}}\right) = x^6y^{-\frac{7}{2}} = \dfrac{x^6}{y^{-\frac{7}{2}}}$

105. $x^{\frac{3}{4}} = \sqrt[4]{x^3}$

107. $3a^{\frac{2}{3}} = 3\sqrt[3]{a^2}$

109. $\sqrt[5]{7x} = (7x)^{\frac{1}{5}}$

111. $\sqrt[3]{27p^3q^9} = 3pq^3$

113. $\sqrt{-49} = \sqrt{49}i = 7i$

115. $\sqrt{-48} = \sqrt{48}i = 4i\sqrt{3}$

117. $(2+3i)+(3-5i) = 5-2i$

119. $(5-3i)-(2+5i) = 5-3i-2-5i = 3-8i$

121. $4i(7-2i) = 4i\cdot 7 - 4i\cdot 2i = 28i - 8i^2$
$= 8 + 28i$

123. $(3-4i)^2 = 9-24i-16 = -7-24i$

125. $\dfrac{5-15i}{5i} = \dfrac{1-3i}{i}\bullet\dfrac{i}{i} = \dfrac{i+3}{-1} = -3-i$

127. $\dfrac{3-2i}{3+2i} = \dfrac{3-2i}{3+2i}\bullet\dfrac{3-2i}{3-2i} = \dfrac{5-12i}{13} = \dfrac{5}{13}-\dfrac{12}{13}i$

Chapter Test 7

1. $\sqrt[3]{9p^7q^5} = p^2q\sqrt[3]{9pq^2}$

3. $\sqrt{6x}\left(\sqrt{18x}-\sqrt{2x}\right)$
$= \sqrt{6x}\bullet\sqrt{18x} - \sqrt{6x}\bullet\sqrt{2x} = 6x\sqrt{3} - 2x\sqrt{3}$
$= 4x\sqrt{3}$

5. $\dfrac{7x}{\sqrt{64y^2}} = \dfrac{7x}{8y}$

7. $\left(16x^4\right)^{\frac{3}{2}} = 4^3\left(x^2\right)^3 = 64x^6$

9. $\sqrt[3]{-27w^6z^9} = -3w^2z^3$

11. $\sqrt{43} \approx 6.6$

13. $\sqrt{x-7} - 2 = 0$
$\sqrt{x-7} = 2$
$x - 7 = 4$
$x = 11$
Check: $\sqrt{(11)-7} - 2 = 0$
The solution set is $\{11\}$.

15. $\dfrac{3}{\sqrt[3]{9x}} = \dfrac{3}{\sqrt[3]{9x}}\bullet\dfrac{\sqrt[3]{3x^2}}{\sqrt[3]{3x^2}} = \dfrac{3\sqrt[3]{3x^2}}{3x} = \dfrac{\sqrt[3]{3x^2}}{x}$

17. $\sqrt[3]{54m^4} + m\sqrt[3]{16m} = 3m\sqrt[3]{2m} + 2m\sqrt[3]{2m}$
$= 5m\sqrt[3]{2m}$

19. $\left(a^7b^3\right)^{\frac{2}{5}} = \sqrt[5]{\left(a^7b^3\right)^2} = \sqrt[5]{a^{14}b^6} = a^2b\sqrt[5]{a^4b}$

21. $5\sqrt{20} + 4\sqrt{45} = 5\sqrt{4\bullet 5} + 4\sqrt{9\bullet 5}$
$= 10\sqrt{5} + 12\sqrt{5} = 22\sqrt{5}$

23. $\left(27m^{\frac{3}{2}}n^{-6}\right)^{\frac{2}{3}} = 9mn^{-4} = \dfrac{9m}{n^4}$

25. $\left(\sqrt{11}-\sqrt{5}\right)^2 = 11 - 2\sqrt{11}\sqrt{5} + 5 = 16 - 2\sqrt{55}$

27. $d = \sqrt{\left[6-(-5)\right]^2 + (0-8)^2} = \sqrt{11^2 + (-8)^2}$
$= \sqrt{185} \approx 13.601$

29. $(-5+3i)(7-6i) = -35 + 30i + 21i - 18i^2$
$= -35 + 51i + 18$
$= -17 + 51i$

Cumulative Review: Chapters 0–7

1. $7x-6(x-1)=2(5+x)+11$
$$7x-6x+6=10+2x+11$$
$$-15=x;\{-15\}$$

3. $6x-4y=18$
$$y=\frac{3}{2}x-\frac{9}{2}$$
$$m=\frac{3}{2}$$

The given line has slope $m=\frac{3}{2}$, as does any line parallel to it.
$$y=\frac{3}{2}x-6$$

5. $5x^2-8x+11-(-3x^2-2x+8)$
$$-(-2x^2-4x+3)$$
$$=5x^2-8x+11+3x^2+2x-8+2x^2+4x-3$$
$$=10x^2-2x$$

7. $2x^3+x^2-3x=x(2x^2+x-3)$
$$=x(2x+3)(x-1)$$

9. $4x^2+8xy-5x-10y$
$$=4x(x+2y)-5(x+2y)$$
$$=(x+2y)(4x-5)$$

11. $4x+3y=7$
$$3y=-4x+7$$
$$y=-\frac{4}{3}x+\frac{7}{3}$$

The slope of the line is $-\frac{4}{3}$.

13. The break-even point occurs when cost equals revenue.
$$30x+500=75x-x^2$$
$$x^2-45x+500=0$$
$$(x-20)(x-25)=0$$
The company breaks even for the x-values 20 and 25.

15. $(\sqrt{3}-5)(\sqrt{2}+3)=\sqrt{3}\sqrt{2}+3\sqrt{3}-5\sqrt{2}-15$
$$=\sqrt{6}+3\sqrt{3}-5\sqrt{2}-15$$

17. $x=-5$

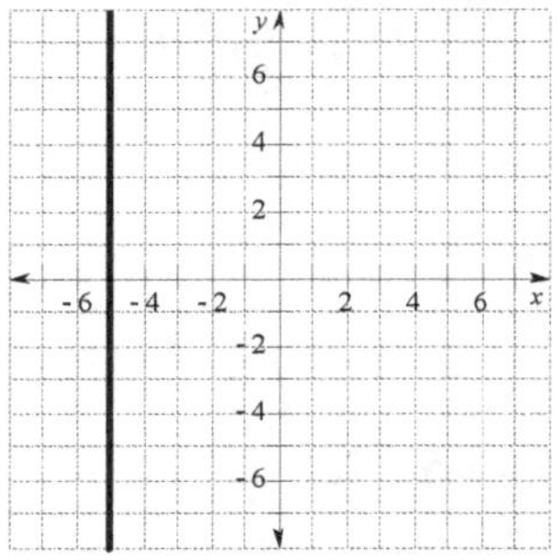

19. $4x-3y=15$
$$x+y=2\rightarrow y=2-x$$
Substitute for y in the first equation.
$$4x-3(2-x)=15 \quad\text{and}\quad y=2-x$$
$$7x-6=15 \qquad =2-(3)=-1$$
$$x=3$$
$\{(3,-1)\}$

21. $\left(\frac{x^2y^{-3}}{x^5y^4}\right)^{-2}=(x^{-3}y^{-7})^{-2}=x^6y^{14}$

23. $\left(\frac{64x^3y^9}{27}\right)^{\frac{1}{3}}=\frac{(64x^3y^9)^{\frac{1}{3}}}{27^{\frac{1}{3}}}=\frac{4xy^3}{3}$

25. $5x-(2-3x)\geq 6+10x$
$$8x-2\geq 6+10x$$
$$-8\geq 2x$$
$$-4\geq x$$
$\{x|x\leq -4\}$

27. Let W be the width of the rectangle and L its length. It follows that

$$L = 2W - 3.$$

$$2W + 2L = 96$$

Using substitution, we get

$$2W + 2(2W - 3) = 96$$

$$2W + 4W - 6 = 96$$

$$6W = 102$$

$$W = \frac{102}{6} = 17$$

The width is 17 inches, and the length is $2(17) - 3$ or 31 inches.

Chapter 8
Quadratic Functions

Exercises 8.1

1. $x^2+9x+14=0$

$(x+7)(x+2)=0$

$\{-7, -2\}$

3. $z^2-2z-35=0$

$(z+5)(z-7)=0$

$\{-5, 7\}$

5. $2x^2-5x-3=0$

$(2x+1)(x-3)=0$

$\left\{-\frac{1}{2}, 3\right\}$

7. $6y^2-y-2=0$

$(2y+1)(3y-2)=0$

$\left\{-\frac{1}{2}, \frac{2}{3}\right\}$

< Objective 1 >

9. $x^2=121$

$x=\pm 11; \{\pm 11\}$

11. $y^2=7$

$y=\pm\sqrt{7}; \{\pm\sqrt{7}\}$

13. $2x^2-12=0$

$x^2=6$

$x=\pm\sqrt{6}; \{\pm\sqrt{6}\}$

15. $2t^2+12=4$

$2t^2=-8$

$t^2=-4$

$t=\pm\sqrt{-4}=\pm 2i$

$\{\pm 2i\}$ nonreal

17. $(x+1)^2=12$

$x+1=\pm\sqrt{12}=\pm 2\sqrt{3}$

$x=-1\pm 2\sqrt{3}; \{-1\pm 2\sqrt{3}\}$

19. $(3z+1)^2-5=0$

$(3z+1)^2=5$

$3z+1=\pm\sqrt{5}$

$3z=-1\pm\sqrt{5}$

$z=\frac{-1\pm\sqrt{5}}{3}; \left\{\frac{-1\pm\sqrt{5}}{3}\right\}$

21. x^2+18x

$\left(\frac{18}{2}\right)^2=81$

23. y^2-8y

$\left(\frac{-8}{2}\right)^2=16$

25. x^2-3x

$\left(\frac{-3}{2}\right)^2=\frac{9}{4}$

27. n^2+n

$\left(\frac{1}{2}\right)^2=\frac{1}{4}$

29. $x^2+\frac{1}{5}x$

$$\left(\frac{1}{2}\bullet\frac{1}{5}\right)^2=\left(\frac{1}{10}\right)^2=\frac{1}{100}$$

31. $x^2-\frac{1}{6}x$

$$\left(\frac{1}{2}\bullet\frac{-1}{6}\right)^2=\frac{1}{144}$$

< Objective 2 >

33.
$$x^2+10x=4$$
$$x^2+10x+25=4+25$$
$$(x+5)^2=29$$
$$x+5=\pm\sqrt{29}$$
$$x=-5\pm\sqrt{29};\{-5\pm\sqrt{29}\}$$

35.
$$y^2-2y=8$$
$$y^2-2y+1=8+1$$
$$(y-1)^2=9$$
$$y-1=\pm 3$$
$$y=1\pm 3;\{-2,4\}$$

37. $x^2-2x-5=0$
$$x^2-2x+1=5+1$$
$$(x-1)^2=6$$
$$x-1=\pm\sqrt{6}$$
$$x=1\pm\sqrt{6};\{1\pm\sqrt{6}\}$$

39. $x^2+10x+13=0$
$$x^2+10x+25=-13+25$$
$$(x+5)^2=12$$
$$x+5=\pm\sqrt{12}$$
$$x=-5\pm 2\sqrt{3};\{-5\pm 2\sqrt{3}\}$$

41. $z^2-5z-7=0$
$$z^2-5z+\frac{25}{4}=7+\frac{25}{4}$$
$$\left(x-\frac{5}{2}\right)^2=\frac{53}{4}$$
$$x-\frac{5}{2}=\pm\frac{\sqrt{53}}{2}$$
$$x=\frac{5}{2}\pm\frac{\sqrt{53}}{2};\left\{\frac{5\pm\sqrt{53}}{2}\right\}$$

43. $m^2-3m-5=0$
$$m^2-3m+\frac{9}{4}=5+\frac{9}{4}$$
$$\left(m-\frac{3}{2}\right)^2=\frac{29}{4}$$
$$m-\frac{3}{2}=\pm\frac{\sqrt{29}}{2}$$
$$m=\frac{3\pm\sqrt{29}}{2};\left\{\frac{3\pm\sqrt{29}}{2}\right\}$$

45.
$$x^2+\frac{1}{2}x=1$$
$$x^2+\frac{1}{2}x+\frac{1}{16}=1+\frac{1}{16}$$
$$\left(x+\frac{1}{4}\right)^2=\frac{17}{16}$$
$$x+\frac{1}{4}=\pm\frac{\sqrt{17}}{4}$$
$$x=-\frac{1}{4}\pm\frac{\sqrt{17}}{4};\left\{\frac{-1\pm\sqrt{17}}{4}\right\}$$

47. $2x^2+2x-1=0$

$$2x^2+2x=1$$
$$x^2+x=\frac{1}{2}$$
$$x^2+x+\frac{1}{4}=\frac{1}{2}+\frac{1}{4}$$
$$\left(x+\frac{1}{2}\right)^2=\frac{3}{4}$$
$$x+\frac{1}{2}=\pm\frac{\sqrt{3}}{2}$$
$$x=-\frac{1}{2}\pm\frac{\sqrt{3}}{2};\left\{\frac{-1\pm\sqrt{3}}{2}\right\}$$

49. $3x^2-8x=2$

$$x^2-\frac{8}{3}x=\frac{2}{3}$$
$$x^2-\frac{8}{3}x+\frac{16}{9}=\frac{2}{3}+\frac{16}{9}$$
$$\left(x-\frac{4}{3}\right)^2=\frac{22}{9}$$
$$x-\frac{4}{3}=\pm\frac{\sqrt{22}}{3}$$
$$x=\frac{4}{3}\pm\frac{\sqrt{22}}{3};\left\{\frac{4\pm\sqrt{22}}{3}\right\}$$

51. $3x^2-2x+12=0$

$$x^2-\frac{2}{3}x=-4$$
$$x^2-\frac{2}{3}x+\frac{1}{9}=-4+\frac{1}{9}$$
$$\left(x-\frac{1}{3}\right)^2=-\frac{35}{9}$$
$$x-\frac{1}{3}=\pm\frac{\sqrt{35}}{3}i$$
$$x=\frac{1}{3}\pm\frac{\sqrt{35}}{3}i$$
$$\left\{\frac{1\pm i\sqrt{35}}{3}\right\}\text{ nonreal}$$

53. $x^2+10x+28=0$

$$x^2+10x+25=-28+25$$
$$(x+5)^2=-3$$
$$x+5=\pm i\sqrt{3}$$
$$x=-5\pm i\sqrt{3}$$
$$\left\{-5\pm i\sqrt{3}\right\}\text{ nonreal}$$

< Objective 3 >

55. Using Pythagorean theorem,

$x^2=20^2+8^2=400+64=464$

$x\approx 21.541$ ft

(rounded to the nearest thousandth)

57. Using Pythagorean theorem,

$15^2=5^2+x^2$

$225=25+x^2$

$200=x^2$

$x\approx 14.142$ ft

(rounded to the nearest thousandth)

59. Using Pythagorean theorem,

$$x^2+(2x)^2=8^2$$
$$x^2+4x^2=64$$
$$5x^2=64$$
$$x^2=\frac{64}{5}$$
$$x\approx 3.578$$
$$2x\approx 7.155$$

The legs have lengths 3.578 m and 7.155 m.

61. Let x be the number.

$$(x+2)^2=64$$
$$x+2=\pm 8$$
$$x=-2\pm 8=6,-10$$

The numbers are 6, −10.

63.

$$x^2+(x+4)^2=8^2$$
$$x^2+x^2+8x+16=64$$
$$2x^2+8x-48=0$$
$$x^2=\frac{-8\pm\sqrt{8^2-4(2)(-48)}}{2(2)}$$
$$x\approx 3.292,-7.292$$
$$x+4\approx 7.292$$

The side lengths are 3.292 in. and 7.292 in.

65.

$$x(x-6)=75$$
$$x^2-6x-75=0$$
$$x=\frac{6\pm\sqrt{(-6)^2-4(1)(-75)}}{2(1)}$$
$$x\approx 12.165,-6.165$$
$$x-6\approx 6.165$$

The dimensions of the rectangle are 6.165 ft by 12.165 ft.

67.

$$160=-16t^2-32t+320$$
$$16t^2+32t-160=0$$
$$16(t^2+2t-10)=0$$
$$t=\frac{-2\pm\sqrt{2^2-4(1)(-10)}}{2(1)}$$
$$t\approx 2.317,-4.317$$

The ball will reach the height after 2.317 s.

69.

$$-200p+35,000=p^2+400p-20,000$$
$$p^2-600p+55,000=0$$
$$p=\frac{600\pm\sqrt{(-600)^2-4(1)(55,000)}}{2(1)}$$
$$p\approx 112.92,487.08$$

The equilibrium price is \$112.92.

71. Let x be the number.

$$(x+3)^2=9$$
$$x+3=\pm 3$$
$$x=-3\pm 3=0,-6$$

The number is 0 or –6.

73.

$$R=x\left(25-\frac{1}{2}x\right)$$
$$294.50=x\left(25-\frac{1}{2}x\right)=25x-\frac{1}{2}x^2$$
$$x^2-50x+589=0$$
$$x^2-50x=-589$$
$$x^2-50x+625=-589+625$$
$$(x-25)^2=36$$
$$x-25=\pm 6$$
$$x=25\pm 6=31,19$$

The number of units sold was 19 or 31

75. True. Not all quadratic expressions are factorable.

77. True. If "completing the square" is being used and the coefficient of x^2 is not 1, divide both sides of the equation by that coefficient.

79. An equation of the form $(x-h)^2 = k$, where k is positive, always has two distinct solutions.

81. When completing the square for $x^2 + bx$, the number added is never negative.

83. $y = x^2 + 12x - 2$

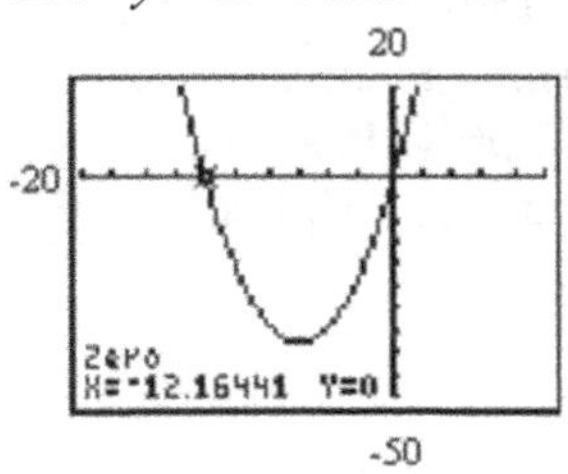

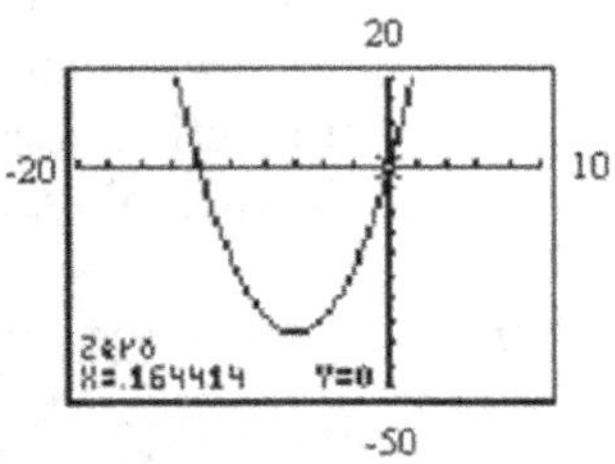

To the nearest tenth, the x-intercepts are (–12.2, 0) and (0.2, 0).

85. $y = x^2 - 2x - 8$

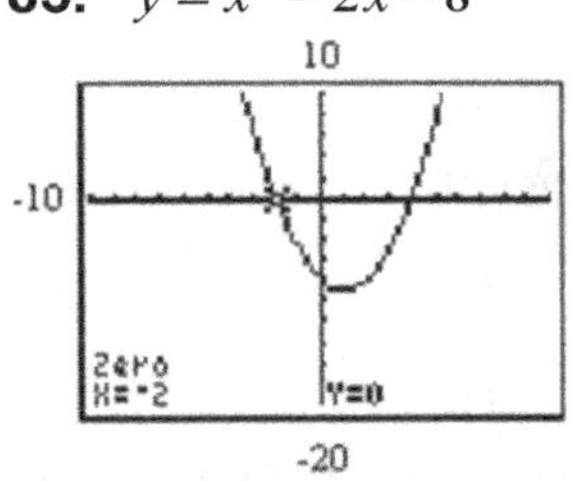

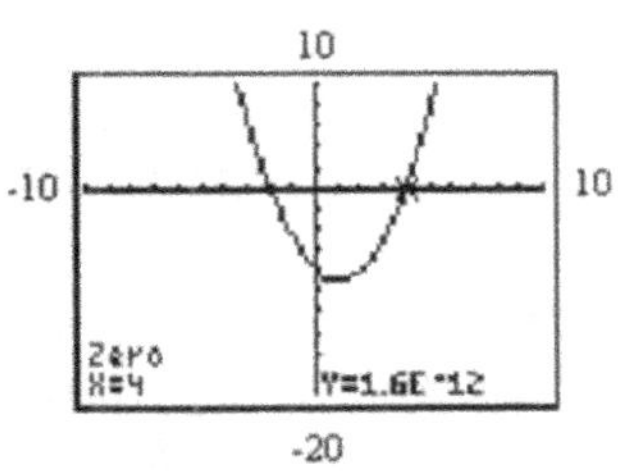

To the nearest tenth, the x-intercepts are (–2.0, 0) and (4.0, 0).

87. $f(x) = x^2 + 1$

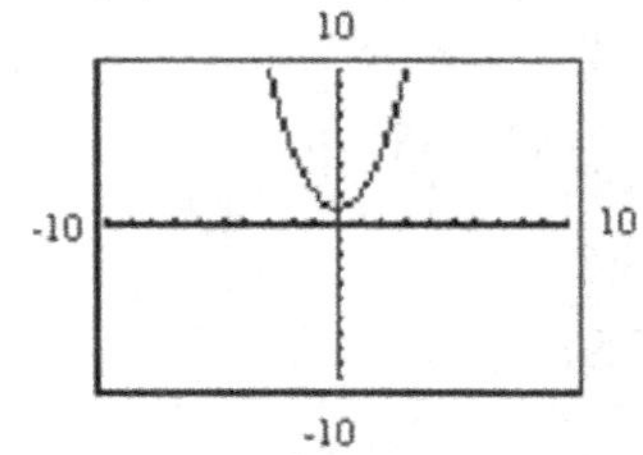

(a) There are no x-intercepts.

(b) $x^2 + 1 = 0$

$$x^2 = -1$$

$$x = \pm i$$

The zeros are $\pm i$.

(c) If the graph of the function has no x-intercepts, then the zeros of the function are not real numbers.

89.

$$M = 2x^2 - 30x$$

$$152 = 2x^2 - 30x$$

$$2x^2 - 30x = 152$$

$$x^2 - 15x = 76$$

$$x^2 - 15x + \frac{225}{4} = 76 + \frac{225}{4}$$

$$\left(x - \frac{15}{2}\right)^2 = \frac{529}{4}$$

$$x - \frac{15}{2} = \pm\sqrt{\frac{529}{4}} = \pm\frac{23}{2}$$

$$x = \frac{15}{2} \pm \frac{23}{2}; \{19, -4\}$$

x is 19 when the moment is 152.

91.

$$V=\frac{1}{350}D^2H-\frac{1}{5}$$
$$70=\frac{1}{350}D^2(48)-\frac{1}{5}$$
$$\frac{350}{5}+\frac{1}{5}=\frac{48}{350}D^2$$
$$\frac{351}{5}=\frac{24}{175}D^2$$
$$\frac{351}{5}\bullet\frac{175}{24}=D^2$$
$$\frac{117}{1}\bullet\frac{35}{8}=D^2$$
$$\frac{4{,}095}{8}=D^2$$
$$511.875=D^2$$
$$\pm\sqrt{511.875}=\sqrt{D^2}$$
$$\pm 22.62=D$$

Since diameter cannot be negative, the diameter, $D=22.62$ in.

93.

$$P=-3x^2+240x-1{,}800$$
$$2{,}325=-3x^2+240x-1{,}800$$
$$-775=x^2-80x+600$$
$$0=x^2-80x+1{,}375=x^2-20x-1{,}125$$
$$=\frac{80\pm\sqrt{80^2-4(1)(1{,}375)}}{2(1)}=\frac{80\pm 30}{2}$$
$$=25,55$$

A total of 25 or 55 appliances may be produced and sold.

95.

$$S=-p^2+23p-11$$
$$\text{and } D=-5p+62$$
$$-5p+62=-p^2+23p-11$$
$$62=p^2+28p-11$$
$$0=p^2+28p-73=p^2-28p+73$$
$$=\frac{28\pm\sqrt{28^2-4(1)(73)}}{2(1)}$$
$$=\frac{28\pm 22.18}{2}=2.91,25.09$$

Since the final answer must be positive in supply and demand, \$2.91 is the answer. The other solution does not provide a positive result when plugged into the original.

97.

$$N=121-4t^2$$
$$0=121-4t^2$$
$$-121=-4t^2$$
$$30.25=t^2$$
$$\pm\sqrt{30.25}=\sqrt{t^2}$$
$$\pm 5.5=t$$

Since time cannot be negative, $t=5.5$ days.

99. x^2+10x

(a) $x^2+5x+5x=x^2+10x$

(b) The missing piece is a square whose side has length 5. Therefore, an area of 25 sq units must be added to the figure to "complete the square."

(c) $x^2+10x+25=(x+5)^2$

101.

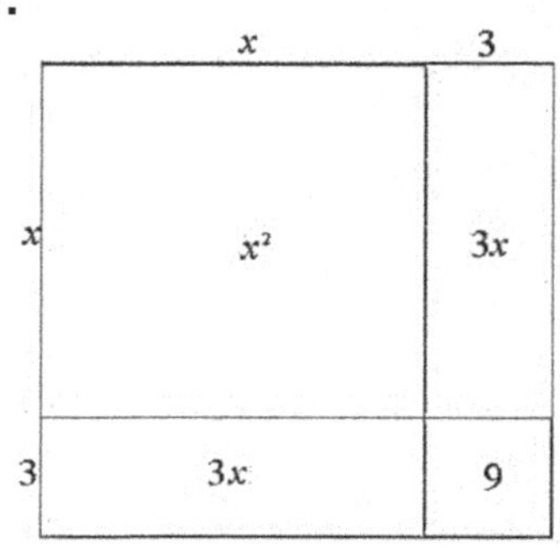

$$x^2+6x+9=(x+3)^2$$

103.

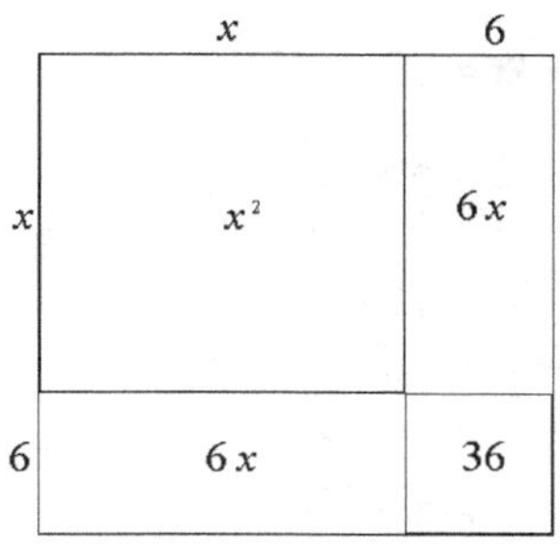

$$x^2+12x+36=(x+6)^2$$

105. Above and Beyond

107. x^2+2ax

$$\left(\frac{2a}{2}\right)^2=a^2$$

109. x^2+3ax

$$\left(\frac{3a}{2}\right)^2=\frac{9a^2}{4}$$

111. a^2x^2+2ax

$$a^2x^2+2ax+1=(ax+1)^2$$

Add 1 to the binomial.
Note: If this were an equation, we would first need to divide both sides by a^2.

113.
$$x^2+2ax=4$$
$$x^2+2ax+a^2=4+a^2$$
$$(x+a)^2=4+a^2$$
$$x+a=\pm\sqrt{4+a^2}$$
$$x=-a\pm\sqrt{4+a^2}$$
$$\left\{-a\pm\sqrt{4+a^2}\right\}$$

Exercises 8.2

< Objective 1 >

1.
$$x^2-5x-14=0$$
$$(x-7)(x+2)=0$$
$$x=7 \text{ or } x=-2;\{-2,7\}$$
$$a=1,\ b=-5,\ c=-14$$
$$x=\frac{-(-5)\pm\sqrt{(-5)^2-4(1)(-14)}}{2(1)}=\frac{5\pm\sqrt{81}}{2}$$
$$=\frac{5\pm9}{2}$$
$$\{-2,7\}$$

3.
$$t^2+8t-65=0$$
$$(t+13)(t-5)=0$$
$$\{-13,5\}$$
$$a=1,\ b=-8,\ c=-65$$
$$t=\frac{-(8)\pm\sqrt{(8)^2-4(1)(-65)}}{2(1)}=\frac{-8\pm\sqrt{324}}{2}$$
$$=\frac{-8\pm18}{2}$$
$$\{-13,5\}$$

5.
$$3x^2+x-10=0$$
$$(3x-5)(x+2)=0$$
$$\left\{-2,\frac{5}{3}\right\}$$
$$x=\frac{-(1)\pm\sqrt{(1)^2-4(3)(-10)}}{2(3)}=\frac{-1\pm\sqrt{121}}{6}$$
$$=\frac{-1\pm11}{6}$$
$$\left\{-2,\frac{5}{3}\right\}$$

7. $16t^2 - 24t + 9 = 0$

$$(4t-3)^2 = 0$$

$$\left\{\frac{3}{4}\right\}$$

$$t = \frac{-(-24) \pm \sqrt{(-24)^2 - 4(16)(9)}}{2(16)} = \frac{24 \pm \sqrt{0}}{32}$$

$$= \frac{3}{4}$$

$$\left\{\frac{3}{4}\right\}$$

9. **(a)** $x^2 - 4x - 7 = 0$

$$x^2 - 4x = 7$$

$$x^2 - 4x + 4 = 7 + 4 = 11$$

$$(x-2)^2 = 11$$

$$x - 2 = \pm\sqrt{11}$$

$$x = 2 \pm \sqrt{11}$$

$$\left\{2 \pm \sqrt{11}\right\}$$

(b) $a = 1,\ b = -4,\ c = -7$

$$x = \frac{-(-4) \pm \sqrt{(-4)^2 - 4(1)(-7)}}{2(1)}$$

$$= \frac{4 \pm \sqrt{44}}{2} = \frac{4 \pm 2\sqrt{11}}{2} = 2 \pm \sqrt{11}$$

$$\left\{2 \pm \sqrt{11}\right\}$$

11. **(a)** $x^2 + 3x - 27 = 0$

$$x^2 + 3x = 27$$

$$x^2 + 3x + \frac{9}{4} = 27 + \frac{9}{4}$$

$$\left(x + \frac{3}{2}\right)^2 = \frac{117}{4}$$

$$x + \frac{3}{2} = \pm\frac{\sqrt{117}}{2}$$

$$x = -\frac{3}{2} \pm \frac{3\sqrt{13}}{2}$$

$$\left\{\frac{-3 \pm 3\sqrt{13}}{2}\right\}$$

(b) $a = 1,\ b = 3,\ c = -27$

$$x = \frac{-(3) \pm \sqrt{(3)^2 - 4(1)(-27)}}{2(1)} = \frac{-3 \pm \sqrt{117}}{2}$$

$$= \frac{-3 \pm 3\sqrt{13}}{2}$$

$$\left\{\frac{-3 \pm 3\sqrt{13}}{2}\right\}$$

13. **(a)**

$$3x^2 - 5x + 1 = 0$$
$$x^2 - \frac{5}{3}x = -\frac{1}{3}$$
$$x^2 - \frac{5}{3}x + \frac{25}{36} = -\frac{1}{3} + \frac{25}{36}$$
$$\left(x - \frac{5}{6}\right)^2 = \frac{13}{36}$$
$$x - \frac{5}{6} = \pm\frac{\sqrt{13}}{6}$$
$$x = \frac{5}{6} \pm \frac{\sqrt{13}}{6}$$
$$\left\{\frac{5 \pm \sqrt{13}}{6}\right\}$$

(b) $a = 3$, $b = -5$, $c = 1$

$$x = \frac{-(-5) \pm \sqrt{(-5)^2 - 4(3)(1)}}{2(3)} = \frac{5 \pm \sqrt{13}}{5}$$
$$\left\{\frac{5 \pm \sqrt{13}}{6}\right\}$$

15. **(a)**

$$2q^2 - 2q - 1 = 0$$
$$q^2 - q = \frac{1}{2}$$
$$q^2 - q + \frac{1}{4} = \frac{1}{2} + \frac{1}{4}$$
$$\left(q - \frac{1}{2}\right)^2 = \frac{3}{4}$$
$$q - \frac{1}{2} = \pm\frac{\sqrt{3}}{2}$$
$$q = \frac{1}{2} \pm \frac{\sqrt{3}}{2}$$
$$\left\{\frac{1 \pm \sqrt{3}}{2}\right\}$$

(b)

$$q = \frac{-(-2) \pm \sqrt{(-2)^2 - 4(2)(-1)}}{2(2)}$$
$$= \frac{2 \pm \sqrt{12}}{4} = \frac{2 \pm 2\sqrt{3}}{4} = \frac{1 \pm \sqrt{3}}{2}$$
$$\left\{\frac{1 \pm \sqrt{3}}{2}\right\}$$

17. **(a)** $3x^2 - x - 2 = 0$

$$x^2 - \frac{x}{3} = \frac{2}{3}$$

$$x^2 - \frac{x}{3} + \frac{1}{36} = \frac{2}{3} + \frac{1}{36}$$

$$\left(x - \frac{1}{6}\right)^2 = \frac{25}{36}$$

$$x - \frac{1}{6} = \pm\frac{5}{6}$$

$$x = \frac{1}{6} \pm \frac{5}{6}$$

$$\left\{-\frac{2}{3}, 1\right\}$$

(b) $x = \dfrac{-(-1) \pm \sqrt{(-1)^2 - 4(3)(-2)}}{2(3)}$

$$= \frac{1 \pm \sqrt{25}}{6} = \frac{1 \pm 5}{6}$$

$$\left\{-\frac{2}{3}, 1\right\}$$

19. **(a)** $2y^2 - y - 5 = 0$

$$y^2 - \frac{y}{2} = \frac{5}{2}$$

$$y^2 - \frac{y}{2} + \frac{1}{16} = \frac{5}{2} + \frac{1}{16}$$

$$\left(y - \frac{1}{4}\right)^2 = \frac{41}{16}$$

$$y - \frac{1}{4} = \pm\frac{\sqrt{41}}{4}$$

$$y = \frac{1}{4} \pm \frac{\sqrt{41}}{4}$$

$$\left\{\frac{1 \pm \sqrt{41}}{4}\right\}$$

(b) $y = \dfrac{1 \pm \sqrt{1 - 4(-10)}}{4} = \dfrac{1 \pm \sqrt{41}}{4}$

$$\left\{\frac{1 \pm \sqrt{41}}{4}\right\}$$

21. $x^2 - 4x + 3 = 0$

$a = 1,\ b = -4,\ c = 3$

$$x = \frac{-(-4) \pm \sqrt{(-4)^2 - 4(1)(3)}}{2(1)}$$

$$= \frac{4 \pm \sqrt{16 - 12}}{2} = \frac{4 \pm \sqrt{4}}{2} = \frac{4 \pm 2}{2} = 2 \pm 1$$

$\{1, 3\}$

23. $p^2 - 8p + 16 = 9$

$a = 1,\ b = -8,\ c = 16$

$$p = \frac{-(-8) \pm \sqrt{(-8)^2 - 4(1)(16)}}{2(1)} = \frac{8 \pm \sqrt{0}}{2} = 4$$

$\{4\}$

25. $-x^2 + 3x + 5 = 0$

$a = -1,\ b = 3,\ c = 5$

$$x = \frac{-3 \pm \sqrt{9 + 20}}{-2} = \frac{3 \pm \sqrt{29}}{2}$$

$$\left\{\frac{3 \pm \sqrt{29}}{2}\right\}$$

27. $-3s^2 + 2s - 1 = 0$

$a = -3,\ b = 2,\ c = -1$

$$s = \frac{-2 \pm \sqrt{4 - 12}}{-6} = \frac{2 \pm \sqrt{-8}}{6} = \frac{2 \pm 2i\sqrt{2}}{6}$$

$$= \frac{1 \pm i\sqrt{2}}{3}$$

$\left\{\dfrac{1 \pm i\sqrt{2}}{3}\right\}$ nonreal

29. $2x^2 - \dfrac{1}{2}x - 5 = 0$

$4x^2 - x - 10 = 0$

$a = 4,\ b = -1,\ c = -10$

$$x = \frac{1 \pm \sqrt{1 + 160}}{8} = \frac{1 \pm \sqrt{161}}{8}$$

$$\left\{\frac{1 \pm \sqrt{161}}{8}\right\}$$

31. $5t^2 - 2t - \frac{2}{3} = 0$

$15t^2 - 6t - 2 = 0$

$a = 15,\ b = -6,\ x = -2$

$t = \frac{6 \pm \sqrt{36+120}}{30} = \frac{6 \pm \sqrt{156}}{30} = \frac{6 \pm 2\sqrt{39}}{30}$

$= \frac{3 \pm \sqrt{39}}{15}$

$\left\{\frac{3 \pm \sqrt{39}}{15}\right\}$

33. $(x-2)(x+3) = 4$

$x^2 + x - 6 = 4$

$x^2 + x - 10 = 0$

$x = \frac{-1 \pm \sqrt{1+40}}{2} = \frac{-1 \pm \sqrt{41}}{2}$

$\left\{\frac{-1 \pm \sqrt{41}}{2}\right\}$

35. $(t+1)(2t-4) - 7 = 0$

$2t^2 - 2t - 11 = 0$

$t = \frac{2 \pm \sqrt{4+88}}{4} = \frac{2 \pm \sqrt{92}}{4}$

$= \frac{2 \pm 2\sqrt{23}}{4} = \frac{1 \pm \sqrt{23}}{2}$

$\left\{\frac{1 \pm \sqrt{23}}{2}\right\}$

37. $2x - 7 = \frac{3}{x}$

$2x^2 - 7x = 3$

$2x^2 - 7x - 3 = 0$

$x = \frac{7 \pm \sqrt{49+24}}{4} = \frac{7 \pm \sqrt{73}}{4}$

$\left\{\frac{7 \pm \sqrt{73}}{4}\right\}$

39. $2t - \frac{3}{t} = 3$

$2t^2 - 3 = 3t$

$2t^2 - 3t - 3 = 0$

$t = \frac{3 \pm \sqrt{9+24}}{4} = \frac{3 \pm \sqrt{33}}{4}$

$\left\{\frac{3 \pm \sqrt{33}}{4}\right\}$

41. $\frac{5}{y^2} + \frac{2}{y} - 1 = 0$

$5 + 2y - y^2 = 0$

$y^2 - 2y - 5 = 0$

$y = \frac{2 \pm \sqrt{4+20}}{2} = \frac{2 \pm 2\sqrt{6}}{2}$

$= 1 \pm \sqrt{6}$

$\{1 \pm \sqrt{6}\}$

43. $2x^2 - 5x = 0$

$b^2 - 4ac = (-5)^2 - 4(2)(0) = 25$

There are 2 real solutions.

45. $m^2 - 18m + 81 = 0$

$b^2 - 4ac = (-18)^2 - 4(1)(81) = 0$

There is 1 real solution.

47. $3x^2 - 7x + 1 = 0$

$b^2 - 4ac = (-7)^2 - 4(3)(1) = 37$

There are 2 real solutions.

49. $2w^2 - 5w + 11 = 0$

$b^2 - 4ac = 25 - 88 = -63$

There are no real solutions.

51. $x^2 - 8x + 16 = 0$

$(x-4)^2 = 0$

$\{4\}$

53. $3t^2 - 7t + 1 = 0$

$$t = \frac{7 \pm \sqrt{49-12}}{6} = \frac{7 \pm \sqrt{37}}{6}$$

$$\left\{\frac{7 \pm \sqrt{37}}{6}\right\}$$

< Objective 2 >

55. $5y^2 - 2y = 0$

$y(5y-2) = 0$

$\left\{0, \frac{2}{5}\right\}$

57. $(x-1)(2x+7) = -6$

$2x^2 + 5x - 1 = 0$

$$x = \frac{-5 \pm \sqrt{25+8}}{4} = \frac{-5 \pm \sqrt{33}}{4}$$

$$\left\{\frac{-5 \pm \sqrt{33}}{4}\right\}$$

59. $x^2 + 9 = 0$

$x^2 = -9$

$x = \pm 3i$

$\{\pm 3i\}$ nonreal

61. $x - 3 - \frac{10}{x} = 0$

$x^2 - 3x - 10 = 0$

$(x-5)(x+2) = 0$

$\{-2, 5\}$

< Objective 3 >

63. $h(t) = 112t - 16t^2$

(a)

$112t - 16t^2 = 112$

$16t^2 - 112t + 112 = 0$

$t^2 - 7t + 7 = 0$

$$t = \frac{7 \pm \sqrt{49-28}}{2} = \frac{7 \pm \sqrt{21}}{2}$$

$\frac{7 - \sqrt{21}}{2} \approx 1.2$

$\frac{7 + \sqrt{21}}{2} \approx 5.8$

The arrow is 112 ft high after 1.2 s (on the way up) and 5.8 s (on the way down).

(b)

$112t - 16t^2 = 144$

$16t^2 - 112t + 144 = 0$

$t^2 - 7t + 9 = 0$

$$t = \frac{7 \pm \sqrt{49-36}}{2} = \frac{7 \pm \sqrt{13}}{2}$$

$\frac{7 - \sqrt{13}}{2} \approx 1.7$

$\frac{7 + \sqrt{13}}{2} \approx 5.3$

The arrow is 144 ft high after 1.7 s (on the way up) and 5.3 s (on the way down).

65. Let x and $x+1$ be consecutive integers.

$x(x+1) = 72$

$x^2 + x - 72 = 0$

$(x+9)(x-8) = 0$

The integers are either –9 and –8, or they are 8 and 9.

67. Let L be the length of the rectangle. Then $L-3$ is its width.

$$L(L-3)=70$$
$$L^2-3L-70=0$$
$$(L-10)(L+7)=0$$

The rectangle is 7 ft by 10 ft.
Note: Length measurements must be nonnegative quantities.

69. Let x be the length of the shorter leg. Then $2x$ is the length of the longer leg.

$$x^2+(2x)^2=6^2$$
$$5x^2=36$$
$$x^2=\frac{36}{5}$$
$$x=\frac{6}{\sqrt{5}}=\frac{6\sqrt{5}}{5}\approx 2.7$$
$$2x\approx 5.4$$

The legs measure 2.7 cm and 5.4 cm.

71. $h(t)=64t-16t^2$

(a) $64t-16t^2=0$

$$t^2-4t=0$$
$$t(t-4)=0$$

It takes 4 s for the ball to return to the ground.

(b) $64t-16t^2=48$

$$t^2-4t+3=0$$
$$(t-1)(t-3)=0$$

The ball reaches 48 ft after 1 s.

73.

$$C(x)=2400-40x+2x^2$$
$$2{,}400-40x+2x^2=5{,}400$$
$$x^2-20x-1{,}500=0$$
$$(x-50)(x+30)=0$$

They can produce 50 chairs for $5,400.

75. Let x be the length of the shortest leg. Then the length of the other leg $x+1$. And the hypotenuse is given by $x+3$.

$$x^2+(x+1)^2=(x+3)^2$$
$$x^2+x^2+2x+1=x^2+6x+9$$
$$x^2-4x-8=0$$
$$x=\frac{4\pm\sqrt{16+32}}{2}=\frac{4\pm 4\sqrt{3}}{2}$$
$$=2\pm 2\sqrt{3}$$
$$2+2\sqrt{3}\approx 5.5$$

The sides are 5.5 in., 6.5 in., and 8.5 in.

77.

$$P(x)=-3x^2+270x$$
$$-3x^2+270x=5{,}100$$
$$x^2-90x=-1{,}700$$
$$x^2-90x+45^2=-1{,}700+45^2$$
$$(x-45)^2=325$$
$$x-45=\pm\sqrt{325}$$
$$x=45\pm 5\sqrt{13}$$
$$45-5\sqrt{13}\approx 27$$
$$45+5\sqrt{13}\approx 63$$

27 or 63 Blu-Ray players must be produced to realize a profit of $5,100.

79.

$$D=S$$
$$-2p+14=-p^2+16p-2$$
$$p^2-18p+16=0$$
$$p=\frac{18\pm\sqrt{(-18)^2-4(16)}}{2}$$
$$=\frac{18\pm\sqrt{260}}{2}=\frac{18\pm 2\sqrt{65}}{2}$$
$$=9\pm\sqrt{65}$$
$$9-\sqrt{65}\approx 0.94$$
$$9+\sqrt{65}\approx 17.06$$

$0.94 and $17.06 are equilibrium prices.

81. $h(t)=70+15t-5t^2$

(a) $70+15t-5t^2=0$

$t^2-3t-14=0$

$$t=\frac{3\pm\sqrt{3^2-4(-14)}}{2}=\frac{3\pm\sqrt{65}}{2}$$

$$\frac{3+\sqrt{65}}{2}\approx 5.5$$

It takes 5.5 s for the ball to return to the ground.

(b) $70+15t-5t^2=80$

$5t^2-15t+10=0$

$t^2-3t+2=0$

$(t-2)(t-1)=0$

The ball is at a height of 80 m after 1 s, and again after 2 s.

83. **(a)** $h(t)=-5t^2+20t+100$

(b) $-5t^2+20t+100=0$

$t^2-4t-20=0$

$$t=\frac{4\pm\sqrt{(-4)^2-4(-20)}}{2}=\frac{4\pm\sqrt{96}}{2}=\frac{4\pm4\sqrt{6}}{2}=2\pm2\sqrt{6}$$

$2+2\sqrt{6}\approx 6.9$

The ball lands on the ground after 6.9 s.

(c) $-5t^2+20t+100=75$

$t^2-4t+5=0$

$(t-5)(t+1)=0$

The ball's height is 75 m after 5 s.

(d) $-5t^2+20t+100=125$

$t^2-4t+5=0$

$b^2-4ac=(-4)^2-4(5)=-4$

There are no real solutions. The ball will not reach 125 m.

85. $(300-2x)(500-2x)=100{,}000$

$150{,}000-1600x+4x^2=100{,}000$

$4x^2-1600x+50{,}000=0$

$4(x^2-400x+12{,}500)=0$

$$x=\frac{400\pm\sqrt{400^2-4(1)(12{,}500)}}{2(1)}$$

$x\approx 34.168, 365.831$ (not possible)

The walkway must be 34.168 ft wide (or less)

87. The quadratic formula can always be used to solve a quadratic equation.

89. The solutions of a quadratic equation are sometimes irrational.

91. **(a)** $x^2-3x-5=0$

$$x=\frac{3\pm\sqrt{9+20}}{2}=\frac{3\pm\sqrt{29}}{2}$$

$$\frac{3-\sqrt{29}}{2}\approx -1.2$$

$$\frac{3+\sqrt{29}}{2}\approx 4.2$$

{−1.2, 4.2}

(b) $f(x)=x^2-3x-5$

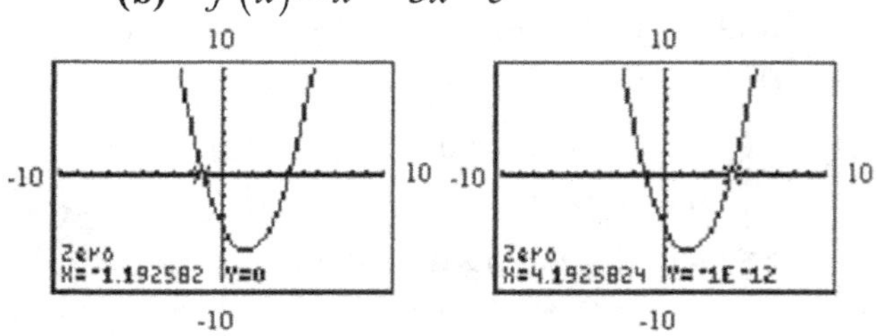

{−1.2, 4.2}

(c) The solutions to the equation are the x-coordinates of the points of intersection of the graphs of f and g.

93. $V = 0.79D^2 - 2D - 4$

$$272 = 0.79D^2 - 2D - 4$$
$$0 = 0.79D^2 - 2D - 276$$
$$D = \frac{-(-2) \pm \sqrt{(-2)^2 - 4(0.79)(-276)}}{2(0.79)}$$
$$= \frac{2 \pm \sqrt{876.16}}{1.58} = 20$$

The diameter of the small end of the log is 20 in.

95.

$$C = -0.0015t^2 + 0.0845t + 0.7170$$
$$0.8 = -0.0015t^2 + 0.0845t + 0.7170$$
$$0 = -0.0015t^2 + 0.0845t - 0.083$$
$$= 0.0015t^2 - 0.0845t + 0.083$$
$$t = \frac{0.0845 \pm \sqrt{(0.0845)^2 - 4(0.0015)(0.083)}}{2(0.0015)}$$
$$= 55.3, 1.0$$

(a) It will take 1 hour for the drug to have an effect.

(b) It will take 55.3 hours for the drug to stop having an effect.

97. Above and Beyond

99. $x^2 + y^2 = z^2$

$$x^2 = z^2 - y^2$$
$$x = \pm\sqrt{z^2 - y^2}$$
$$\left\{\pm\sqrt{z^2 - y^2}\right\}$$

101. $x^2 - 36a^2 = 0$

$$x^2 = 36a^2$$
$$x = \pm 6a$$

{±6a}

103. $2x^2 + 5ax - 3a^2 = 0$

$$(2x - a)(x + 3a) = 0$$
$$\left\{-3a, \frac{a}{2}\right\}$$

105. $2x^2 - ax - 2a^2 = 0$

$$x = \frac{-(a) \pm \sqrt{(a)^2 - 4(2)(-2a^2)}}{2(2)} = \frac{-a \pm \sqrt{17a^2}}{4}$$
$$= \frac{-a \pm |a|\sqrt{17}}{4}$$
$$\left\{\frac{-a \pm |a|\sqrt{17}}{4}\right\}$$

107.

$$\begin{array}{r} x^2 + 2x - 5 \\ x - 5 \overline{) x^3 - 3x^2 - 15x + 25} \\ \underline{x^3 - 5x^2} \\ 2x^2 - 15x + 25 \\ \underline{2x^2 - 10x} \\ -5x + 25 \\ \underline{-5x + 25} \\ 0 \end{array}$$

$$x^3 - 3x^2 - 15x + 25 = (x - 5)(x^2 + 2x - 5)$$

Therefore, if $x^2 + 2x - 5 = 0$, then $x^3 - 3x^2 - 15x + 25 = 0$.

$$x = \frac{-2 \pm \sqrt{2^2 - 4(-5)}}{2} = \frac{-2 \pm \sqrt{24}}{2}$$
$$= \frac{-2 \pm 2\sqrt{6}}{2} = -1 \pm \sqrt{6}$$
$$\left\{-1 \pm \sqrt{6}\right\}$$

109. $f(x) = x^3 + 1 = (x + 1)(x^2 - x + 1)$

One of the zeros is -1.

$$x^2 - x + 1 = 0$$
$$x = \frac{1 \pm \sqrt{1 - 4}}{2} = \frac{1 \pm \sqrt{-3}}{2} = \frac{1 \pm i\sqrt{3}}{2}$$

$\left\{-1, \frac{1 \pm i\sqrt{3}}{2}\right\}$ two nonreal zeros

111. $0=x^6-1=(x^3+1)(x^3-1)$
$=(x+1)(x^2-x+1)(x-1)(x^2+x+1)$
From exercises 107 and 108, we have $\left\{\pm1,\frac{1\pm i\sqrt{3}}{2},\frac{-1\pm i\sqrt{3}}{2}\right\}$, four nonreal solutions.

113. Above and Beyond

115. Above and Beyond

Exercises 8.3

1. **(f)** $y=-2x+1$

3. **(g)** $y=x^2+2x-3=(x+3)(x-1)$

5. **(h)** $y=-x^2+6x-8=-(x-2)(x-4)$

7. **(c)** $y=2x+1$

9. $y=x^2-3$: **(a)**, **(c)**

11. $y=x^2-3x-4=(x-4)(x+1)$: **(a)**, **(c)**

13. $y=-x^2-3x+10=-(x+5)(x-2)$: **(b)**, **(c)**

15. $M=\left(\frac{-2+1}{2},\frac{6+7}{2}\right)=\left(-\frac{1}{2},\frac{13}{2}\right)$

17. The point (6, 5) is a horizontal distance of 2 units from $x=4$, so (2, 5) is the symmetric point, the same distance from the axis of symmetry.

19. The point (–1, 7) is a horizontal distance of 4 units from $x=3$, so (7, –7) is the symmetric point, the same distance from the axis of symmetry.

21. $x=\frac{-1+5}{2}=\frac{4}{2}=2$

23. $x=\frac{-6+(-1)}{2}=\frac{-7}{2}$

< Objectives 1 and 2 >

25. $f(x)=-x^2+4=-(x+2)(x-2)$
$a=-1,\ b=0,\ c=4$
Axis of Symmetry: $x=-\frac{b}{2a}=-\frac{(0)}{2(-1)}=0$
Vertex: $y_v=-(0)^2+4=4$; (0, 4)
x-intercepts: (–2, 0) and (2, 0)

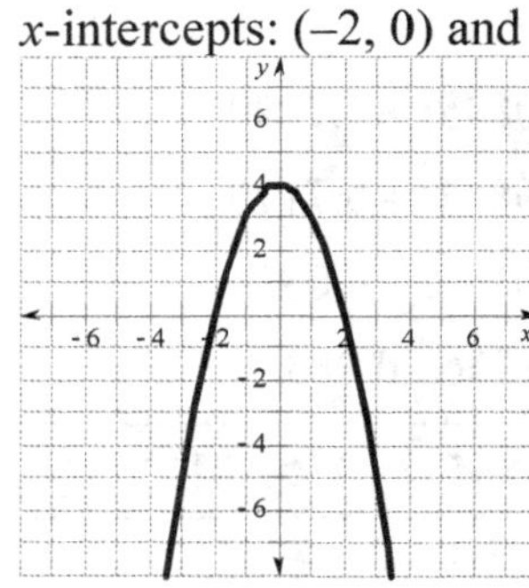

27. $f(x)=-x^2-2x=-x(x+2)$
$a=-1,\ b=-2,\ c=0$
Axis of Symmetry: $x=-\frac{b}{2a}=-\frac{(-2)}{2(-1)}=-1$
Vertex: (–1, 1)
x-intercepts: (–2, 0) and (0, 0)

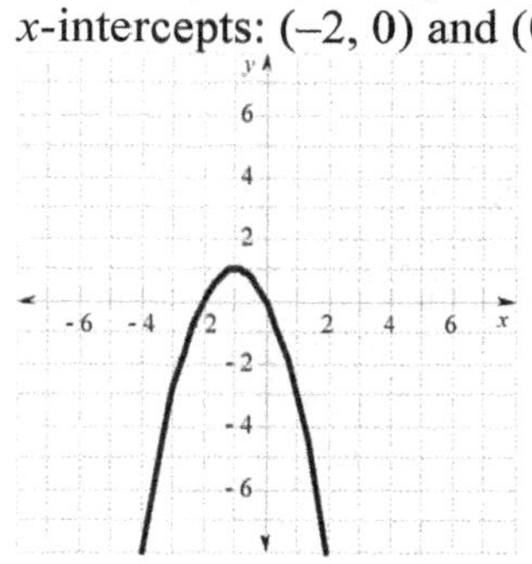

29. $f(x)=x^2-6x+5=(x-1)(x-5)$
$a=1,\ b=-6,\ c=5$
Axis of Symmetry: $x=3$
Vertex: (3, –4)
x-intercepts: (1, 0) and (5, 0)

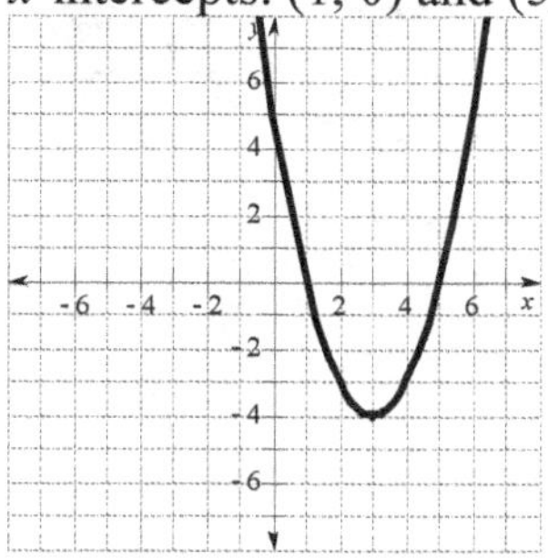

31. $f(x)=x^2-5x+6=(x-2)(x-3)$
$a=1,\ b=-5,\ c=6$
Axis of Symmetry: $x=\frac{5}{2}$
Vertex: $\left(\frac{5}{2},-\frac{1}{4}\right)$
x-intercepts: (2, 0) and (3, 0)

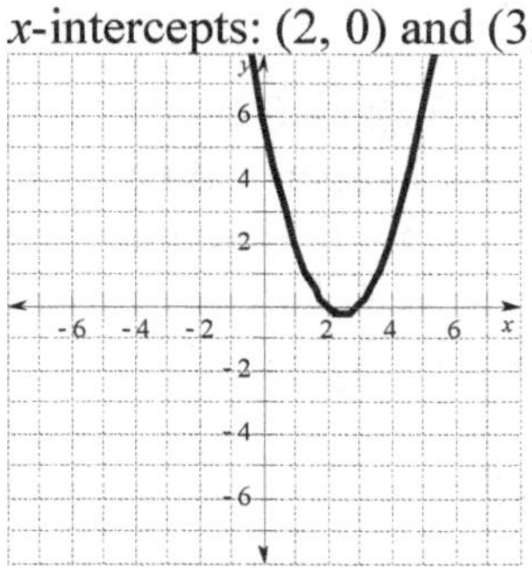

33. $f(x)=x^2-6x+8=(x-2)(x-4)$
$a=1,\ b=-6,\ c=8$
Axis of Symmetry: $x=3$
Vertex: (3, –1)
x-intercepts: (2, 0) and (4, 0)

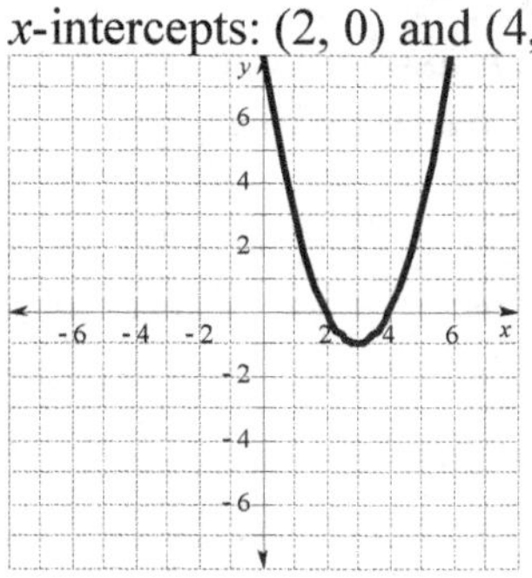

35. $f(x)=-x^2-6x-5=-(x+5)(x+1)$
$a=-1,\ b=-6,\ c=-5$
Axis of Symmetry: $x=-3$
Vertex: (–3, 4)
x-intercepts: (–5, 0) and (–1, 0)

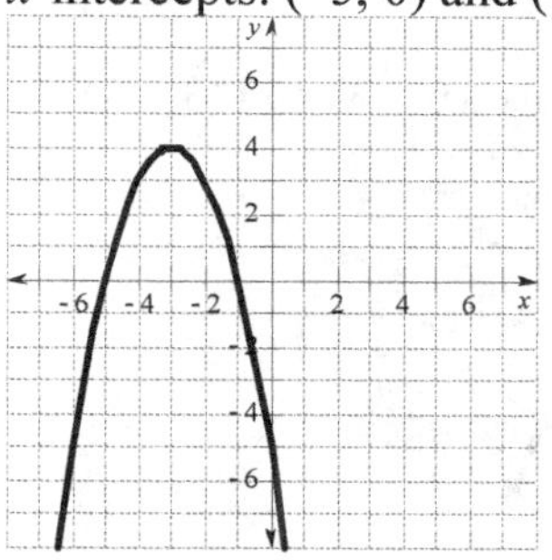

37. $f(x)=x^2-2x-1$

Axis of Symmetry: $x=-\frac{b}{2a}=-\frac{(-2)}{2(1)}=1$

Vertex: $y_v=(1)^2-2(1)-1=-2$; (1, –2)
Symmetric Points: (0, –1) and (2, –1)

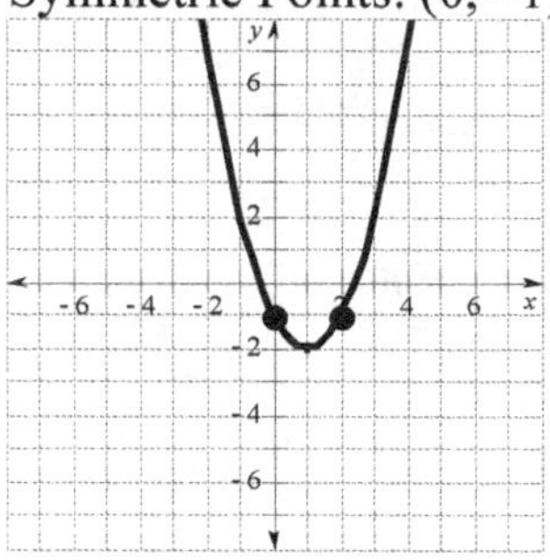

39. $f(x)=x^2-4x-1$

Axis of Symmetry: $x=-\frac{b}{2a}=-\frac{(-4)}{2(1)}=2$

Vertex: $y_v=(2)^2-4(2)-1=-5$; (2, –5)
Symmetric Points: (0, –1) and (4, –1)

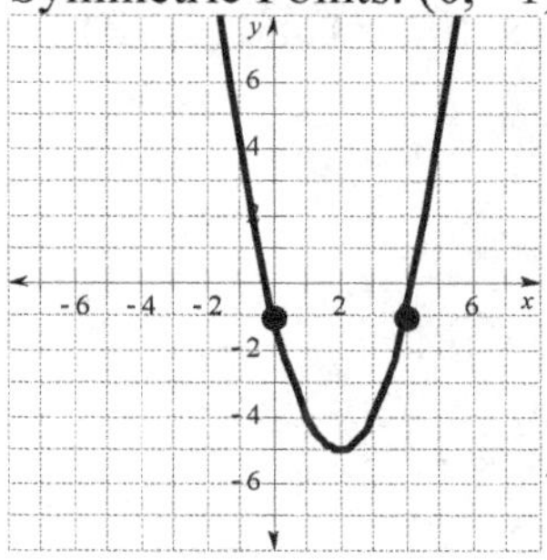

41. $f(x)=-x^2+3x-3$

Axis of Symmetry: $x=\frac{3}{2}$

Vertex: $\left(\frac{3}{2},-\frac{3}{4}\right)$

Symmetric Points: (0, –3) and (3, –3)

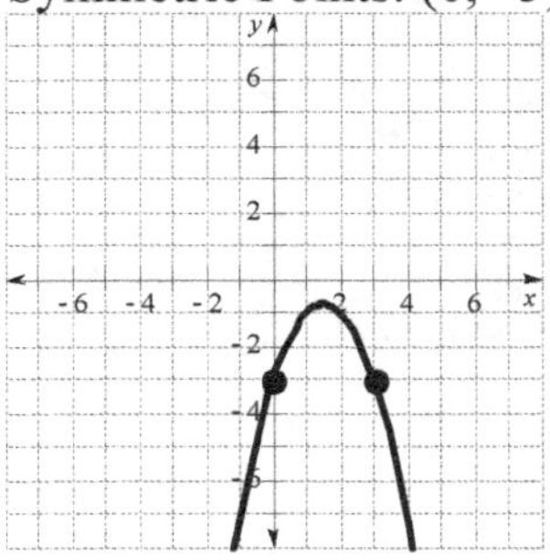

43. $f(x)=2x^2+4x-1$

Axis of Symmetry: $x=-1$

Vertex: (–1, –3)

Symmetric Points: (0, –1) and (–2, –1)

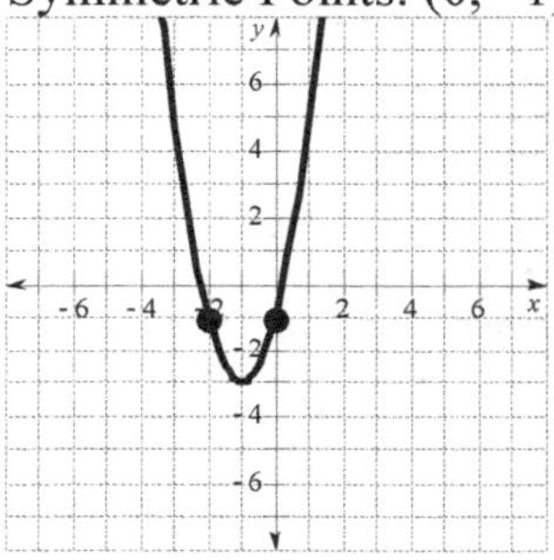

45. $f(x)=-\frac{1}{3}x^2+x-3$

Axis of Symmetry: $x=\frac{3}{2}$

Vertex: $\left(\frac{3}{2},-\frac{9}{4}\right)$

Symmetric Points: (0, –3) and (3, –3)

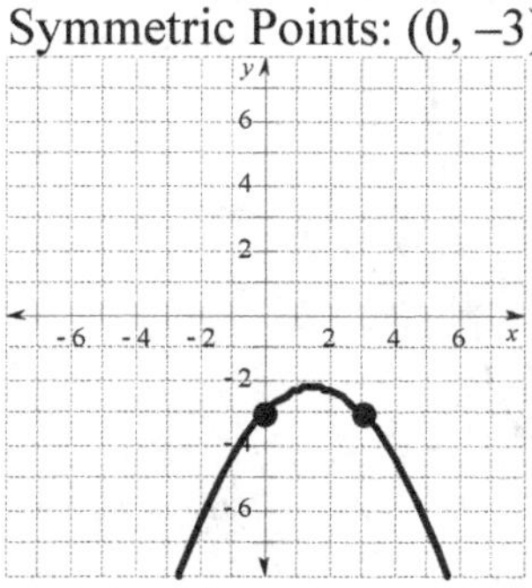

47. $f(x)=3x^2+12x+5$

Axis of Symmetry: $x=-2$

Vertex: (–2, –7)

Symmetric Points: (0, 5) and (–4, 5)

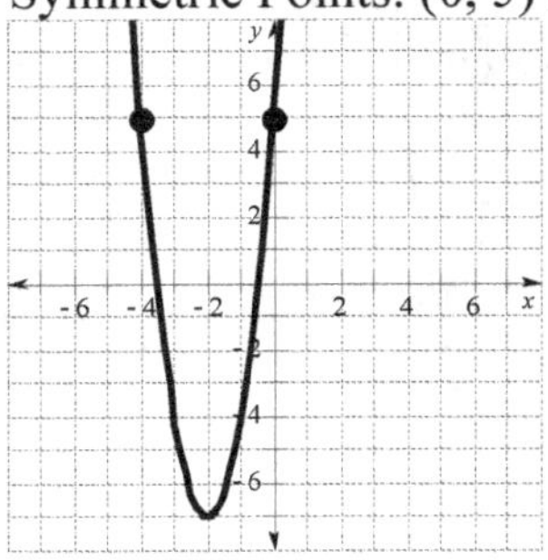

49. The vertex of a parabola is always located on the axis of symmetry.

51. The graph of $y=ax^2+bx+c$ sometimes intersects the *x*-axis.

53. The graph of $y=ax^2+bx+c$ always intersects the *y*-axis.

55. $C=-0.0015t^2+0.0845t+0.7170$

Note that the parabola opens down, hence to find the drug's maximum, find the vertex.

$$\frac{-b}{2a}=\frac{-0.0845}{2(-0.0015)}=\frac{-0.0845}{-0.003}=28.2$$

$$f\left(\frac{-b}{2a}\right)=f(28.2)$$

$$=-0.0015(28.2)^2+0.0845(28.2)+0.7170$$

$$=-1.19286+2.3829+0.7170=1.91$$

(a) The drug's concentration will be at maximum at 28.2 hr.

(b) The drug's maximum concentration in the bloodstream will be 1.91 ng/mL.

Exercises 8.4

< Objective 2 >

1. $P(x) = -0.3x^2 + 60x - 400$

$$x_v = -\frac{60}{2(-0.3)} = 100$$

$$y_v = -0.3(100)^2 + 60(100) - 400$$
$$= 2,600$$

A maximum profit of \$2,600 is achieved by selling 100 items.

3. $2L + 2W = 2,000$

$$L + W = 1,000$$
$$W = 1,000 - L$$
$$A(L) = L(1,000 - L) = 1,000L - L^2$$
$$L_v = \frac{-1,000}{-2} = 500$$
$$W_v = 1,000 - (500) = 500$$
$$A_v = 1,000(500) - 500(500)$$
$$= 250,000$$

A 500-ft square encloses the maximum area of 250,000 sq ft.

5. $h(t) = -16t^2 + 96t$

$$t_v = -\frac{96}{-32} = 3$$

$$h(2) = -16(3)^2 + 96(3) = 144$$

144 ft is the maximum height the ball will attain.

< Objective 3 >

7. $x^4 - 14x^2 + 45 = 0$

Let $u = x^2$. Then $u^2 = x^4$.

$$u^2 - 14u + 45 = 0$$
$$(u-9)(u-5) = 0$$
$$u = 9 \quad \text{or} \quad u = 5$$
$$x^2 = 9 \qquad x^2 = 5$$
$$x = \pm 3 \qquad x = \pm\sqrt{5}$$
$$\{\pm 3, \pm\sqrt{5}\}$$

9. $6x^4 - 7x^2 + 2 = 0$

Let $u = x^2$. Then $u^2 = x^4$.

$$6u^2 - 7u + 2 = 0$$
$$(2u-1)(3u-2) = 0$$
$$2u = 1 \quad \text{or} \quad 3u = 2$$
$$u = \frac{1}{2} \qquad u = \frac{2}{3}$$
$$x^2 = \frac{1}{2} \qquad x^2 = \frac{2}{3}$$
$$x = \pm\sqrt{\frac{1}{2}} \qquad x = \pm\sqrt{\frac{2}{3}}$$
$$= \pm\frac{1}{\sqrt{2}} \cdot \frac{\sqrt{2}}{\sqrt{2}} \qquad = \pm\frac{\sqrt{2}}{\sqrt{3}} \cdot \frac{\sqrt{3}}{\sqrt{3}}$$
$$= \pm\frac{\sqrt{2}}{2} \qquad = \pm\frac{\sqrt{6}}{3}$$
$$\left\{\pm\frac{\sqrt{2}}{2}, \pm\frac{\sqrt{6}}{3}\right\}$$

11. $x^4 + x^2 - 20 = 0$

Let $u = x^2$. Then $u^2 = x^4$.

$$u^2 + u - 20 = 0$$
$$(u-4)(u+5) = 0$$
$$u = 4 \quad \text{or} \quad u = -5$$
$$x^2 = 4 \qquad x^2 = -5$$
$$x = \pm 2 \qquad x = \pm i\sqrt{5}$$

$\{\pm 2, \pm i\sqrt{5}\}$ two nonreal solutions

13. $x^4+11x^2+28=0$

Let $u=x^2$. Then $u^2=x^4$.

$u^2+11u+28=0$

$(u+4)(u+7)=0$

$u=-4$ or $u=-7$

$x^2=-4$ $\quad x^2=-7$

$x=\pm 2i$ $\quad x=\pm i\sqrt{7}$

$\{\pm 2i, \pm i\sqrt{7}\}$ four nonreal solutions

15. $x-8\sqrt{x}+15=0$

Let $u=\sqrt{x}$. Then $u^2=x$.

$u^2-8u+15=0$

$(u-3)(u-5)=0$

$u=3$ or $u=5$

$\sqrt{x}=3$ $\quad \sqrt{x}=5$

$x=9$ $\quad x=25$

{9, 25}

17. $x-4\sqrt{x}-21=0$

Let $u=\sqrt{x}$. Then $u^2=x$.

$u^2-4u-21=0$

$(u-7)(u+3)=0$

$u=7$ or $u=-3$

$\sqrt{x}=7$ $\quad \sqrt{x}\neq -3$

$x=49$

{49}

19. $x^4-7x^2-4=0$

Let $u=x^2$. Then $u^2=x^4$.

$u^2-7u-4=0$

$$u=\frac{7\pm\sqrt{(-7)^2-4(1)(-4)}}{2(1)}$$

$$u=\frac{7\pm\sqrt{49+16}}{2}=\frac{7\pm\sqrt{65}}{2}$$

$u=\frac{7+\sqrt{65}}{2}$ or $u=\frac{7-\sqrt{65}}{2}$

$x^2=\frac{7+\sqrt{65}}{2}$ $\quad \frac{7-\sqrt{65}}{2}<0$

$$x=\pm\sqrt{\frac{7+\sqrt{65}}{2}}$$

$=\pm 2.744$

{±2.744}

21. $2x^4-7x^2+4=0$

Let $u=x^2$. Then $u^2=x^4$.

$2u^2-7u+4=0$

$$u=\frac{7\pm\sqrt{(-7)^2-4(2)(4)}}{2(2)}$$

$$u=\frac{7\pm\sqrt{49-32}}{4}=\frac{7\pm\sqrt{17}}{4}$$

$u=\frac{7+\sqrt{17}}{4}$ or $u=\frac{7-\sqrt{17}}{4}$

$x^2=\frac{7+\sqrt{17}}{4}$ $\quad x^2=\frac{7-\sqrt{17}}{4}$

$x=\pm\sqrt{\frac{7+\sqrt{17}}{2}}$ $\quad x=\pm\sqrt{\frac{7-\sqrt{17}}{2}}$

$=\pm 1.668$ $\quad =\pm 0.848$

{±1.668, ±0.848}

< Objective 1 >

23. $0 = x^2 + x - 12$

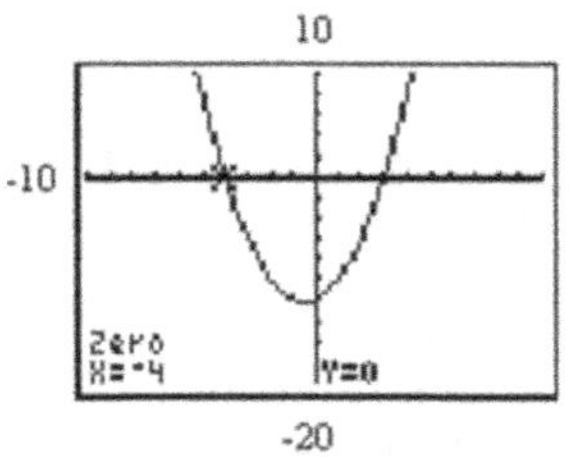

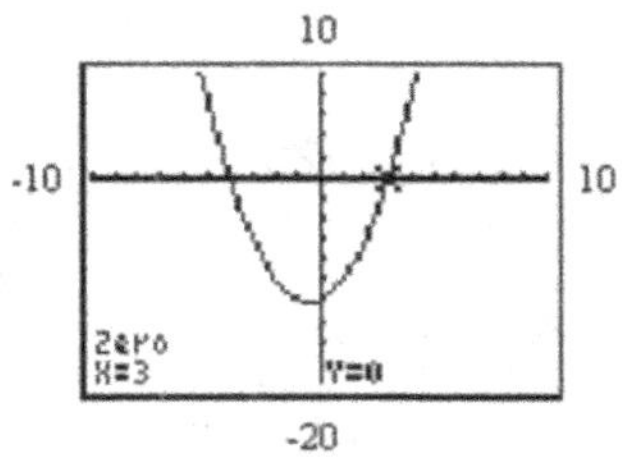

$\{-4, 3\}$

25. $0 = 6x^2 - 19x$

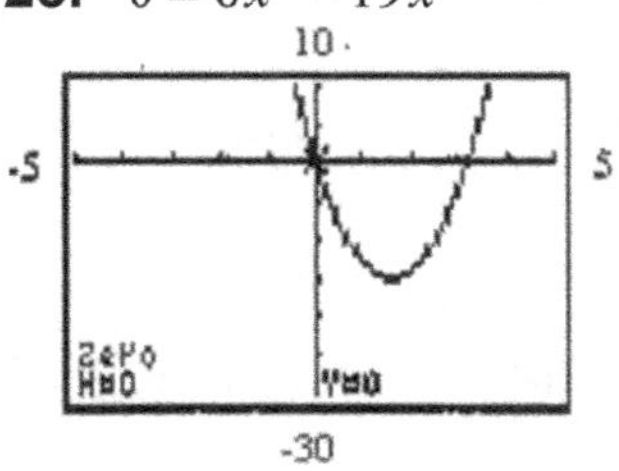

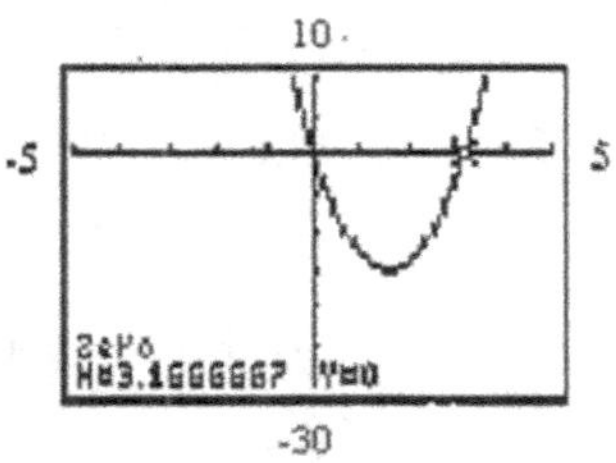

$\{0, 3.167\}$

27. $0 = 9x^2 + 12x - 7$

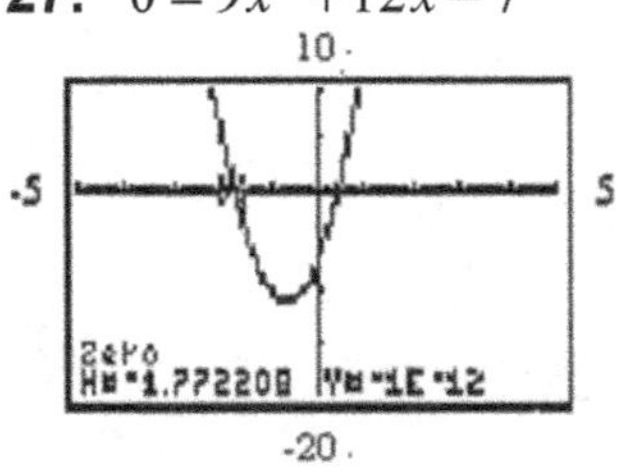

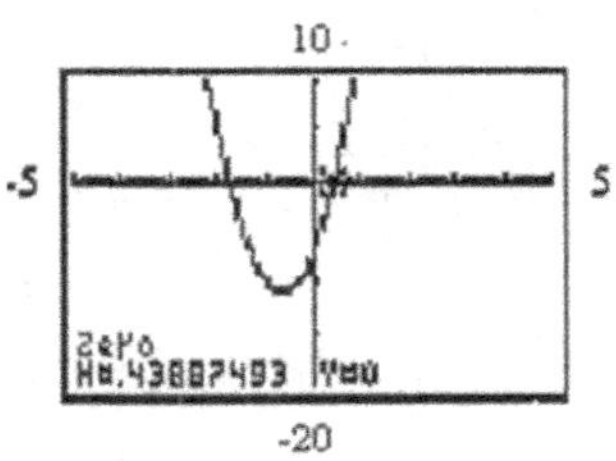

$\{-1.772, 0.439\}$

29. $0 = x^2 + 2x - 7$

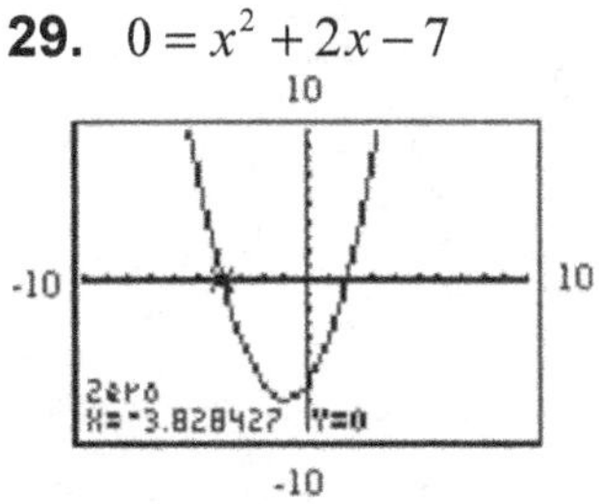

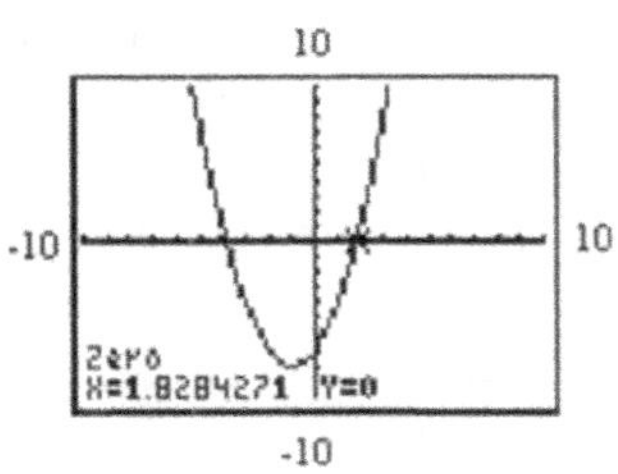

$\{-3.828, 1.828\}$

31. $f(x) = 2(x-3)^2 + 1$

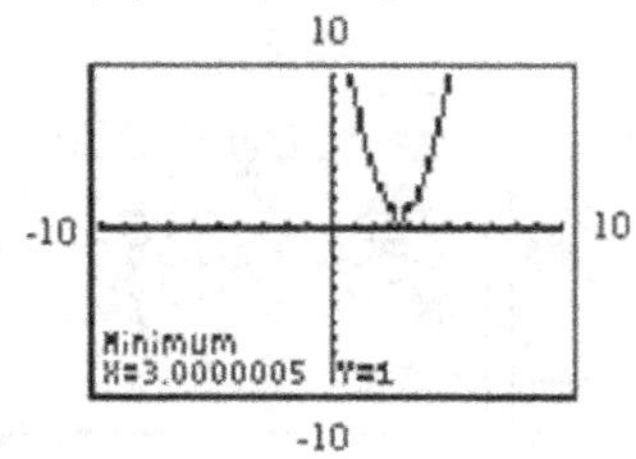

(a) $(3, 1)$

(b) $R = \{y \mid y \ge 1\}$

33. $f(x) = -(x-1)^2 + 2$

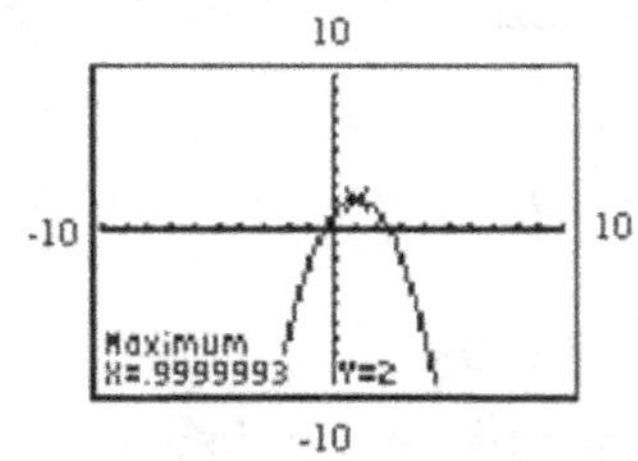

(a) $(1, 2)$

(b) $R = \{y \mid y \le 2\}$

35. $f(x) = 3(x+1)^2 - 2$

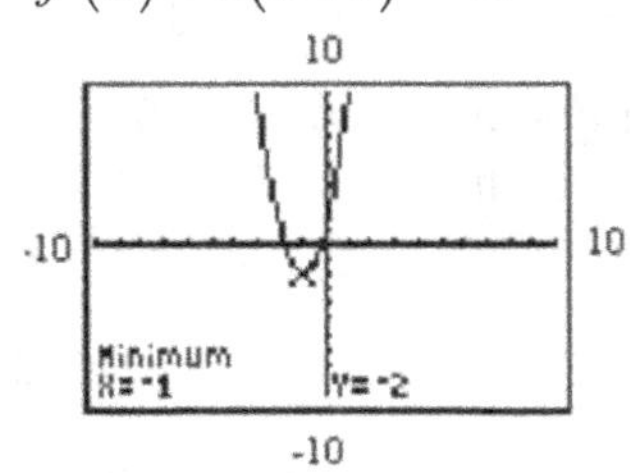

(a) $(-1, -2)$

(b) $\{y \mid y \ge -2\}$

37. $y = -0.008x^2 + 0.946x - 1.174$

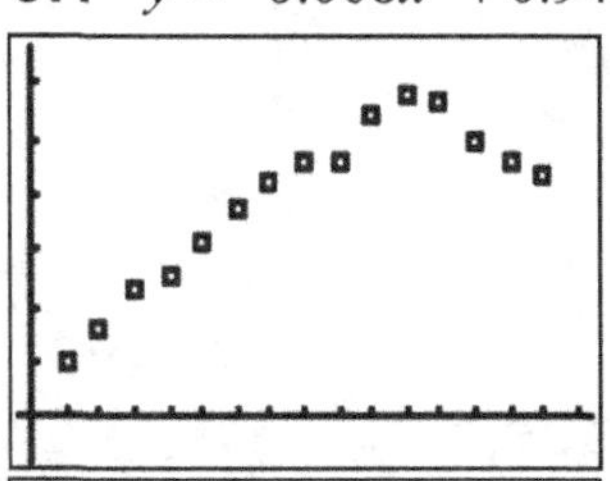
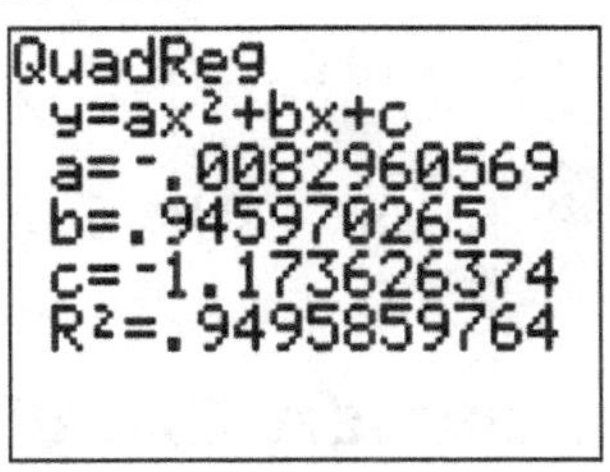

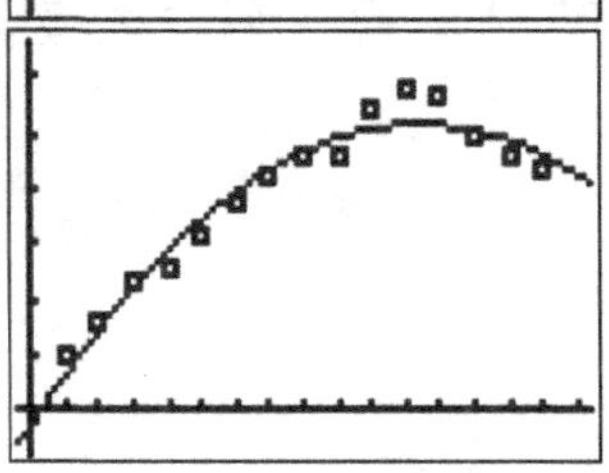

39. $y = -0.010x^2 + 0.899x + 5.424$

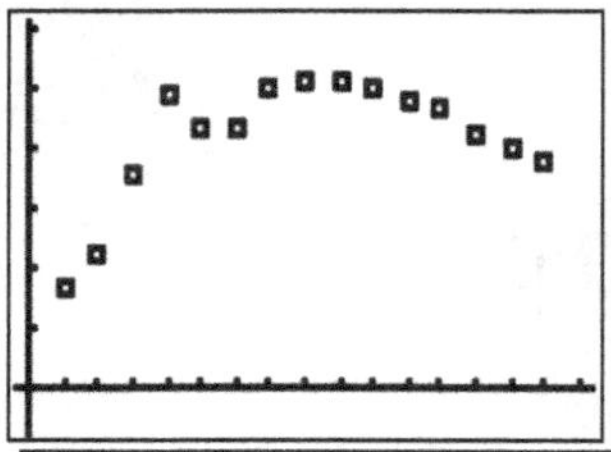

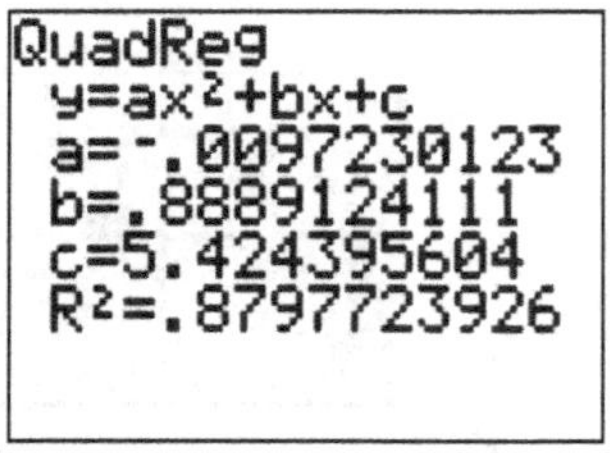

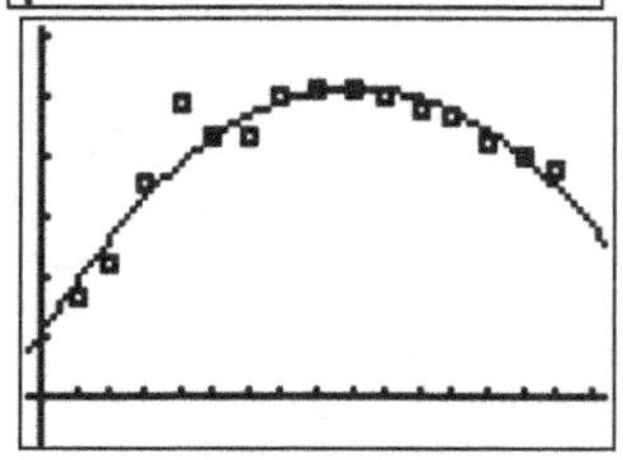

41. $P = -t^2 + 120t + 20$

$t_v = -\frac{120}{-2} = 60$

The maximum number of people will be sick 60 days after the outbreak.

43. $T = 0.4t^2 - 2.6t + 103$

$t_v = -\frac{-2.6}{0.8} = 3.25$

The patient's temperature will reach its minimum after 3.25 h.

45. $f(x) = 3x^2 - 25$

$-3 \le x \le 3$

$-25 \le y \le 0$

47. $f(x) = -2x^2 + 5x - 7$

$-2 \le x \le 4$

$-10 \le y \le 0$

49. Above and Beyond

Summary Exercises

1. $x^2 - 12 = 0$

$x^2 = 12$

$x = \pm\sqrt{12} = \pm 2\sqrt{3}$

$\{\pm 2\sqrt{3}\}$

3. $(x-2)^2 = 20$

$x - 2 = \pm\sqrt{20}$

$x = 2 \pm 2\sqrt{5}$

$\{2 \pm 2\sqrt{5}\}$

5. $x^2 - 14x$

$\left(\frac{(-14)}{2}\right)^2 = 49$

$x^2 - 14x + 49 = (x-7)^2$

7. $x^2 - 4x - 5 = 0$

$x^2 - 4x = 5$

$x^2 - 4x + 4 = 5 + 4$

$(x-2)^2 = 9$

$x - 2 = \pm 3$

$x = 2 \pm 3$

$\{-1, 5\}$

9. $w^2 - 10w - 3 = 0$

$w^2 - 10w = 3$

$w^2 - 10w + 25 = 3 + 25$

$(w-5)^2 = 28$

$w - 5 = \pm\sqrt{28}$

$w = 5 \pm 2\sqrt{7}$

$\{5 \pm 2\sqrt{7}\}$

11. $2x^2-6x+1=0$

$$2x^2-6x=-1$$
$$x^2-3x=-\frac{1}{2}$$
$$x^2-3x+\frac{9}{4}=-\frac{1}{2}+\frac{9}{4}$$
$$\left(x-\frac{3}{2}\right)^2=\frac{7}{4}$$
$$x-\frac{3}{2}=\pm\frac{\sqrt{7}}{2}$$
$$x=\frac{3}{2}\pm\frac{\sqrt{7}}{2}$$
$$\left\{\frac{3\pm\sqrt{7}}{2}\right\}$$

13. $x^2-5x-24=0$

$a=1,,\ b=-5,\ c=-24$

$$x=\frac{-(-5)\pm\sqrt{(-5)^2-4(1)(-24)}}{2(1)}=\frac{5\pm\sqrt{121}}{2}$$
$$=\frac{5\pm 11}{2}$$

$\{-3, 8\}$

15. $x^2=5x-2$

$$x^2-5x+2=0$$

$a=1,\ b=-5,\ c=2$

$$x=\frac{-(-5)\pm\sqrt{(-5)^2-4(1)(2)}}{2(1)}=\frac{5\pm\sqrt{17}}{2}$$
$$\left\{\frac{5\pm\sqrt{17}}{2}\right\}$$

17. $3y^2+4y=1$

$$3y^2+4y-1=0$$
$$y=\frac{-(4)\pm\sqrt{(4)^2-4(3)(-1)}}{2(3)}$$
$$=\frac{-4\pm\sqrt{28}}{6}=\frac{-4\pm 2\sqrt{7}}{6}$$
$$=\frac{-2\pm\sqrt{7}}{3}$$
$$\left\{\frac{-2\pm\sqrt{7}}{3}\right\}$$

19. $(x-5)(x+3)=13$

$$x^2-2x-28=0$$
$$x=\frac{2\pm\sqrt{4+112}}{2}=\frac{2\pm 2\sqrt{29}}{2}$$
$$=1\pm\sqrt{29}$$
$$\left\{1\pm\sqrt{29}\right\}$$

21. $3x^2+2x+5=0$

$$x=\frac{-2\pm\sqrt{4-60}}{6}=\frac{-2\pm\sqrt{-56}}{6}$$
$$=\frac{-2\pm 2i\sqrt{14}}{6}=\frac{-1\pm i\sqrt{14}}{3}$$

$\left\{\frac{-1\pm i\sqrt{14}}{3}\right\}$ nonreal

23. $x^2-3x+3=0$

$$b^2-4ac=(-3)^2-4(1)(3)=-3$$

There are no real solutions.

25. $4x^2-12x+9=0$

$$b^2-4ac=(-12)^2-4(4)(9)=0$$

There is 1 real solution.

27. Let x and $12-x$ be the integers.

$$x(12-x)=32$$
$$12x-x^2=32$$
$$x^2-12x+32=0$$
$$(x-4)(x-8)=0$$

The integers are 4 and 8.

29. Let x be a positive integer.

$$2x^2=8x+10$$
$$x^2=4x+5$$
$$x^2-4x-5=0$$
$$(x-5)(x+1)=0$$

The positive integer is 5.

31. Let x and $2x-3$ be the dimensions of the rectangle.

$$x(2x-3)=35$$
$$2x^2-3x-35=0$$
$$(x-5)(2x+7)=0$$

The rectangle is 5 cm by 7 cm.

33.

$$P(x)=-3x^2+240x$$
$$-3x^2+240x\ge 4{,}500$$
$$3x^2-240x+4{,}500\le 0$$
$$x^2-80x+1{,}500\le 0$$
$$(x-30)(x-50)\le 0$$

Selling between 30 and 50 patio chairs (inclusive) gives a profit of at least $4,500.

35. Let x and $2x+1$ be the dimensions of the rectangle.

$$2x(2x+1)=x(2x+1)+36$$
$$4x^2+2x=2x^2+x+36$$
$$2x^2+x-36=0$$
$$(x-4)(2x+9)=0$$

The rectangle is 4 cm by 9 cm.

37. The dimensions of a right triangle are shown in the figure.

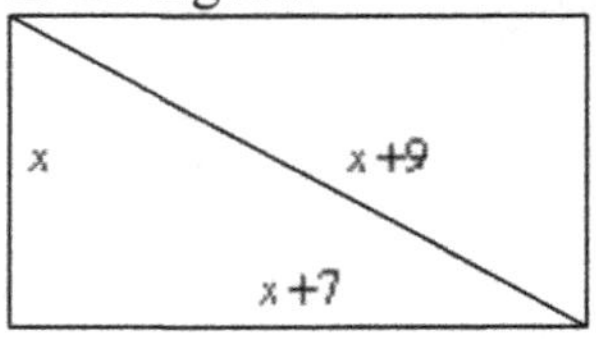

$$x^2+(x+7)^2=(x+9)^2$$
$$x^2+x^2+14x+49=x^2+18x+81$$
$$x^2-4x-32=0$$
$$(x-8)(x+4)=0$$

The rectangle is 8 ft by 15 ft.

39. Let x and $x+2$ be the length of the legs of the triangle.

$$x^2+(x+2)^2=8^2$$
$$x^2+x^2+4x+4=64$$
$$x^2+2x-30=0$$
$$x=\frac{-2\pm\sqrt{4+120}}{2}=\frac{-2\pm 2\sqrt{31}}{2}$$
$$=-1\pm\sqrt{31}$$
$$-1+\sqrt{31}\approx 4.6$$
$$\left(-1+\sqrt{31}\right)+2=1+\sqrt{31}\approx 6.6$$

The triangle's legs are 4.6 m and 6.6 m.

41.

$$f(x)=x^2-x-2$$
$$x^2-x-2=0$$
$$(x+1)(x-2)=0$$

The zeros are –1 and 2.

43.

$$f(x)=-2x^2-7x-6$$
$$-2x^2-7x-6=0$$
$$2x^2+7x+6=0$$
$$(x+2)(2x+3)=0$$

The zeros are –2 and $-\frac{3}{2}$.

45. $f(x)=x^2$

$a=1,\ b=0,\ c=0$

Axis of Symmetry: $x=-\dfrac{b}{2a}=-\dfrac{(0)}{2(1)}=0$

Vertex: $y_v=f(x_v)=f(0)=(0)^2=0$;
(0, 0)

47. $f(x)=x^2-5$

$a=1,\ b=0,\ c=-5$

Axis of Symmetry: $x=-\dfrac{b}{2a}=-\dfrac{(0)}{2(1)}=0$

Vertex: $y_v=f(x_v)=f(0)=(0)^2-5=-5$;
(0, –5)

49. $f(x)=(x+2)^2=x^2+4x+4$

$a=1,\ b=4,\ c=4$

Axis of Symmetry: $x=-2$

Vertex: (–2, 0)

51. $f(x)=(x+3)^2+1=x^2+6x+10$

$a=1,\ b=6,\ c=10$

Axis of Symmetry: $x=-3$

Vertex: (–3, 1)

53. $f(x)=-(x-5)^2-2=-x^2+10x-27$

$a=-1,\ b=10,\ c=-27$

Axis of Symmetry: $x=5$

Vertex: (5, –2)

55. $f(x)=-x^2+2x$

$a=-1,\ b=2,\ c=0$

Axis of Symmetry: $x=1$

Vertex: (1, 1)

57. $f(x)=-x^2-x+6$

$a=-1,\ b=-1,\ c=6$

Axis of Symmetry: $x=-\dfrac{1}{2}$

Vertex: $\left(-\dfrac{1}{2},\dfrac{25}{4}\right)$

59. $f(x)=-x^2-6x+4$

$a=-1,\ b=-6,\ c=4$

Axis of Symmetry: $x=-3$

Vertex: (–3, 13)

61. $0=2x^2+5x-9$

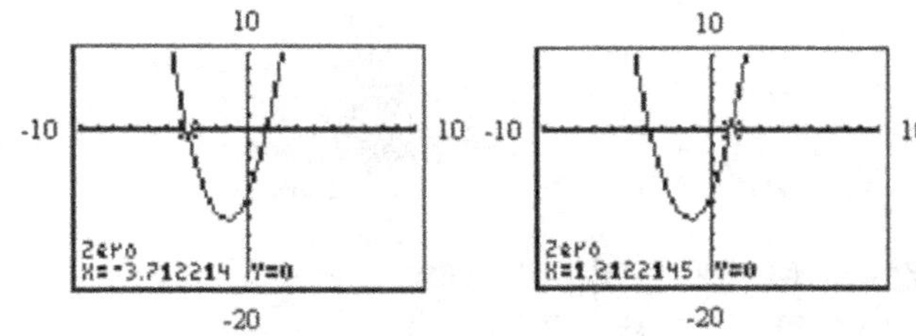

{–3.712, 1.212}

63. $0=x^2-7x+5$

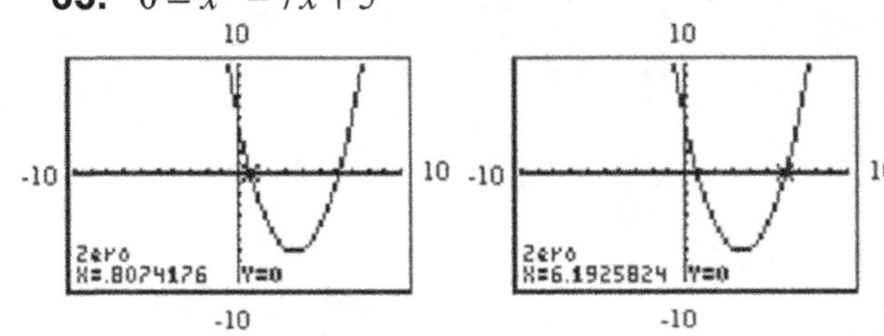

{0.807, 6.193}

65. $f(x)=x^2$

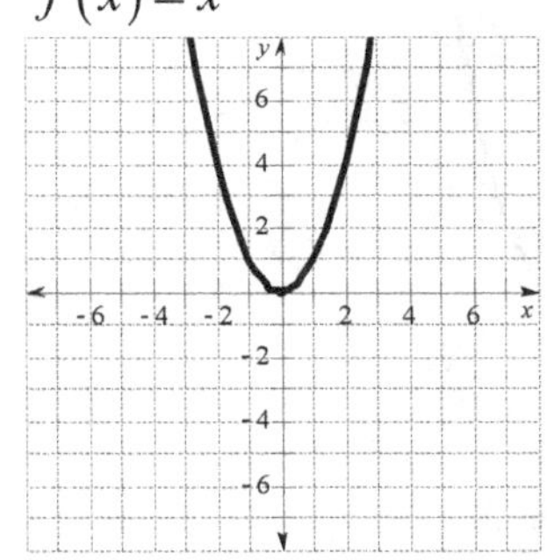

67. $f(x)=x^2-5$

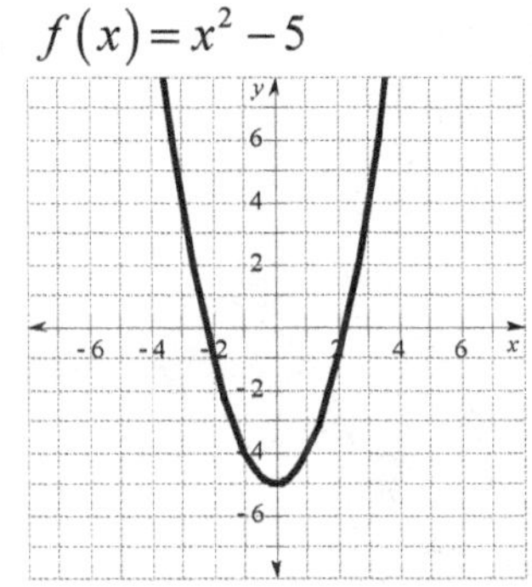

69. $f(x)=(x+2)^2$

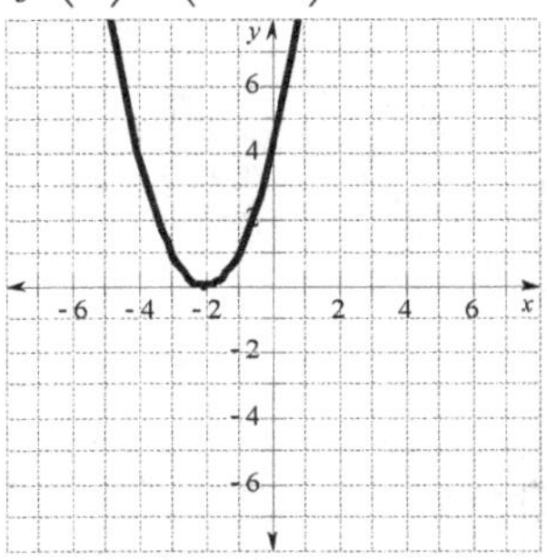

71. $f(x)=(x+3)^2+1$

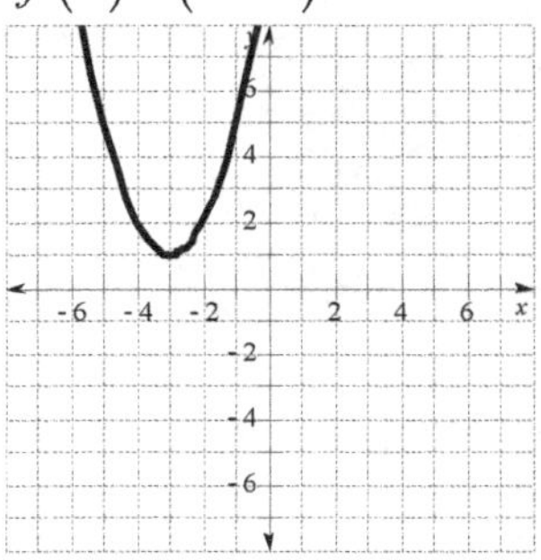

73. $f(x)=x^2-4x$

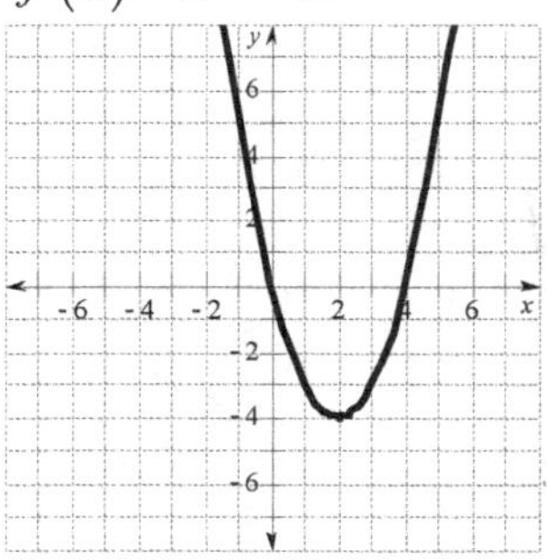

75. $f(x)=x^2+2x-3$

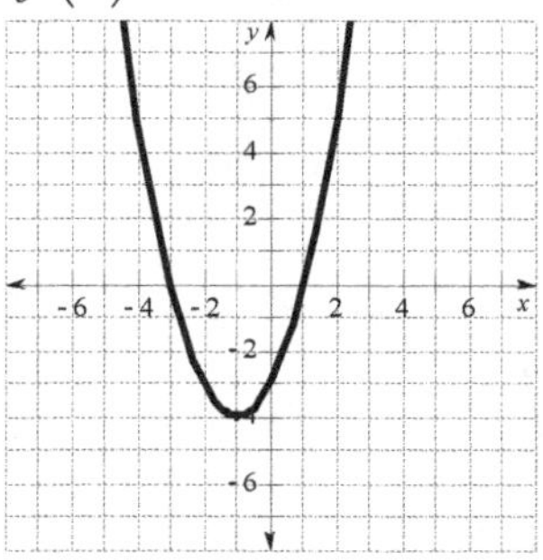

77. $f(x)=-x^2-x+6$

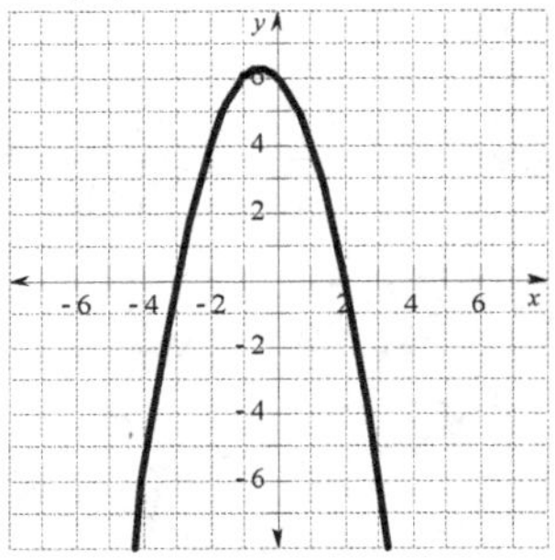

79. $f(x)=x^2+4x+5$

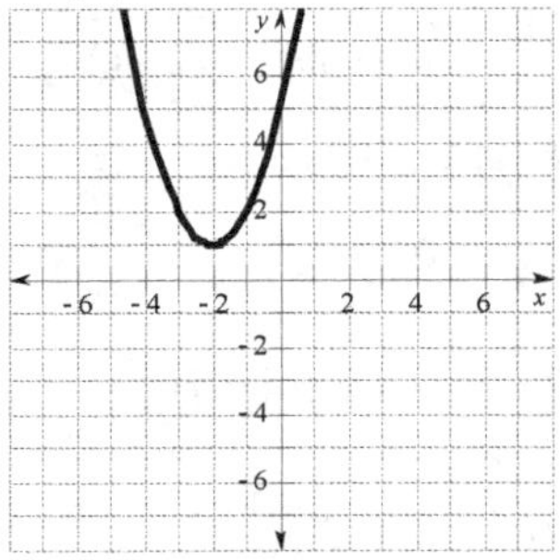

81. $f(x)=x^2-2x+4$

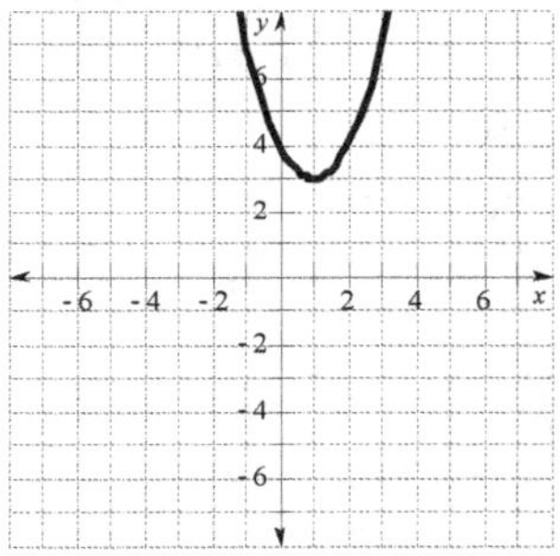

83. $f(x)=2x^2-4x+1$

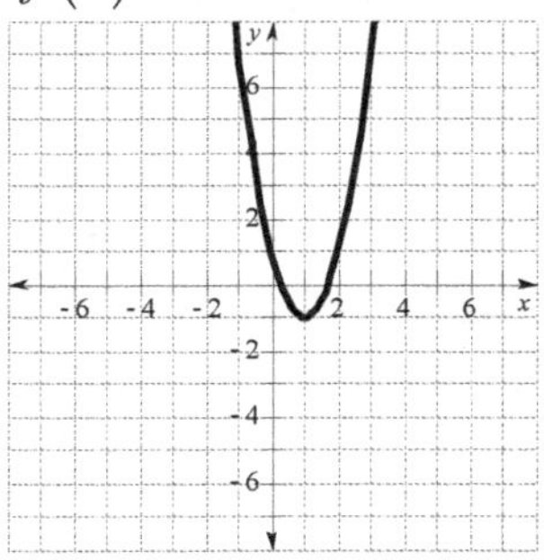

85. $x^4-13x^2+40=0$

Let $u=x^2$. Then $u^2=x^4$.

$u^2-13u+40=0$

$(u-8)(u-5)=0$

$u=8$ or $u=5$

$x^2=8$ $\quad x^2=5$

$x=\pm\sqrt{8}=\pm 2\sqrt{2}$ $\quad x=\pm\sqrt{5}$

$\{\pm 2\sqrt{2},\pm\sqrt{5}\}$

87. $x^4-5x^2-36=0$

Let $u=x^2$. Then $u^2=x^4$.

$u^2-5u-36=0$

$(u-9)(u+4)=0$

$u=9$ or $u=-4$

$x^2=9$ $\quad x^2=-4$

$x=\pm 3$ $\quad x=\pm 2i$

{±3, ±2i} two nonreal solutions

89. $x-7\sqrt{x}+12=0$

Let $u=\sqrt{x}$. Then $u^2=x$.

$u^2-7u+12=0$

$(u-3)(u-4)=0$

$u=3$ or $u=4$

$\sqrt{x}=3$ $\quad \sqrt{x}=4$

$x=9$ $\quad x=16$

{9, 16}

91. $x^4-6x^2-10=0$

Let $u=x^2$. Then $u^2=x^4$.

$u^2-6u-10=0$

$$u=\frac{6\pm\sqrt{(-6)^2-4(1)(-10)}}{2(1)}$$

$$u=\frac{6\pm\sqrt{36+40}}{2}=\frac{6\pm\sqrt{76}}{2}$$

$u=\dfrac{6+\sqrt{76}}{2}$ or $u=\dfrac{6-\sqrt{76}}{2}$

$x^2=\dfrac{6+\sqrt{76}}{2}$ $\quad \dfrac{6-\sqrt{76}}{2}<0$

$$x=\pm\sqrt{\frac{6+\sqrt{76}}{2}}$$

$=\pm 2.713$

{±2.713}

Chapter Test 8

1. $y=-3(x+2)^2+1=-3x^2-12x-11$

Axis of Symmetry: $-\dfrac{b}{2a}=-\dfrac{(-12)}{2(-3)}=-2$

Vertex: $y_v=f(-2)=-3\left[(-2)+2\right]^2+1=1$;

(–2, 1)

3. $y=-2x^2+6x-3$

Axis of Symmetry: $-\dfrac{b}{2a}=-\dfrac{(6)}{2(-2)}=\dfrac{3}{2}$

Vertex: $y_v=f\left(\dfrac{3}{2}\right)=-2\left(\dfrac{3}{2}\right)^2+6\left(\dfrac{3}{2}\right)-3=\dfrac{3}{2}$

; $\left(\dfrac{3}{2},\dfrac{3}{2}\right)$

5. $y=x^2-6x+2$

Axis of Symmetry: $-\dfrac{b}{2a}=-\dfrac{(-6)}{2(1)}=3$

Vertex: $y_v=f(3)=(3)^2-6(3)+2=-7$;

(3, –7)

7. $2x^2-10x+3=0$

$$x^2-5x=-\frac{3}{2}$$

$$x^2-5x+\frac{25}{4}=-\frac{3}{2}+\frac{25}{4}$$

$$\left(x-\frac{5}{2}\right)^2=\frac{19}{4}$$

$$x-\frac{5}{2}=\pm\frac{\sqrt{19}}{2}$$

$$x=\frac{5}{2}\pm\frac{\sqrt{19}}{2}$$

$$\left\{\frac{5\pm\sqrt{19}}{2}\right\}$$

9. $0=x^2+3x-7$

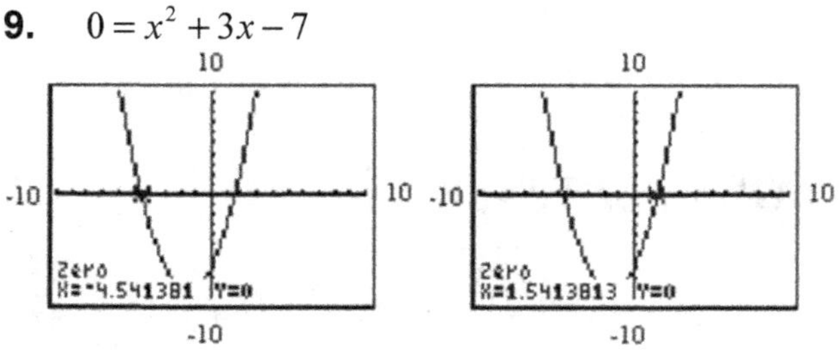

{–4.541, 1.541}

11. $f(x)=(x-5)^2$

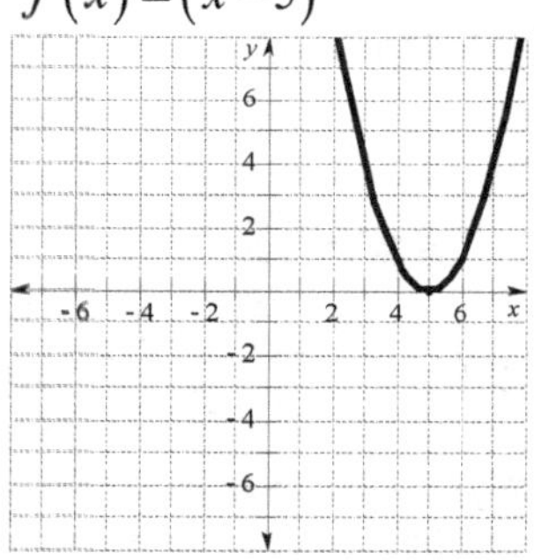

13. $f(x)=-2(x-3)^2-1$

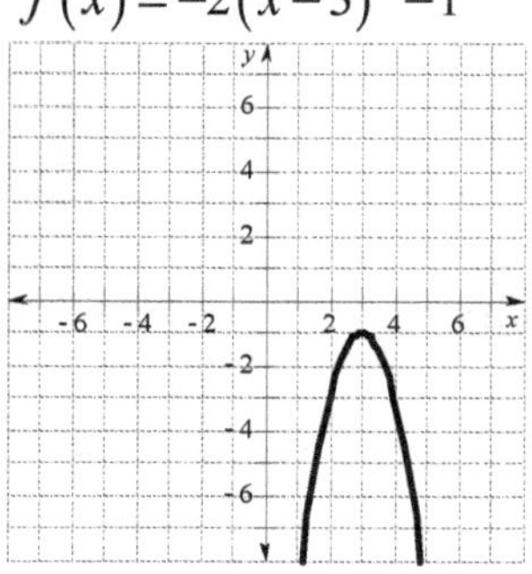

15. $2x^2+7x+3=0$

$$(x+3)(2x+1)=0$$

$$\left\{-3,-\frac{1}{2}\right\}$$

17. $4x^3-9x=0$

$$x(4x^2-9)=0$$

$$x(2x+3)(2x-3)=0$$

$$\left\{-\frac{3}{2},0,\frac{3}{2}\right\}$$

19. $h(t)=-16t^2+32t+32$

$$-16t^2+32t+32=0$$

$$t^2-2t-2=0$$

$$t=\frac{2\pm\sqrt{4+8}}{2}=1\pm\sqrt{3}$$

$$1+\sqrt{3}\approx 2.7$$

The ball lands after 2.7 s.

21. $x^2+4x=7$

$$x^2+4x-7=0$$

$a=1$, $b=4$, $c=-7$

$$x=\frac{-(4)\pm\sqrt{(4)^2-4(1)(-7)}}{2(1)}=\frac{-4\pm\sqrt{44}}{2}$$

$$=\frac{-4\pm2\sqrt{11}}{2}=-2\pm\sqrt{11}$$

$$\left\{-2\pm\sqrt{11}\right\}$$

23. $2x^2+2x+5=0$

$a=2$, $b=2$, $c=5$

$$x=\frac{-(2)\pm\sqrt{(2)^2-4(2)(5)}}{2(2)}$$

$$=\frac{-2\pm\sqrt{4-40}}{4}=\frac{-2\pm\sqrt{-36}}{4}=\frac{-2\pm6i}{4}$$

$$=\frac{-1\pm3i}{2}$$

$\left\{\frac{-1\pm3i}{2}\right\}$ nonreal

25. $x^4 - 4x^2 - 32 = 0$

Let $u = x^2$. Then $u^2 = x^4$.

$u^2 - 4u - 32 = 0$

$(u-8)(u+4) = 0$

$u = 8$ or $u = -4$

$x^2 = 8$ $\quad x^2 = -4$

$x = \pm\sqrt{8} = \pm 2\sqrt{2}$ $\quad x = \pm 2i$

$\{\pm 2\sqrt{2}, \pm 2i\}$ two nonreal solutions

27. $4w^2 - 20 = 0$

$4w^2 = 20$

$w^2 = 5$

$w = \pm\sqrt{5}$

$\{\pm\sqrt{5}\}$

29. $4(x-1)^2 = 23$

$(x-1)^2 = \frac{23}{4}$

$x - 1 = \pm\frac{\sqrt{23}}{2}$

$x = 1 \pm \frac{\sqrt{23}}{2}$

$\left\{\frac{2 \pm \sqrt{23}}{2}\right\}$

Cumulative Review: Chapters 0–8

1. $2x - 3y = 6$

$y = \frac{2}{3}x - 2$

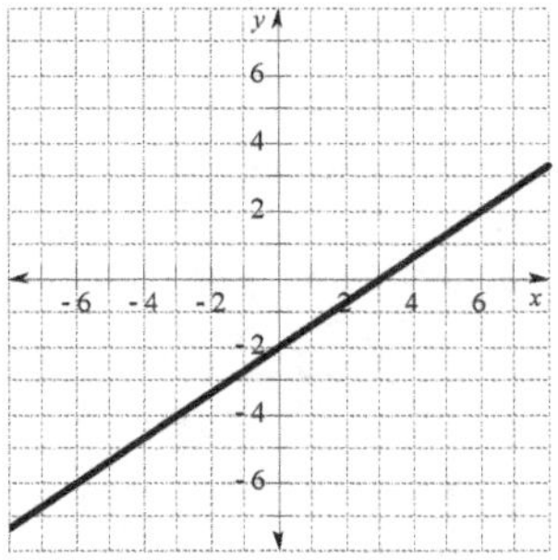

3. $y = 4$

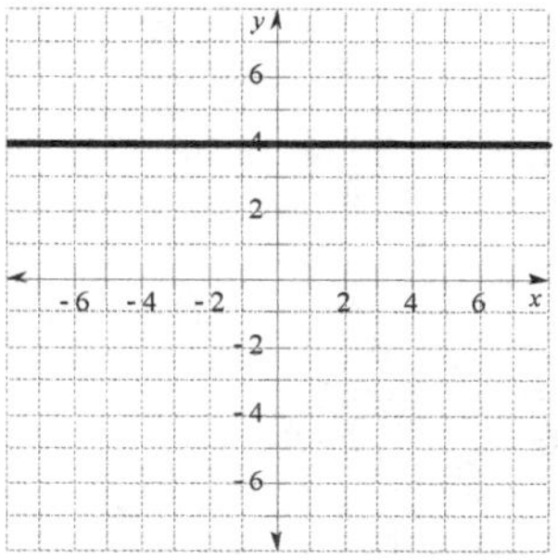

5. $m = \frac{-1-3}{-5-(-2)} = \frac{4}{3}$

7. $f(x) = (x^2 - 1)(x + 3) = x^3 + 3x^2 - x - 3$

9. $\sqrt{\frac{2}{3}} + 7\sqrt{6} = \frac{\sqrt{2}}{\sqrt{3}} + 7\sqrt{6} = \frac{\sqrt{2}}{\sqrt{3}} \cdot \frac{\sqrt{3}}{\sqrt{3}} + 7\sqrt{6}$

$= \frac{\sqrt{6}}{3} + 7\sqrt{6} \cdot \frac{3}{3} = \frac{\sqrt{6}}{3} + \frac{21\sqrt{6}}{3}$

$= \frac{22\sqrt{6}}{3}$

11. $2x - 7 = 0$

$2x = 7$

$x = \frac{7}{2}; \left\{\frac{7}{2}\right\}$

13. $0 = (x-3)(x+5)$

$\{-5, 3\}$

15. $x^2 + 7x - 30 = 0$

$(x+10)(x-3) = 0$

$\{-10, 3\}$

17. $(x-3)^2 = 5$

$x - 3 = \pm\sqrt{5}$

$x = 3 \pm \sqrt{5}$

$\{3 \pm \sqrt{5}\}$

19.

$$\frac{x}{3}-\frac{4}{9}=\frac{5}{18}$$
$$18\bullet\left(\frac{x}{3}-\frac{4}{9}\right)=18\bullet\frac{5}{18}$$
$$6x-8=5$$
$$x=\frac{13}{6}$$
$$\left\{\frac{13}{6}\right\}$$

21. $x-2\le 7$
$$x\le 9$$
$$\{x|x\le 9\}$$

23. Let x be the number.
$$5x-7=-72$$
$$5x=-65$$
$$x=-13$$
The number is –13.

25.

$$P=-4x^2+320x$$
$$-4x^2+320x\ge 4{,}956$$
$$x^2-80x+1{,}239\le 0$$
$$(x-21)(x-59)\le 0$$
Selling 21 or 59 receivers guarantees at least \$4,956 in profit.

Chapter 9
Rational Expressions

Exercises 9.1

< Objective 1 >

1. Substitute 5 for x.

$$\frac{3(5)}{2(5)-1}=\frac{15}{9}=\frac{5}{3}$$

3. Substitute -4 for x.

$$\frac{3(-4)+10}{(-4)+2}=\frac{-2}{-2}=1$$

5. Substitute -2 for x.

$$\frac{(-2)^2+(-2)}{(-2)^2+2(-2)}=\frac{2}{0}$$

The expression is undefined for $x=-2$.

7. Substitute -1 for x.

$$\frac{2-3(-1)}{(-1)^2-4}=\frac{5}{-3}=-\frac{5}{3}$$

< Objective 2 >

9. $\frac{x}{x-3}$ is undefined if $x=3$.

11. $\frac{x+5}{3}$ is never undefined.

13. $\frac{2x-3}{2x-1}$ is undefined if $x=\frac{1}{2}$.

15. $\frac{2x+5}{x}$ is undefined if $x=0$.

17. $\frac{x(x+1)}{x+2}$ is undefined if $x=-2$.

19. $\frac{5-3x}{2x}$ is undefined if $x=0$.

< Objective 3 >

21. $\frac{14}{21}=\frac{2}{3}$

23. $\frac{4x^5}{6x^2}=\frac{2x^3}{3}$

25. $\frac{10x^2y^5}{25xy^2}=\frac{2xy^3}{5}$

27. $\frac{-36x^5y^3}{21x^2y^5}=-\frac{12x^3}{7y^2}$

29. $\frac{28a^5b^3c^2}{84a^2bc^4}=\frac{a^3b^2}{3c^2}$

31. $\frac{6x-24}{x^2-16}=\frac{6\cancel{(x-4)}}{(x+4)\cancel{(x-4)}}=\frac{6}{x+4}$

33. $\frac{x^2+2x+1}{6x+6}=\frac{(x+1)^2}{6(x+1)}=\frac{x+1}{6}$

35. $\frac{x^2-13x+36}{x^2-81}=\frac{(x-4)(x-9)}{(x+9)(x-9)}=\frac{x-4}{x+9}$

37. $\frac{3b^2-7b-6}{b-3}=\frac{(3b+2)(b-3)}{b-3}=3b+2$

39. $$\frac{2y^2+3yz-5z^2}{2y^2+11yz+15z^2}=\frac{(y-z)(2y+5z)}{(y+3z)(2y+5z)}=\frac{y-z}{y+3z}$$

41. $$\frac{x^3-64}{x^2-16}=\frac{(x-4)(x^2+4x+16)}{(x+4)(x-4)}=\frac{x^2+4x+16}{x+4}$$

43. $$\frac{a^4-81}{a^2+5a+6}=\frac{(a^2+9)\cancel{(a+3)}(a-3)}{(a+2)\cancel{(a+3)}}=\frac{(a^2+9)(a-3)}{a+2}$$

45. $$\frac{xy-2x+3y-6}{x^2+8x+15}=\frac{(x+3)(y-2)}{(x+3)(x+5)}=\frac{y-2}{x+5}$$

47. $$\frac{x^2+3x-18}{x^3-3x^2-2x+6}=\frac{(x+6)(x-3)}{(x^2-2)(x-3)}=\frac{x+6}{x^2-2}$$

49. $$\frac{2m-10}{25-m^2}=\frac{2(m-5)}{(5+m)(5-m)}=-\frac{2(m-5)}{(5+m)(m-5)}=-\frac{2}{m+5}$$

51. $$\frac{121-x^2}{2x^2-21x-11}=\frac{\overset{-1}{\cancel{(11-x)}}(11+x)}{(2x+1)\cancel{(x-11)}}=-\frac{x+11}{2x+1}$$

53. $$\frac{2(x+h)-2x}{(x+h)-x}=\frac{2x+2h-2x}{h}=\frac{2h}{h}=2$$

55. $$\frac{3(x+h)-3-(3x-3)}{(x+h)-x}=\frac{3x+3h-3-3x+3}{h}=\frac{3h}{h}=3$$

57. $$\frac{(x+h)^2-x^2}{(x+h)-x}=\frac{x^2+2xh+h^2-x^2}{h}=\frac{2xh+h^2}{h}=\frac{h(2x+h)}{h}=2x+h$$

59. $$\frac{x^2-16y^2}{3x^2-12xy}\div(x^2+4xy)=\frac{x^2-16y^2}{3x^2-12xy}\bullet\frac{1}{x^2+4xy}=\frac{(x+4y)(x-4y)}{3x(x-4y)\bullet x(x+4y)}=\frac{1}{3x^2}$$

< Objective 4 >

61. $f(x)=-7x^2+2x-5$ is a rational function.

63. $f(x)=\dfrac{x^2-x-1}{x+2}$ is a rational function.

65. $f(x)=5x^2-\sqrt[3]{x}$ is not a rational function.

67. $$f(x)=\frac{x^2-x-2}{x+1}=\frac{(x-2)\cancel{(x+1)}}{\cancel{x+1}}=x-2;\ x\neq-1$$

69. $$f(x)=\frac{3x^2+5x-2}{x+2}=\frac{(3x-1)\cancel{(x+2)}}{\cancel{x+2}}=3x-1;x\neq-2$$

71. $$f(x)=\frac{x^2+4x+4}{5(x+2)}=\frac{(x+2)\cancel{(x+2)}}{5\cancel{(x+2)}}=\frac{x+2}{5};x\neq-2$$

73. $$f(x)=\frac{x^2-2x-8}{x^2-x-6}=\frac{\cancel{(x+2)}(x-4)}{\cancel{(x+2)}(x-3)}=\frac{x-4}{x-3};x\neq-2,3$$

75. $f(x)=\frac{x^2+4x+3}{x^2+7x+6}=\frac{(x+3)\cancel{(x+1)}}{(x+6)\cancel{(x+1)}}$

$=\frac{x+3}{x+6}; x \neq -6,-1$

77. $f(x)=\frac{x^2-4x+3}{x^2-1}=\frac{\cancel{(x-1)}(x-3)}{(x+1)\cancel{(x-1)}}$

$=\frac{x-3}{x+1}; x \neq -1,1$

< Objective 5 >

79. **(a)** The total cost is given by $8.75x+3,500$; therefore, the per unit cost is given by

$\bar{C}(x)=\frac{8.75x+3500}{x}$.

(b) $\frac{8.75(50)+3500}{50}=78.75$

When producing 50 units, the average cost per unit is \$78.75.

$R(x)=\frac{300x^2}{x^2+9}$

81. $\frac{x-7}{2x+6} \div \frac{21-3x}{x^2+3x}=\frac{x-7}{2x+6} \bullet \frac{x^2+3x}{21-3x}$

$=\frac{(x-7) \bullet x(x+3)}{2(x+3) \bullet (-3)(x-7)}$

$=\frac{-x}{6}$

83. True.

85. A rational expression is always the ratio of two polynomials.

87. $f(x)=\frac{1}{x-3}$

(a) The function is undefined if $x=3$.

(b)

x	*f*(*x*)
4	1
3.1	10
3.01	100
3.001	1,000
3.0001	10,000

(c) Answers will vary.

(d)

x	*f*(*x*)
2	−1
2.9	−10
2.99	−100
2.999	−1,000
2.9999	−10,000

(e) Answers will vary.

(f) Answers will vary; graph provided.

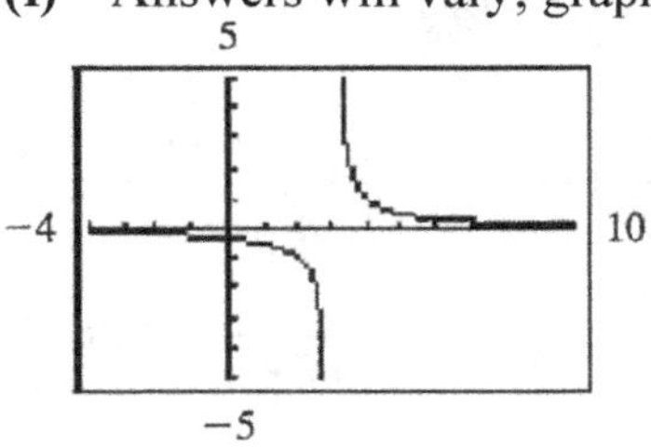

89. $p=\frac{6whs+6wh}{3s^2+6s+3}=\frac{6wh(s+1)}{3(s^2+2s+1)}$

$=\frac{6wh(s+1)}{3(s+1)(s+1)}=\frac{2wh}{(s+1)}$

91. Let *x* be the younger child's age. Then $x+5$ is the older child's age. The ratio of the younger child's age to the older child's age is $\frac{x}{x+5}$.

93. Above and Beyond

95. Above and Beyond

97. Above and Beyond

99. Above and Beyond

101. Above and Beyond

Exercises 9.2

< Objectives 1 and 2 >

1. $\frac{x^2}{3}\bullet\frac{6x}{x^4}=\frac{6x^3}{3x^4}=\frac{2}{x}$

3. $\frac{x}{8x^4}\div\frac{x^5}{24}=\frac{x}{8x^4}\bullet\frac{24}{x^5}=\frac{24x}{8x^9}=\frac{3}{x^8}$

5. $\frac{4xy^2}{15x^3}\bullet\frac{25xy}{16y^3}=\frac{100x^2y^3}{240x^3y^3}=\frac{5}{12x}$

7. $\frac{8b^3}{15ab}\div\frac{2ab^2}{20ab^3}=\frac{8b^3}{15ab}\bullet\frac{20ab^3}{2ab^2}=\frac{160ab^6}{30a^2b^3}$
$=\frac{16b^3}{3a}$

9. $\frac{m^3n}{2mn}\bullet\frac{6mn^2}{m^3n}\div\frac{3mn}{5m^2n}=\frac{m^3n}{2mn}\bullet\frac{6mn^2}{m^3n}\bullet\frac{5m^2n}{3mn}$
$=\frac{30m^6n^4}{6m^5n^3}=5mn$

11. $\frac{6x+18}{4x}\bullet\frac{16x^3}{3x+9}=\frac{6\cancel{(x+3)}\bullet16x^3}{4x\bullet3\cancel{(x+3)}}=8x^2$

13. $\frac{3b-15}{6b}\div\frac{4b-20}{9b^2}=\frac{3\cancel{(b-5)}}{6b}\bullet\frac{9b^2}{4\cancel{(b-5)}}$
$=\frac{9b}{8}$

15. $\frac{x^2-3x-10}{5x}\bullet\frac{15x^2}{3x-15}$
$=\frac{\cancel{(x-5)}(x+2)\bullet15x^2}{5x\bullet3\cancel{(x-5)}}=\frac{15x^2(x+2)}{15x}$
$=x(x+2)=x^2+2x$

17. $\frac{c^2+2c-8}{6c}\div\frac{5c+20}{18c}$
$=\frac{\cancel{(c+4)}(c-2)}{\cancel{6c}}\bullet\frac{\overset{3}{\cancel{18c}}}{5\cancel{(c+4)}}=\frac{3(c-2)}{5}$
$=\frac{3c-6}{5}$

19. $\frac{x^2-x-12}{3x-12}\bullet\frac{15x^3}{x^2-9}$
$=\frac{\cancel{(x+3)}\cancel{(x-4)}\bullet\overset{5}{\cancel{15}}x^3}{\cancel{3}\cancel{(x-4)}\bullet\cancel{(x+3)}(x-3)}=\frac{5x^3}{x-3}$

21. $\frac{d^2-3d-18}{16d-96}\div\frac{d^2-9}{20d}$
$=\frac{\cancel{(d-6)}\cancel{(d+3)}}{\underset{4}{\cancel{16}}\cancel{(d-6)}}\bullet\frac{\overset{5}{\cancel{20}}d}{\cancel{(d+3)}(d-3)}$
$=\frac{5d}{4(d-3)}=\frac{5d}{4d-12}$

23. $\frac{2x^2-x-3}{3x^2+7x+4}\bullet\frac{3x^2-11x-20}{4x^2-9}$
$=\frac{\cancel{(2x-3)}\cancel{(x+1)}\bullet\cancel{(3x+4)}(x-5)}{\cancel{(3x+4)}\cancel{(x+1)}\bullet(2x+3)\cancel{(2x-3)}}$
$=\frac{x-5}{2x+3}$

25. $\frac{a^2-9}{2a^2-6a}\div\frac{2a^2+5a-3}{4a^2-1}$
$=\frac{(a+3)(a-3)}{2a(a-3)}\bullet\frac{(2a+1)(2a-1)}{(2a-1)(a+3)}=\frac{2a+1}{2a}$

27. $\frac{2w-6}{w^2+2w}\bullet\frac{3w}{3-w}=\frac{2(w-3)\bullet3w}{w(w+2)\bullet(3-w)}$
$=-\frac{6}{w+2}$

29. $\dfrac{a-7}{2a+6} \div \dfrac{21-3a}{a^2+3a} = \dfrac{a-7}{2(a+3)} \bullet \dfrac{a(a+3)}{3(7-a)}$

$= -\dfrac{a}{6}$

31. $\dfrac{x^2-9y^2}{2x^2-xy-15y^2} \cdot \dfrac{4x+10y}{x^2+3xy}$

$= \dfrac{(x+3y)(x-3y)\cdot 2(2x+5y)}{(2x+5y)(x-3y)\cdot x(x+3y)}$

$= \dfrac{2}{x}$

33. $\dfrac{3m^2-5mn+2n^2}{9m^2-4n^2} \div \dfrac{m^3-m^2n}{9m^2+6mn}$

$= \dfrac{(3m-2n)(m-n)}{(3m+2n)(3m-2n)} \bullet \dfrac{3m(3m+2n)}{m^2(m-n)} = \dfrac{3}{m}$

35. $\dfrac{x^3+8}{x^2-4} \bullet \dfrac{5x-10}{x^3-2x^2+4x}$

$= \dfrac{(x+2)(x^2-2x+4)\bullet 5(x-2)}{(x+2)(x-2)\bullet x(x^2-2x+4)} = \dfrac{5}{x}$

< Objective 3 >

37. $f(x) = \dfrac{x^2-3x-4}{x+2} = \dfrac{(x-4)(x+1)}{x+2}$

$g(x) = \dfrac{x^2-2x-8}{x-4} = \dfrac{(x-4)(x+2)}{x-4} = x+2;$

$x \neq 4$

(a) $f(0)\bullet g(0) = (-2)\bullet(2) = -4$

(b) $f(4)\bullet g(4)$ is undefined.

(c) $h(x) = f(x)\bullet g(x)$

$= \dfrac{(x-4)(x+1)}{x+2}\bullet(x+2)$

$= (x-4)(x+1) = x^2-3x-4$

$x \neq -2; x \neq 4$

(d) $h(0) = (1)(-4) = -4$

(e) $h(4)$ is undefined.

39. $f(x) = \dfrac{2x^2-3x-5}{x+2} = \dfrac{(2x-5)(x+1)}{x+2}$

$g(x) = \dfrac{3x^2+5x-2}{x+1} = \dfrac{(3x-1)(x+2)}{x+1}$

(a) $f(1)\bullet g(1) = (-2)(3) = -6$

(b) $f(-2)\bullet g(-2)$ is undefined.

(c)

$h(x) = f(x)\bullet g(x)$

$= \dfrac{(2x-5)(x+1)}{x+2}\bullet\dfrac{(3x-1)(x+2)}{x+1}$

$= (2x-5)(3x-1) = 6x^2-17x+5$

$x \neq -2; x \neq -1$

(d) $h(1) = (-3)(2) = -6$

(e) $h(-2)$ is undefined.

41. $f(x) = \dfrac{3x^2+x-2}{x-2} = \dfrac{(3x-2)(x+1)}{x-2}$

$g(x) = \dfrac{x^2-4x-5}{x+4} = \dfrac{(x-5)(x+1)}{x+4}$

(a) $\dfrac{f(0)}{g(0)} = \dfrac{(1)}{\left(-\dfrac{5}{4}\right)} = -\dfrac{4}{5}$

(b) $\dfrac{f(1)}{g(1)} = \dfrac{(-2)}{\left(-\dfrac{8}{5}\right)} = \dfrac{5}{4}$

(c) $h(x) = \dfrac{f(x)}{g(x)} = \dfrac{\left[\dfrac{(3x-2)(x+1)}{x-2}\right]}{\left[\dfrac{(x-5)(x+1)}{x+4}\right]}$

$= \dfrac{(3x-2)(x+1)}{x-2} \div \dfrac{(x-5)(x+1)}{x+4}$

$= \dfrac{(3x-2)(x+1)}{x-2} \bullet \dfrac{x+4}{(x-5)(x+1)}$

$= \dfrac{(3x-2)(x+4)}{(x-2)(x-5)}$

$x \neq -1; x \neq -4$

(d) $h(x)$ is undefined if $x \in \{-4, -1, 2, 5\}$.

43. $A = l \cdot w = \left(\frac{3x-2}{x-2}\right)\left(\frac{2x-4}{x-1}\right)$

$= \frac{3x-2}{\cancel{x-2}} \cdot \frac{2\cancel{(x-2)}}{x-1} = \frac{2(3x-2)}{x-1}$

The area of the room is $\frac{2(3x-2)}{x-1}$.

45. True.

47. True. The product of three negative values is negative.

49. False. Division by zero results in a quotient of zero.

51. $\frac{x^3-3x^2+2x-6}{x^2-9} \cdot \frac{5x^2+15x}{20x}$

$= \frac{(x^2+2)(x-3) \cdot 5x(x+3)}{(x+3)(x-3) \cdot 20x} = \frac{x^2+2}{4}$

53. $\frac{x^4-16}{x^2+x-6} \div (x^3+4x)$

$= \frac{(x^2+4)(x+2)(x-2)}{(x+3)(x-2)} \cdot \frac{1}{x(x^2+4)}$

$= \frac{x+2}{x(x+3)} = \frac{x+2}{x^2+3x}$

55. $\frac{\frac{2}{3}}{\frac{1}{4}} = \frac{2}{3} \div \frac{1}{4} = \frac{2}{3} \cdot \frac{4}{1} = \frac{8}{3}$

The ratio of herbicides to insecticides used in the U.S. is $\frac{8}{3}$.

57. $\frac{x^2+5x}{3x-6} \cdot \frac{x^2-4}{3x^2+15x} \cdot \frac{6x}{x^2+6x+8}$

$= \frac{x(x+5)(x+2)(x-2) \cdot 6x}{3(x-2) \cdot 3x(x+5)(x+2)(x+4)}$

$= \frac{2x}{3(x+4)}$

59. $\frac{x^2-2x-8}{2x-8} \cdot \frac{x^2+5x}{x^2+5x+6} \div \frac{x^2+2x-15}{x^2-9}$

$= \frac{x^2-2x-8}{2x-8} \cdot \frac{x^2+5x}{x^2+5x+6} \cdot \frac{x^2-9}{x^2+2x-15}$

$= \frac{(x-4)(x+2) \cdot x((x+5)(x+3)(x-3)}{2(x-4)(x+2)(x+3)(x+5)(x-3)}$

$= \frac{x}{2}$

Exercises 9.3

< Objective 1 >

1. $\frac{9}{4x^3} + \frac{3}{4x^3} = \frac{12}{4x^3} = \frac{3}{x^3}$

3. $\frac{5}{3a+7} + \frac{2}{3a+7} = \frac{7}{3a+7}$

5. $\frac{2x}{x-3} - \frac{6}{x-3} = \frac{2x-6}{x-3} = \frac{2(x-3)}{x-3} = 2$

7. $\frac{y^2}{2y+8} + \frac{3y-4}{2y+8} = \frac{y^2+3y-4}{2y+8}$

$= \frac{(y+4)(y-1)}{2(y+4)} = \frac{y-1}{2}$

9. $\frac{5m-2}{m-6} - \frac{3m+10}{m-6} = \frac{5m-2-(3m+10)}{m-6}$

$= \frac{5m-2-3m-10}{m-6}$

$= \frac{2m-12}{m-6} = \frac{2(m-6)}{m-6} = 2$

11. $\frac{x-7}{x^2-x-6} + \frac{2x-2}{x^2-x-6} = \frac{3x-9}{x^2-x-6}$

$= \frac{3(x-3)}{(x-3)(x+2)}$

$= \frac{3}{x+2}$

< Objectives 2 and 3 >

13. $\frac{3}{2x}+\frac{4}{5x}=\frac{5}{5}\left(\frac{3}{2x}\right)+\frac{2}{2}\left(\frac{4}{5x}\right)=\frac{15}{10x}+\frac{8}{10x}$
$=\frac{23}{10x}$

15. $\frac{6}{a}+\frac{3}{a^2}=\frac{a}{a}\left(\frac{6}{a}\right)+\frac{3}{a^2}=\frac{6a}{a^2}+\frac{3}{a^2}=\frac{6a+3}{a^2}$
$=\frac{3(2a+1)}{a^2}$

17. $\frac{2}{m}-\frac{2}{n}=\frac{n}{n}\left(\frac{2}{m}\right)-\frac{m}{m}\left(\frac{2}{n}\right)=\frac{2n-2m}{mn}$
$=\frac{2(n-m)}{mn}$

19. $\frac{3}{4b^2}-\frac{5}{3b^3}=\frac{3b}{3b}\left(\frac{3}{4b^2}\right)-\frac{4}{4}\left(\frac{5}{3b^3}\right)$
$=\frac{9b}{12b^3}-\frac{20}{12b^3}=\frac{9b-20}{12b^3}$

21. $\frac{3}{b}-\frac{1}{b-3}=\left(\frac{b-3}{b-3}\right)\left(\frac{3}{b}\right)-\frac{b}{b}\left(\frac{1}{b-3}\right)$
$=\frac{3b-9}{b(b-3)}-\frac{b}{b(b-3)}=\frac{2b-9}{b(b-3)}$

23. $\frac{2}{x+1}+\frac{3}{x+2}$
$=\left(\frac{x+2}{x+2}\right)\left(\frac{2}{x+1}\right)+\left(\frac{x+1}{x+1}\right)\left(\frac{3}{x+2}\right)$
$=\frac{(2x+4)+(3x+3)}{(x+1)(x+2)}=\frac{5x+7}{(x+1)(x+2)}$

25. $\frac{5}{y-3}-\frac{1}{y+1}=\frac{5(y+1)-(y-3)}{(y-3)(y+1)}$
$=\frac{5y+5-y+3}{(y-3)(y+1)}$
$=\frac{4y+8}{(y-3)(y+1)}$
$=\frac{4(y+2)}{(y-3)(y+1)}$

27. $\frac{3w}{w-6}+\frac{4w}{w-2}=\frac{3w(w-2)+4w(w-6)}{(w-6)(w-2)}$
$=\frac{3w^2-6w+4w^2-24w}{(w-6)(w-2)}$
$=\frac{7w^2-30w}{(w-6)(w-2)}$
$=\frac{w(7w-30)}{(w-6)(w-2)}$

29. $\frac{3x}{3x-2}-\frac{2x}{2x+1}=\frac{3x(2x+1)-2x(3x-2)}{(3x-2)(2x+1)}$
$=\frac{6x^2+3x-6x^2+4x}{(3x-2)(2x+1)}$
$=\frac{7x}{(3x-2)(2x+1)}$

31. $\frac{5}{x-9}+\frac{4}{9-x}=\frac{5}{x-9}-\frac{4}{x-9}=\frac{1}{x-9}$

33. $\frac{3}{x^2-16}+\frac{2}{x-4}=\frac{3}{(x+4)(x-4)}+\frac{2}{x-4}$
$=\frac{3+2(x+4)}{(x+4)(x-4)}$
$=\frac{2x+11}{(x+4)(x-4)}$

35. $\dfrac{4m}{m^2-3m+2}-\dfrac{1}{m-2}$

$=\dfrac{4m}{(m-1)(m-2)}-\dfrac{1}{m-2}=\dfrac{4m-(m-1)}{(m-1)(m-2)}$

$=\dfrac{3m+1}{(m-1)(m-2)}$

37. $\dfrac{5x+5}{x^2+3x+2}-\dfrac{x-3}{x^2+5x-6}$

$=\dfrac{5(-4)+5}{(-4)^2+3(-4)+2}-\dfrac{(-4)-3}{(-4)^2+5(-4)-6}$

$=\dfrac{-15}{6}-\dfrac{-7}{-10}=-\dfrac{16}{5}$

39. $\dfrac{2m+2n}{m^2-n^2}+\dfrac{m-2n}{m^2+2mn+n^2}$

$=\dfrac{2(3)+2(2)}{(3)^2-(2)^2}+\dfrac{(3)-2(2)}{(3)^2+2(3)(2)+(2)^2}$

$=2+\dfrac{-1}{25}=\dfrac{49}{25}$

41. $\dfrac{1}{a-3}-\dfrac{1}{a+3}+\dfrac{2a}{a^2-9}$

$=\dfrac{1}{(4)-3}-\dfrac{1}{(4)+3}+\dfrac{2(4)}{(4)^2-9}=1-\dfrac{1}{7}+\dfrac{8}{7}$

$=2$

43. $\dfrac{3w^2+16w-8}{w^2+2w-8}+\dfrac{w}{w+4}-\dfrac{w-1}{w-2}$

$=\dfrac{3(3)^2+16(3)-8}{(3)^2+2(3)-8}+\dfrac{(3)}{(3)+4}-\dfrac{(3)-1}{(3)-2}$

$=\dfrac{67}{7}+\dfrac{3}{7}-2=8$

45. $\dfrac{a^2-9}{2a^2-5a-3}\bullet\left(\dfrac{1}{a-2}+\dfrac{1}{a+3}\right)$

$=\dfrac{(-3)^2-9}{2(-3)^2-5(-3)-3}$

$\bullet\left(\dfrac{1}{(-3)-2}+\dfrac{1}{(-3)+3}\right)$

$=\dfrac{0}{33}\bullet\left(-\dfrac{1}{5}+\dfrac{1}{0}\right)$

This expression is undefined when $a=-3$.

< Objective 4 >

47. $f(x)=\dfrac{3x}{x+1}$

$g(x)=\dfrac{2x}{x-3}$

(a) $f(1)+g(1)=\left(\dfrac{3}{2}\right)+\left(\dfrac{2}{-2}\right)=\dfrac{1}{2}$

(b) $h(x)=f(x)+g(x)=\dfrac{3x}{x+1}+\dfrac{2x}{x-3}$

$=\dfrac{3x(x-3)+2x(x+1)}{(x+1)(x-3)}$

$=\dfrac{3x^2-9x+2x^2+2x}{(x+1)(x-3)}$

$=\dfrac{5x^2-7x}{(x+1)(x-3)}$

(c) $h(1)=\dfrac{5(1)^2-7(1)}{[(1)+1][(1)-3]}=\dfrac{1}{2}$

$(1,h(1))=\left(1,\dfrac{1}{2}\right)$

49. $f(x)=\dfrac{x}{x+1}$

$g(x)=\dfrac{1}{x^2+2x+1}=\dfrac{1}{(x+1)^2}$

(a) $f(1)+g(1)=\left(\dfrac{1}{2}\right)+\left(\dfrac{1}{4}\right)=\dfrac{3}{4}$

(b) $h(x)=f(x)+g(x)=\dfrac{x}{x+1}+\dfrac{1}{(x+1)^2}$

$=\dfrac{x(x+1)+1}{(x+1)^2}=\dfrac{x^2+x+1}{(x+1)^2}$

(c) $h(1)=\dfrac{(1)^2+(1)+1}{[(1)+1]^2}=\dfrac{3}{4}$

$(1,h(1))=\left(1,\dfrac{3}{4}\right)$

51. $f(x)=\dfrac{x+5}{x-5}$

$g(x)=\dfrac{x-5}{x+5}$

(a) $f(1)-g(1)=\left(-\dfrac{3}{2}\right)-\left(-\dfrac{2}{3}\right)=-\dfrac{5}{6}$

(b) $h(x)=f(x)-g(x)=\dfrac{x+5}{x-5}-\dfrac{x-5}{x+5}$

$=\dfrac{(x+5)(x+5)-(x-5)(x-5)}{(x+5)(x-5)}$

$=\dfrac{x^2+10x+25-x^2+10x-25}{(x+5)(x-5)}$

$=\dfrac{20x}{(x+5)(x-5)}$

(c) $h(1)=\dfrac{20(1)}{[(1)+5][(1)-5]}=-\dfrac{5}{6}$

$(1,h(1))=\left(1,-\dfrac{5}{6}\right)$

53. $f(x)=\dfrac{x+9}{4x-36}=\dfrac{x+9}{4(x-9)}$

$g(x)=\dfrac{x-9}{x^2-18x+81}=\dfrac{x-9}{(x-9)^2}=\dfrac{1}{x-9}$

(a) $f(1)-g(1)=\left(-\dfrac{5}{16}\right)-\left(-\dfrac{1}{8}\right)=-\dfrac{3}{16}$

(b) $h(x)=f(x)-g(x)=\dfrac{x+9}{4(x-9)}-\dfrac{1}{x-9}$

$=\dfrac{x+9-4}{4(x+9)}=\dfrac{x+5}{4(x-9)}$

(c) $h(1)=\dfrac{6}{-32}=-\dfrac{3}{16}$

$(1,h(1))=\left(1,-\dfrac{3}{16}\right)$

55. $\dfrac{2x}{x+3}+\dfrac{2x}{x+3}+\dfrac{6}{x+3}+\dfrac{6}{x+3}$

$=\dfrac{2x+2x+6+6}{x+3}=\dfrac{4x+12}{x+3}=\dfrac{4(x+3)}{x+3}=4$

57. Let x represent the first even integer. Then $x+2$ represents the next even integer.

$\dfrac{1}{x}+\dfrac{1}{x+2}=\dfrac{1\cdot(x+2)}{x\cdot(x+2)}+\dfrac{1\cdot x}{(x+2)\cdot x}$

$=\dfrac{x+2}{x(x+2)}+\dfrac{x}{x(x+2)}=\dfrac{x+2+x}{x(x+2)}$

$=\dfrac{2x+2}{x(x+2)}$

59. The perimeter is two times the length plus two times the width.

$$\frac{2(2x+1)}{5}+\frac{2(4)}{3x+1}$$
$$=\frac{2(2x+1)\bullet(3x+1)}{5\bullet(3x+1)}+\frac{2(4)\bullet 5}{(3x+1)\bullet 5}$$
$$=\frac{2(2x+1)(3x+1)}{5(3x+1)}+\frac{40}{5(3x+1)}$$
$$=\frac{2(6x^2+5x+1)}{5(3x+1)}+\frac{40}{5(3x+1)}$$
$$=\frac{12x^2+10x+2+40}{5(3x+1)}=\frac{12x^2+10x+42}{5(3x+1)}$$
$$=\frac{2(6x^2+5x+21)}{5(3x+1)}$$

61. True

63. False

65. always

67. sometimes

69. $\dfrac{6y}{y^2-8y+15}+\dfrac{9}{y-3}$

$$=\frac{6y}{(y-3)(y-5)}+\frac{9}{y-3}=\frac{6y+9(y-5)}{(y-3)(y-5)}$$
$$=\frac{15y-45}{(y-3)(y-5)}=\frac{15(y-3)}{(y-3)(y-5)}=\frac{15}{y-5}$$

71. $\dfrac{6x}{x^2-10x+24}-\dfrac{18}{x-6}$

$$=\frac{6x}{(x-4)(x-6)}-\frac{18}{x-6}=\frac{6x-18(x-4)}{(x-4)(x-6)}$$
$$=\frac{-12x+72}{(x-4)(x-6)}=\frac{-12(x-6)}{(x-4)(x-6)}=-\frac{12}{x-4}$$

73. $\dfrac{2}{z^2-4}+\dfrac{3}{z^2+2z-8}$

$$=\frac{2}{(z+2)(z-2)}+\frac{3}{(z-2)(z+4)}$$
$$=\frac{2(z+4)+3(z+2)}{(z+2)(z-2)(z+4)}$$
$$=\frac{2z+8+3z+6}{(z+2)(z-2)(z+4)}$$
$$=\frac{5z+14}{(z+2)(z-2)(z+4)}$$

Exercises 9.4

< Objectives 1 and 2 >

1. $\dfrac{\frac{2}{3}}{\frac{3}{4}}=\dfrac{2}{3}\div\dfrac{3}{4}=\dfrac{2}{3}\bullet\dfrac{4}{3}=\dfrac{8}{9}$

3. $\dfrac{\frac{3}{4}+\frac{1}{3}}{\frac{1}{6}+\frac{3}{4}}=\dfrac{12\left(\frac{3}{4}+\frac{1}{3}\right)}{12\left(\frac{1}{6}+\frac{3}{4}\right)}=\dfrac{9+4}{2+9}=\dfrac{13}{11}$

5. $\dfrac{2+\frac{1}{3}}{3-\frac{1}{5}}=\dfrac{\frac{7}{3}}{\frac{14}{5}}=\dfrac{7}{3}\div\dfrac{14}{5}=\dfrac{7}{3}\bullet\dfrac{5}{14}=\dfrac{5}{6}$

7. $\dfrac{\frac{x}{8}}{\frac{x^2}{4}}=\dfrac{x}{8}\div\dfrac{x^2}{4}=\dfrac{x}{8}\bullet\dfrac{4}{x^2}=\dfrac{1}{2x}$

9. $\dfrac{\frac{3}{m}}{\frac{6}{m^2}}=\dfrac{3}{m}\div\dfrac{6}{m^2}=\dfrac{3}{m}\bullet\dfrac{m^2}{6}=\dfrac{m}{2}$

11. $$\frac{\frac{y+1}{y}}{\frac{y-1}{2y}}=\frac{y+1}{y}\div\frac{y-1}{2y}=\frac{y+1}{y}\bullet\frac{2y}{y-1}$$
$$=\frac{2(y+1)}{y-1}$$

13. $$\frac{\frac{a+2b}{3a}}{\frac{a^2+2ab}{9b}}=\frac{a+2b}{3a}\div\frac{9b}{a(a+2b)}=\frac{3b}{a^2}$$

15. $$\frac{\frac{x-3}{x^2-25}}{\frac{x^2+x-12}{x^2+5x}}$$
$$=\frac{x-3}{x^2-25}\div\frac{x^2+x-12}{x^2+5x}$$
$$=\frac{x-3}{(x+5)(x-5)}\bullet\frac{x(x+5)}{(x-3)(x+4)}$$
$$=\frac{x}{(x-5)(x+4)}$$

17. $$\frac{2-\frac{1}{x}}{2+\frac{1}{x}}=\frac{x\left(2-\frac{1}{x}\right)}{x\left(2+\frac{1}{x}\right)}=\frac{2x-1}{2x+1}$$

19. $$\frac{\frac{1}{x}-\frac{1}{y}}{\frac{1}{xy}}=\frac{xy\left(\frac{1}{x}-\frac{1}{y}\right)}{xy\left(\frac{1}{xy}\right)}=\frac{y-x}{1}=y-x$$

21. $$\frac{\frac{x^2}{y^2}-1}{\frac{x}{y}+1}=\frac{y^2\left(\frac{x^2}{y^2}-1\right)}{y^2\left(\frac{x}{y}+1\right)}=\frac{x^2-y^2}{xy+y^2}$$
$$=\frac{(x+y)(x-y)}{y(x+y)}=\frac{x-y}{y}$$

23. $$\frac{1+\frac{3}{a}-\frac{4}{a^2}}{1+\frac{2}{a}-\frac{3}{a^2}}=\frac{a^2\left(1+\frac{3}{a}-\frac{4}{a^2}\right)}{a^2\left(1+\frac{2}{a}-\frac{3}{a^2}\right)}=\frac{a^2+3a-4}{a^2+2a-3}$$
$$=\frac{(a+4)(a-1)}{(a+3)(a-1)}=\frac{a+4}{a+3}$$

25. $$\frac{\frac{x^2}{y}+2x+y}{\frac{1}{y^2}-\frac{1}{x^2}}=\frac{x^2y^2\left(\frac{x^2}{y}+2x+y\right)}{x^2y^2\left(\frac{1}{y^2}-\frac{1}{x^2}\right)}$$
$$=\frac{x^4y+2x^3y^2+x^2y^3}{x^2-y^2}$$
$$=\frac{x^2y(x+y)^2}{(x+y)(x-y)}=\frac{x^2y(x+y)}{(x-y)}$$

27. $$\frac{2-\frac{2}{x+1}}{2+\frac{2}{x+1}}=\frac{(x+1)\left(2-\frac{2}{x+1}\right)}{(x+1)\left(2+\frac{2}{x+1}\right)}=\frac{2(x+1)-2}{2(x+1)+2}$$
$$=\frac{2x}{2x+4}=\frac{x}{x+2}$$

29. $$\frac{1-\frac{1}{y-1}}{y-\frac{8}{y+2}}=\frac{(y-1)(y+2)\left(1-\frac{1}{y-1}\right)}{(y-1)(y+2)\left(y-\frac{8}{y+2}\right)}$$
$$=\frac{(y-1)(y+2)-(y+2)}{y(y-1)(y+2)-8(y-1)}$$
$$=\frac{y^2-4}{(y-1)(y^2-6y+8)}$$
$$=\frac{(y+2)(y-2)}{(y-1)(y-2)(y+4)}$$
$$=\frac{y+2}{(y-1)(y+4)}$$

31. $$\frac{\dfrac{1}{x-3}+\dfrac{1}{x+3}}{\dfrac{1}{x-3}-\dfrac{1}{x+3}}$$
$$=\frac{(x-3)(x+3)\left(\dfrac{1}{x-3}+\dfrac{1}{x+3}\right)}{(x-3)(x+3)\left(\dfrac{1}{x-3}-\dfrac{1}{x+3}\right)}$$
$$=\frac{x+3+x-3}{x+3-(x-3)}=\frac{2x}{6}=\frac{x}{3}$$

33. $$\frac{\dfrac{x}{x+1}+\dfrac{1}{x-1}}{\dfrac{x}{x-1}-\dfrac{1}{x+1}}=\frac{(x+1)(x-1)\left(\dfrac{x}{x+1}+\dfrac{1}{x-1}\right)}{(x+1)(x-1)\left(\dfrac{x}{x-1}-\dfrac{1}{x+1}\right)}$$
$$=\frac{x(x-1)+(x+1)}{x(x+1)-(x-1)}=\frac{x^2+1}{x^2+1}=1$$

35. $$\frac{\dfrac{a+1}{a-1}-\dfrac{a-1}{a+1}}{\dfrac{a+1}{a-1}+\dfrac{a-1}{a+1}}=\frac{(a+1)(a-1)\left(\dfrac{a+1}{a-1}-\dfrac{a-1}{a+1}\right)}{(a+1)(a-1)\left(\dfrac{a+1}{a-1}+\dfrac{a-1}{a+1}\right)}$$
$$=\frac{(a+1)^2-(a-1)^2}{(a+1)^2+(a-1)^2}$$
$$=\frac{a^2+2a+1-(a^2-2a+1)}{a^2+2a+1+a^2-2a+1}$$
$$=\frac{4a}{2a^2+2}=\frac{2a}{a^2+1}$$

37. $$1+\frac{1}{1+\dfrac{1}{x}}=1+\frac{x}{x+1}=\frac{x+1}{x+1}+\frac{x}{x+1}=\frac{2x+1}{x+1}$$

39. $$1+\frac{1}{1+\dfrac{1}{1+\dfrac{1}{x}}}=1+\frac{1}{1+\dfrac{x}{x+1}}=1+\frac{x+1}{(x+1)+x}$$
$$=1+\frac{x+1}{2x+1}=\frac{2x+1}{2x+1}+\frac{x+1}{2x+1}$$
$$=\frac{3x+2}{2x+1}$$

41. $$\frac{f(x+h)-f(x)}{h}=\frac{\dfrac{5}{x+h}-\dfrac{5}{x}}{h}\bullet\frac{x(x+h)}{x(x+h)}$$
$$=\frac{5x-5(x+h)}{hx(x+h)}$$
$$=\frac{-5h}{hx(x+h)}=\frac{-5}{x(x+h)}$$

43. $$\frac{f(x+h)-f(x)}{h}=\frac{-\dfrac{1}{x+h}--\dfrac{1}{x}}{h}\bullet\frac{x(x+h)}{x(x+h)}$$
$$=\frac{-x+(x+h)}{hx(x+h)}$$
$$=\frac{h}{hx(x+h)}=\frac{1}{x(x+h)}$$

< Objective 3 >

45. The area of a rectangle is the product of the two sides.

$$w=\frac{A}{l}=\frac{\dfrac{2}{3}}{\dfrac{2x+6}{12x-15}}=\frac{2}{3}\bullet\frac{12x-15}{2x+6}$$
$$=\frac{2\bullet3(4x-5)}{3\bullet2(x+3)}=\frac{4x-5}{x+3}$$

47. True.

49. The expressions are almost equivalent.

x	−3	−2	−1	0	1	2	3
$\dfrac{1-\frac{2}{x}}{1-\frac{4}{x^2}}$	3	Error	−1	Error	0.333	Error	0.6
$\dfrac{x}{x+2}$	3	Error	−1	0	0.333	0.5	0.6

51. $C_T = \dfrac{1}{\dfrac{1}{C_L} + \dfrac{1}{C_{CW}}} = \dfrac{1 \cdot C_L C_{CW}}{\left(\dfrac{1}{C_L} + \dfrac{1}{C_{CW}}\right) \cdot C_L C_{CW}}$

$= \dfrac{C_L C_{CW}}{C_{CW} + C_L}$

53. $R_{eq} = \left(R_1^{-1} + R_2^{-1}\right)^{-1} = \left(\dfrac{1}{R_1} + \dfrac{1}{R_2}\right)^{-1}$

$= \dfrac{1}{\dfrac{1}{R_1} + \dfrac{1}{R_2}} = \dfrac{1 \cdot R_1 R_2}{\left(\dfrac{1}{R_1} + \dfrac{1}{R_2}\right) \cdot R_1 R_2}$

$= \dfrac{R_1 R_2}{R_2 + R_1} = \dfrac{R_1 R_2}{R_1 + R_2}$

55. $R = \dfrac{2}{\dfrac{1}{R_1} + \dfrac{1}{R_2}} = \dfrac{2}{\dfrac{1}{40} + \dfrac{1}{50}} = \dfrac{2}{\dfrac{9}{200}} = \dfrac{400}{9}$

$= 44\dfrac{4}{9}$

57. $R = \dfrac{2}{\dfrac{1}{R_1} + \dfrac{1}{R_2}} = \dfrac{2}{\dfrac{1}{50} + \dfrac{1}{60}} = \dfrac{2}{\dfrac{11}{300}} = \dfrac{600}{11}$

$= 54\dfrac{6}{11}$ mi/hr

59. Above and Beyond.

61. The reciprocal of $x + \dfrac{6}{x-1}$ is

$\dfrac{1}{x + \dfrac{6}{x-1}} = \dfrac{x-1}{x(x-1)+6} = \dfrac{x-1}{x^2 - x + 6}$.

63. $\dfrac{\dfrac{1}{x} + \dfrac{1}{x-1}}{2} = \dfrac{x(x-1)\left(\dfrac{1}{x} + \dfrac{1}{x-1}\right)}{2x(x-1)} = \dfrac{x-1+x}{2x(x-1)}$

$= \dfrac{2x-1}{2x(x-1)}$

65. Answers will vary.

67. Above and Beyond

Exercises 9.5

< Objective 1 >

1. $f(x) = \dfrac{7}{x}$; Domain $\{x \mid x \neq 0\}$

$f(0) = \dfrac{7}{0}$ undefined; no y-intercept

$0 = \dfrac{7}{x}$ has no solution; no x-intercept

3. $f(x) = \dfrac{3}{x-9}$; Domain $\{x \mid x \neq 9\}$

$f(0) = \dfrac{3}{0-9} = -\dfrac{1}{3}$; y-intercept $\left(0, -\dfrac{1}{3}\right)$

$0 = \dfrac{3}{x-9}$ has no solution; no x-intercept

5. $f(x) = \dfrac{8}{2-x}$; Domain $\{x \mid x \neq 2\}$

$f(0) = \dfrac{8}{2-0} = 4$; y-intercept (0, 4)

$0 = \dfrac{8}{2-x}$ has no solution; no x-intercept

7. $f(x) = \dfrac{x-6}{x-4}$; Domain $\{x \mid x \neq 4\}$

$f(0) = \dfrac{0-6}{0-4} = \dfrac{3}{2}$; y-intercept $\left(0, \dfrac{3}{2}\right)$

$0 = \dfrac{x-6}{x-4}$ has solution $x = 6$, so the x-intercept is (6, 0).

9. $f(x) = \dfrac{2-x}{x-4}$; Domain $\{x \mid x \neq 4\}$

$f(0) = \dfrac{2-0}{0-4} = -\dfrac{1}{2}$; y-intercept $\left(0, -\dfrac{1}{2}\right)$

$0 = \dfrac{2-x}{x-4}$ has solution $x = 2$, so the x-intercept is (2, 0).

11. $f(x)=\dfrac{2x-10}{x+4}$; Domain $\{x|x\neq -4\}$

$f(0)=\dfrac{2(0)-10}{0+4}=-\dfrac{5}{2}$; y-intercept $\left(0,\dfrac{3}{2}\right)$

$0=\dfrac{2x-10}{x+4}=2x-10$

$10=2x$

$5=x$

The x-intercept is (5, 0).

13. $f(x)=\dfrac{2-4x}{x-3}$; Domain $\{x|x\neq 3\}$

$f(0)=\dfrac{2-4(0)}{0-3}=-\dfrac{2}{3}$; y-intercept $\left(0,-\dfrac{2}{3}\right)$

$0=\dfrac{2-4x}{x-3}=2-4x$

$4x=2$

$x=\dfrac{2}{4}=\dfrac{1}{2}$

The x-intercept is $\left(\dfrac{1}{2},0\right)$.

< Objective 3 >

15. $f(x)=\dfrac{7}{x}$ has vertical asymptote $x=0$ since the function is undefined at that value. The horizontal asymptote is $y=0$ since the y-values get closer and closer to 0 when larger and larger x-values are tried.

17. $f(x)=\dfrac{3}{x-9}$ has vertical asymptote $x=9$ since the function is undefined at that value. The horizontal asymptote is $y=0$ since the y-values get closer and closer to 0 when larger and larger x-values are tried.

19. $f(x)=\dfrac{8}{2-x}$ has vertical asymptote $x=2$ since the function is undefined at that value. The horizontal asymptote is $y=0$ since the y-values get closer and closer to 0 when larger and larger x-values are tried.

21. $f(x)=\dfrac{x-6}{x-4}$ has vertical asymptote $x=4$ since the function is undefined at that value. The horizontal asymptote is $y=1$ since the y-values get closer and closer to 1 when larger and larger x-values are tried.

23. $f(x)=\dfrac{2-x}{x-4}$ has vertical asymptote $x=4$ since the function is undefined at that value. The horizontal asymptote is $y=-1$ since the y-values get closer and closer to –1 when larger and larger x-values are tried.

25. $f(x)=\dfrac{2x-10}{x+4}$ has vertical asymptote $x=-4$ since the function is undefined at that value. The horizontal asymptote is $y=2$ since the y-values get closer and closer to 2 when larger and larger x-values are tried.

27. $f(x)=\dfrac{2-4x}{x-3}$ has vertical asymptote $x=3$ since the function is undefined at that value. The horizontal asymptote is $y=-4$ since the y-values get closer and closer to –4 when larger and larger x-values are tried.

< Objective 2 >

29. $f(x)=\dfrac{6}{x}$

The domain is $\{x|x\neq 0\}$. The vertical asymptote is $x=0$. The horizontal asymptote is $y=0$.

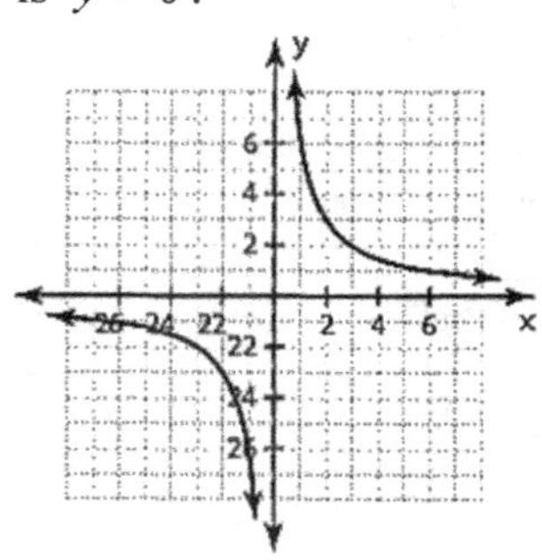

31. $f(x) = -\frac{8}{x}$

The domain is $\{x | x \neq 0\}$. The vertical asymptote is $x = 0$. The horizontal asymptote is $y = 0$.

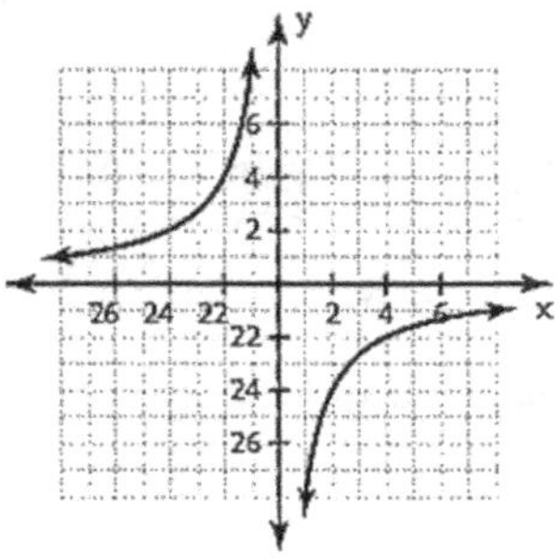

33. $f(x) = \frac{10}{x-2}$

The domain is $\{x | x \neq 2\}$. The vertical asymptote is $x = 2$. The horizontal asymptote is $y = 0$.

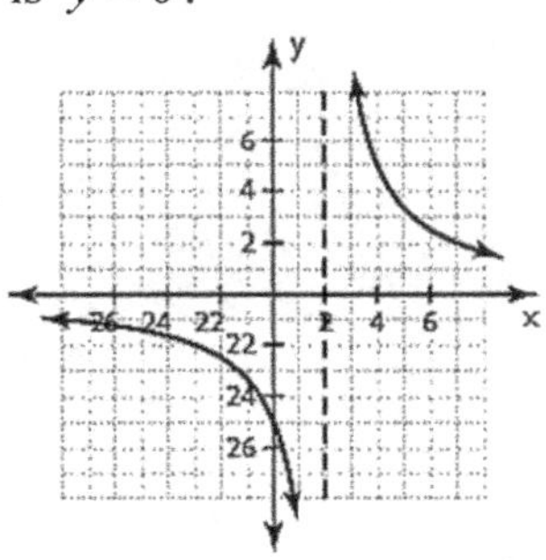

35. $f(x) = \frac{6}{3-x}$

The domain is $\{x | x \neq 3\}$. The vertical asymptote is $x = 3$. The horizontal asymptote is $y = 0$.

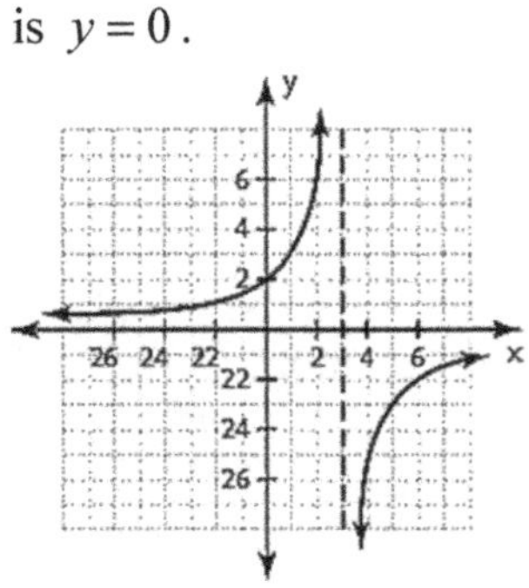

37. $f(x) = \frac{x+5}{x-2}$

The domain is $\{x | x \neq 2\}$. The vertical asymptote is $x = 2$. The horizontal asymptote is $y = 1$.

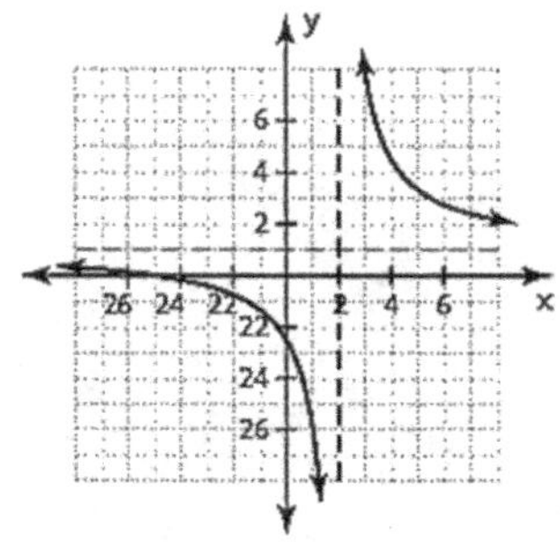

39. $f(x) = \frac{4-x}{x+2}$

The domain is $\{x | x \neq -2\}$. The vertical asymptote is $x = -2$. The horizontal asymptote is $y = -1$.

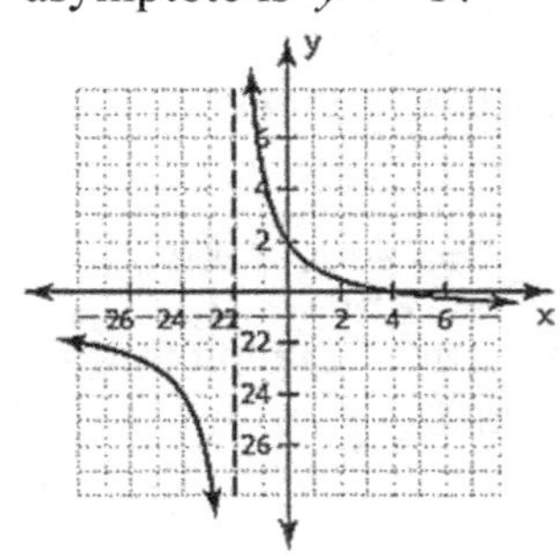

41. $f(x) = \frac{2x-6}{x+4}$

The domain is $\{x | x \neq -4\}$. The vertical asymptote is $x = -4$. The horizontal asymptote is $y = 2$.

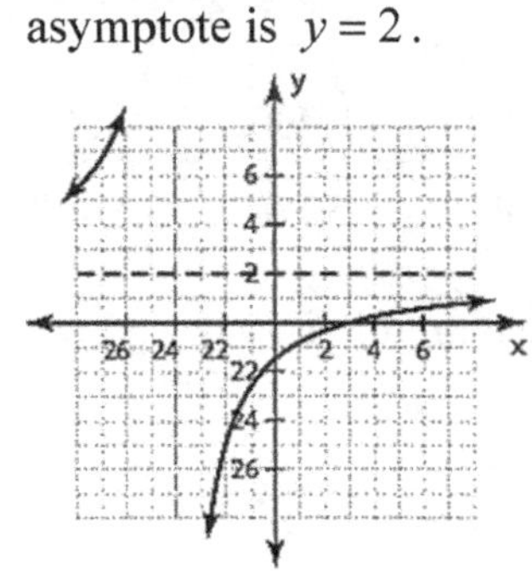

43. $f(x)=\dfrac{2-4x}{x-3}$

The domain is $\{x|x\neq 3\}$. The vertical asymptote is $x=3$. The horizontal asymptote is $y=-4$.

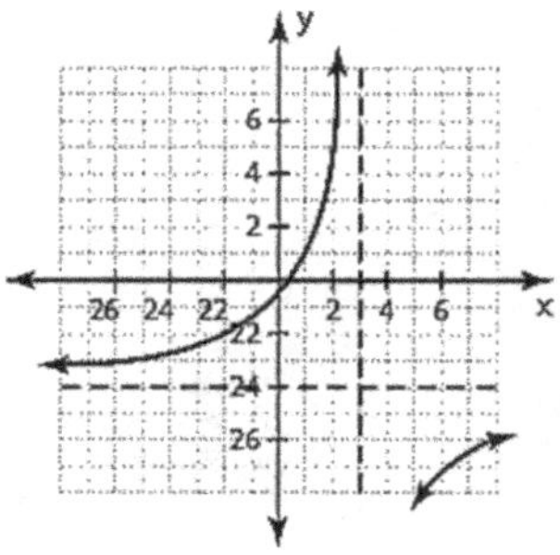

45. The graph of $f(x)=\dfrac{a}{x}$, $a\neq 0$, <u>never</u> has an x-intercept.

47. The graph of $f(x)=\dfrac{a(x-b)}{x-c}$, $a\neq 0$, $b\neq 0$, and $b\neq c$, <u>always</u> has an x-intercept.

49. The graph of $f(x)=\dfrac{a}{x-c}$, $a\neq 0$, <u>sometimes</u> has a vertical asymptote which lies exactly on the y-axis.

51. $f(x)=\dfrac{7x-5}{2x+3}$

x	100	1,000	10,000
$f(x)$	3.4236	3.4923	3.4992

x	−100	−1,000	−10,000
$f(x)$	3.5787	3.5078	3.5008

The horizontal asymptote is $y=\dfrac{7}{2}$.

53. $f(x)=\dfrac{3x+7}{8-9x}$

x	100	1,000	10,000
$f(x)$	−0.3442	−0.3344	−0.3334

x	−100	−1,000	−10,000
$f(x)$	−0.3227	−0.3323	−0.3332

The horizontal asymptote is $y=-\dfrac{1}{3}$.

55. The horizontal asymptote is $y=\dfrac{a}{c}$.

57. Above and Beyond

59. Above and Beyond

Exercises 9.6

1. $\dfrac{x}{4}-\dfrac{x}{5}=2$ is an equation.

$$\frac{x}{4}-\frac{x}{5}=2$$
$$20\left(\frac{x}{4}-\frac{x}{5}\right)=20\bullet 2$$
$$5x-4x=40$$
$$x=40;\{40\}$$

3. $\dfrac{x}{2}-\dfrac{x}{5}$ is an expression.

$$\frac{x}{2}-\frac{x}{5}=\frac{3x}{10}$$

5. $\dfrac{3x+1}{4}=x-1$ is an equation.

$$\frac{3x+1}{4}=x-1$$
$$3x+1=4x-4$$
$$5=x;\{5\}$$

7. $\frac{x}{4}=\frac{x}{12}+\frac{1}{2}$ is an equation.

$$\frac{x}{4}=\frac{x}{12}+\frac{1}{2}$$
$$3x=x+6$$
$$x=3;\{3\}$$

< Objective 1 >

9. LCD = 6

$$\frac{x}{3}+\frac{3}{2}=\frac{x}{6}+\frac{7}{3}$$
$$6\left(\frac{x}{3}+\frac{3}{2}\right)=6\left(\frac{x}{6}+\frac{7}{3}\right)$$
$$2x+9=x+14$$
$$x=5;\{5\}$$

11. LCD = $4x$

$$\frac{4}{x}+\frac{3}{4}=\frac{10}{x}$$
$$4x\left(\frac{4}{x}+\frac{3}{4}\right)=4x\bullet\frac{10}{x}$$
$$16+3x=40$$
$$x=8;\{8\}$$

13. LCD = $4x$

$$\frac{5}{4x}-\frac{1}{2}=\frac{1}{2x}$$
$$4x\left(\frac{5}{4x}-\frac{1}{2}\right)=4x\bullet\frac{1}{2x}$$
$$5-2x=2$$
$$x=\frac{3}{2};\left\{\frac{3}{2}\right\}$$

15. LCD = $(x+3)(x+5)$

$$\frac{4}{x+5}=\frac{3}{x+3}$$
$$\frac{4(x+3)(x+5)}{x+5}=\frac{3(x+3)(x+5)}{x+3}$$
$$4x+12=3x+15$$
$$x=3;\{3\}$$

17. LCD = $x(x+3)$

$$\frac{9}{x}+2=\frac{2x}{x+3}$$
$$9(x+3)+2x(x+3)=2x\bullet x$$
$$2x^2+15x+27=2x^2$$
$$15x+27=0$$
$$x=-\frac{9}{5};\left\{-\frac{9}{5}\right\}$$

19. LCD = $x(x+2)$

$$\frac{3}{x+2}-\frac{5}{x}=\frac{13}{x+2}$$
$$3x-5(x+2)=13x$$
$$-2x-10=13x$$
$$x=-\frac{2}{3};\left\{-\frac{2}{3}\right\}$$

21. LCD = $2(x-2)$

$$\frac{3}{2}+\frac{2}{2x-4}=\frac{1}{x-2}$$
$$\frac{3}{2}+\frac{1}{x-2}=\frac{1}{x-2}$$
$$\frac{3}{2}=0$$

False. No solutions; $\varnothing$

23. LCD = $3(x+4)$

$$\frac{x}{3x+12}+\frac{x-1}{x+4}=\frac{5}{3}$$
$$x+3(x-1)=5(x+4)$$
$$4x-3=5x+20$$
$$x=-23;\{-23\}$$

25. LCD = $15(x+6)$

$$\frac{16-3x}{x+6}+\frac{x+3}{5}=\frac{3x-2}{15}$$
$$15(16-3x)+3(x+6)(x+3)=(x+6)(3x-2)$$
$$3x^2-18x+294=3x^2+16x-12$$
$$-34x=-306$$
$$x=9;\{9\}$$

27. LCD $=(x-2)(x+2)$

$$\frac{1}{x-2}-\frac{2}{x+2}=\frac{2}{x^2-4}$$
$$(x+2)-2(x-2)=2$$
$$-x+6=2$$
$$x=4;\{4\}$$

29. LCD $=(x-5)(x+5)$

$$\frac{7}{x+5}-\frac{1}{x-5}=\frac{x}{x^2-25}$$
$$7(x-5)-(x+5)=x$$
$$6x-40=x$$
$$5x=40$$
$$x=8;\{8\}$$

31. LCD $=(x-3)(x+2)$

$$\frac{11}{x+2}-\frac{5}{x^2-x-6}=\frac{1}{x-3}$$
$$11(x-3)-5=x+2$$
$$11x-38=x+2$$
$$10x=40$$
$$x=4;\{4\}$$

33. LCD $=(x-2)(x+3)$

$$\frac{5}{x-2}-\frac{3}{x+3}=\frac{24}{x^2+x-6}$$
$$5(x+3)-3(x-2)=24$$
$$2x+21=24$$
$$2x=3$$
$$x=\frac{3}{2};\left\{\frac{3}{2}\right\}$$

35. LCD $=(x-3)$

$$\frac{x}{x-3}-2=\frac{3}{x-3}$$
$$x-2(x-3)=3$$
$$-x+6=3$$
$$x=3$$

No solutions; $\varnothing$

Note: 3 is an excluded value.

37. LCD $=x(x-3)(x+2)$

$$\frac{2}{x^2-3x}-\frac{1}{x^2+2x}=\frac{2}{x^2-x-6}$$
$$2(x+2)-(x-3)=2x$$
$$x+7=2x$$
$$x=7;\{7\}$$

39. LCD $=(x-3)(x+3)(x-1)$

$$\frac{2}{x^2-4x+3}-\frac{3}{x^2-9}=\frac{2}{x^2+2x-3}$$
$$2(x+3)-3(x-1)=2(x-3)$$
$$-x+9=2x-6$$
$$-3x=-15$$
$$x=5;\{5\}$$

41. LCD $=(x-5)(x+5)$

$$\frac{7}{x-5}-\frac{3}{x+5}=\frac{40}{x^2-25}$$
$$7(x+5)-3(x-5)=40$$
$$4x+50=40$$
$$4x=-10$$
$$x=-\frac{5}{2};\left\{-\frac{5}{2}\right\}$$

43. LCD $=(x-3)(x-5)$

$$\frac{2x}{x-3}+\frac{2}{x-5}=\frac{3x}{x^2-8x+15}$$
$$2x(x-5)+2(x-3)=3x$$
$$2x^2-8x-6=3x$$
$$2x^2-11x-6=0$$
$$(2x+1)(x-6)=0$$
$$\left\{-\frac{1}{2},6\right\}$$

45. LCD $=(x-3)(x+2)$

$$\frac{2x}{x+2}=\frac{5}{x^2-x-6}-\frac{1}{x-3}$$
$$2x(x-3)=5-(x+2)$$
$$2x^2-6x=3-x$$
$$2x^2-5x-3=0$$
$$(2x+1)(x-3)=0$$
$$\left\{-\frac{1}{2}\right\}$$

Note: 3 is an excluded value.

47. LCD $=(x-2)(x+3)$

$$\frac{7}{x-2}+\frac{16}{x+3}=3$$
$$7(x+3)+16(x-2)=3(x+3)(x-2)$$
$$23x-11=3x^2+3x-18$$
$$3x^2-20x-7=0$$
$$(3x+1)(x-7)=0$$
$$\left\{-\frac{1}{3},7\right\}$$

49. LCD $=(x-3)(x+3)$

$$\frac{11}{x-3}-1=\frac{10}{x+3}$$
$$11(x+3)-(x-3)(x+3)=10(x-3)$$
$$11x+33-x^2+9=10x-30$$
$$-x^2+x+72=0$$
$$x^2-x-72=0$$
$$(x-9)(x+8)=0$$

{–8, 9}

< Objective 2 >

51.
$$\frac{x}{11}=\frac{12}{33}$$
$$33x=132$$
$$x=4$$

53.
$$\frac{5}{8}=\frac{20}{x}$$
$$5x=160$$
$$x=32$$

55.
$$\frac{x+1}{5}=\frac{20}{25}$$
$$25(x+1)=100$$
$$25x+25=100$$
$$25x=75$$
$$x=3$$

57.
$$\frac{3}{5}=\frac{x-1}{20}$$
$$60=5(x-1)=5x-5$$
$$65=5x$$
$$13=x \text{ or } x=13$$

59.
$$\frac{x}{6}=\frac{x+5}{16}$$
$$16x=6(x+5)=6x+30$$
$$10x=30$$
$$x=3$$

61.
$$\frac{x}{x+7}=\frac{10}{17}$$
$$17x=10(x+7)=10x+70$$
$$7x=70$$
$$x=10$$

63.
$$\frac{2}{x-1}=\frac{6}{x+9}$$
$$2(x+9)=6(x-1)$$
$$2x+18=6x-6$$
$$24=4x$$
$$6=x \text{ or } x=6$$

65. $\frac{1}{x+3}=\frac{7}{x^2-9}=\frac{7}{(x-3)(x+3)}$

The LCD is $(x-3)(x+3)$.

$$(x-3)(x+3)\bullet\frac{1}{x+3}$$
$$=(x-3)(x+3)\bullet\frac{7}{(x-3)(x+3)}$$
$$x-3=7$$
$$x=10$$

67.
$$\frac{x}{x+5}=\frac{3x}{5x+9}$$
$$x(5x+9)=3x(x+5)$$
$$5x^2+9x=3x^2+15x$$
$$2x^2-6x=0$$
$$2x(x-3)=0$$
$$x=3;3x=9$$
$$x+5=8;5x+9=24$$

3 and 9 are the shown sides of the first triangle. The sides of the second are 8 and 24.

< Objective 3 >

69. $\text{LCD}=abx$
$$\frac{1}{x}=\frac{1}{a}-\frac{1}{b}$$
$$ab=bx-ax=x(b-a)$$
$$x=\frac{ab}{b-a}$$

71.
$$LCD=RR_1R_2$$
$$\frac{1}{R}=\frac{1}{R_1}+\frac{1}{R_2}$$
$$R_1R_2=RR_2+RR_1$$
$$R_1R_2-RR_1=RR_2$$
$$R_1(R_2-R)=RR_2$$
$$R_1=\frac{RR_2}{R_2-R}$$

73.
$$y=\frac{x+1}{x-1}$$
$$y(x-1)=x+1$$
$$xy-y=x+1$$
$$xy-x=y+1$$
$$x(y-1)=y+1$$
$$x=\frac{y+1}{y-1}$$

< Objective 4 >

75. Let x be the (first) number. Then $4x$ is the second number.
$$\frac{1}{x}+\frac{1}{4x}=\frac{5}{24}$$
$$24+6=5x$$
$$30=5x$$
$$x=6$$
$$4x=24$$

The numbers are 6 and 24.

77. Let x be the (first) number.
$$\frac{11-x}{15-x}=\frac{1}{3}$$
$$3(11-x)=15-x$$
$$33-3x=15-x$$
$$-2x=-18$$
$$x=9$$

The number is 9.

79. Let c be the rate of the river's current.

	Distance	Rate	Time
Downstream	3	$20+c$	$\frac{3}{20+c}$
Upstream	2	$20-c$	$\frac{2}{20-c}$

$$\frac{3}{20+c}=\frac{2}{20-c}$$
$$3(20-c)=2(20+c)$$
$$60-3c=40+2c$$
$$20=5c$$
$$c=4$$

The rate of river's current is 4 mi/hr.

81. Let r be the plane's airspeed.

	Distance	Rate	Time
With Wind	720	$r+30$	$\frac{720}{r+30}$
Against wind	720	$r-30$	$\frac{720}{r-30}$

$$\frac{720}{r+30}+\frac{720}{r-30}=10$$
$$\frac{72}{r+30}+\frac{72}{r-30}=1$$
$$72(r-30)+72(r+30)=(r-30)(r+30)$$
$$r^2-144r-900=0$$
$$(r-150)(r+6)=0$$

The plane's airspeed was 150 mi/hr.

83. Let t be the amount of time (in minutes) that it would take the two printers, operating together.

$$\frac{t}{60}+\frac{t}{90}=1$$
$$180\left(\frac{t}{60}+\frac{t}{90}\right)=180\bullet 1$$
$$3t+2t=180$$
$$5t=180$$
$$t=36$$

It would take 36 min with the two printers, operating together, to print the checks.

85. Let r be Po Ling's rate on a bicycle.

	Distance	Rate	Time
Bicycle	75	r	$\frac{75}{r}$
Drive	165	$r+30$	$\frac{165}{r+30}$

$$\frac{75}{r}=\frac{165}{r+30}$$
$$75(r+30)=165r$$
$$75r+2250=165r$$
$$2250=90r$$
$$25=r$$

Po Ling's bicycling rate is 25 mi/hr and her driving rate is 55 mi/hr.

87. Let t be the plane's time (in hours) on the first leg of the trip.

	Distance	Rate	Time
First leg	540	$\frac{540}{t}$	t
Second leg	360	$\frac{360}{t-1}$	$t-1$

$$\frac{540}{t}=\frac{360}{t-1}$$
$$540(t-1)=360t$$
$$540t-540=360t$$
$$180t=540$$
$$t=3$$

The plane's time for the first leg of the trip was 3 hr, and the time for the second leg was 2 hr.

89. Let t be the amount of time (in hours) that it would take both crews working together.

$$\frac{t}{15}+\frac{t}{10}=1$$
$$30\left(\frac{t}{15}+\frac{t}{10}\right)=30\bullet 1$$
$$2t+3t=30$$
$$5t=30$$
$$t=6$$

If the two crews worked together, they would complete the section of highway in 6 hr.

91. Let t be the amount of time (in hours) that it would take an experienced roofer to shingle a new section of roof working alone. Her assistant would then take $2t$ hours to shingle the same section, working alone.

$$\frac{4}{t}+\frac{4}{2t}=1$$
$$2t\left(\frac{4}{t}+\frac{4}{2t}\right)=2t\bullet 1$$
$$8+4=2t$$
$$12=2t$$
$$6=t$$

The experienced roofer, working alone, would take 6 hr to shingle a new section of roof.

93. Let x be the speed of the plane in ft/s.

$$\frac{60}{88}=\frac{150}{x}$$
$$60x=13,200$$
$$x=220$$

The plane's speed is 220 ft/s.

95. Let x be the amount of gasoline.

$$\frac{5}{160}=\frac{x}{384}$$
$$1920=160x$$
$$12=x$$

The 384-mi trip will take 12 gal of gasoline.

97. Let x be the amount Sveta earns in 1 year (52 weeks).

$$\frac{13,500}{20}=\frac{x}{52}$$
$$702,000=20x$$
$$35,100=x$$

She will earn \$35,100 in one year.

99. Let x be number of ladybug beetles. Then $110-x$ is the number of praying mantises.

$$\frac{x}{110-x}=\frac{7}{4}$$
$$4x=770-7x$$
$$11x=770$$
$$x=70;110-x=40$$

There are 70 ladybugs and 40 praying mantises.

101. Let x be the amount the brother receives. Then $12,000-x$ is the amount the sister receives.

$$\frac{x}{12,000-x}=\frac{2}{3}$$
$$3x=2(12,000-x)=24,000-2x$$
$$5x=24,000$$
$$x=4,800;$$
$$12,000-x=12,000-4,800=7,200$$

The brother receives \$4,800 and the sister receives \$7,200.

103. False; when solving a rational equation, we need to multiply both sides of the equation by the least common denominator of all the rational expressions appearing in the equation.

105. We must always check a possible solution to a rational equation, in case we have obtained an extraneous solution.

107. The value of the term $\frac{5}{x}$, $x\neq 0$ is <u>sometimes</u> equal to one.

109. The value of the term $\frac{5}{x}$, $x \neq 0$ is <u>never</u> equal to zero.

111.

$$\frac{1}{R_{eq}} = \frac{1}{R_1} + \frac{1}{R_2}$$

$$\frac{1}{0.41} = \frac{1}{R_1} + \frac{1}{0.92}$$

$$\frac{1}{0.41} \bullet (0.3772R_1) = \left(\frac{1}{R_1} + \frac{1}{0\ 92}\right) \bullet (0.3772R_1)$$

$$0.92R_1 = 0.3772 - 0.41R_1$$

$$0.51R_1 = 0.3772$$

$$R_1 \approx 0.74$$

To the nearest hundredth, the resistance is 0.74 Ω.

113. Above and Beyond.

Summary Exercises

1. $\frac{x}{2}$ is never undefined.

3. $\frac{2}{x-5}$ is undefined if $x = 5$.

5. $\frac{18x^5}{24x^3} = \frac{3x^2}{4}$

7. $\frac{7y-49}{y-7} = \frac{7(y-7)}{y-7} = 7$

9. $\frac{9-x^2}{x^2+2x-15} = \frac{(3+x)\overset{-1}{(3-x)}}{(x+5)(x-3)} = -\frac{x+3}{x+5}$

11. $\frac{6a^2-ab-b^2}{9a^2-b^2} = \frac{(2a-b)(3a+b)}{(3a+b)(3a-b)} = \frac{2a-b}{3a-b}$

13. $f(x) = \frac{x^2-3x-4}{x+1} = x-4; x \neq -1$

There is a hole at (–1, –5).

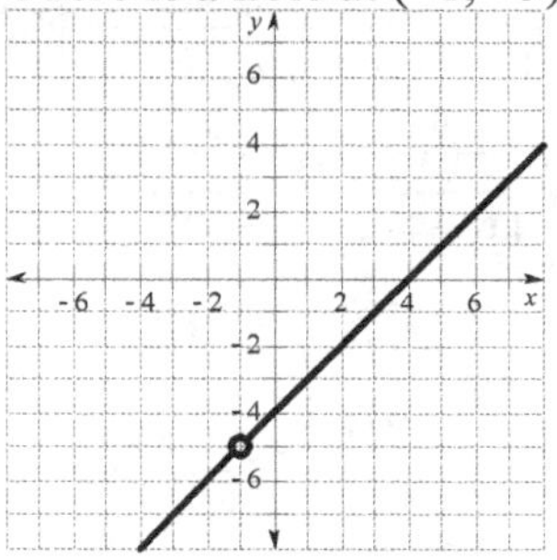

15. $\frac{x^7}{36} \bullet \frac{24}{x^4} = \frac{24x^7}{36x^4} = \frac{2x^3}{3}$

17. $\frac{6y-18}{9y} \bullet \frac{10}{5y-15} - \frac{60(y-3)}{45y(y-3)} = \frac{4}{3y}$

19. $\frac{a^2-2a}{a^2-4} \div \frac{2a^2}{3a+6} = \frac{a(a-2)}{(a+2)(a-2)} \bullet \frac{3(a+2)}{2a^2}$

$$= \frac{3}{2a}$$

21. $\frac{x^2-2xy-3y^2}{x^2-xy-2y^2} \bullet \frac{x^2-4y^2}{x^2-8xy+15y^2}$

$$= \frac{(x-3y)(x+y) \bullet (x+2y)(x-2y)}{(x-2y)(x+y) \bullet (x-3y)(x-5y)}$$

$$= \frac{x+2y}{x-5y}$$

23. $f(x) = \frac{x^2-16}{x-5} = \frac{(x+4)(x-4)}{x-5}$

$$g(x) = \frac{x^2-25}{x+4} = \frac{(x+5)(x-5)}{x+4}$$

(a) $f(3) \bullet g(3) = \left(\frac{7}{2}\right) \bullet \left(-\frac{16}{7}\right) = -8$

(b) $h(x) = f(x) \bullet g(x)$

$$= \frac{(x+4)(x-4)}{x-5} \bullet \frac{(x+5)(x-5)}{x+4}$$

$$= (x-4)(x+5) = x^2 + x - 20$$

(c) $h(3) = -8$

25. $\dfrac{5x+7}{x+4}-\dfrac{2x-5}{x-4}$

$=\left(\dfrac{x-4}{x-4}\right)\left(\dfrac{5x+7}{x+4}\right)-\left(\dfrac{x+4}{x+4}\right)\left(\dfrac{2x-5}{x-4}\right)$

$=\dfrac{(x-4)(5x+7)-(x+4)(2x-5)}{(x+4)(x-4)}$

$=\dfrac{5x^2-13x-28-\left(2x^2+3x-20\right)}{(x+4)(x-4)}$

$=\dfrac{3x^2-16x-8}{(x+4)(x-4)}$

27. $\dfrac{2}{x-5}-\dfrac{1}{x}=\dfrac{2x-(x-5)}{x(x-5)}=\dfrac{x+5}{x(x-5)}$

29. $\dfrac{2}{3m-3}-\dfrac{5}{2m-2}=\dfrac{2}{3(m-1)}-\dfrac{5}{2(m-1)}$

$=\dfrac{2\cdot 2-5\cdot 3}{6(m-1)}=-\dfrac{11}{6(m-1)}$

31. $\dfrac{6}{3x+3}-\dfrac{6}{5x-5}=\dfrac{6}{3(x+1)}-\dfrac{6}{5(x-1)}$

$=\dfrac{6\cdot 5(x-1)-6\cdot 3(x+1)}{3\cdot 5\cdot (x+1)(x-1)}$

$=\dfrac{12x-48}{15(x+1)(x-1)}$

$=\dfrac{4(x-4)}{5(x+1)(x-1)}$

33. $\dfrac{2}{s-1}-\dfrac{6s}{s^2+s-2}=\dfrac{2}{s-1}-\dfrac{6s}{(s-1)(s+2)}$

$=\dfrac{2(s+2)-6s}{(s-1)(s+2)}$

$=\dfrac{-4s+4}{(s-1)(s+2)}$

$=-\dfrac{4(s-1)}{(s-1)(s+2)}=-\dfrac{4}{s+2}$

35. $\dfrac{x^2-14x-8}{x^2-2x-8}+\dfrac{2x}{x-4}-\dfrac{3}{x+2}$

$=\dfrac{x^2-14x-8}{(x-4)(x+2)}+\dfrac{2x(x+2)}{(x-4)(x+2)}$

$-\dfrac{3(x-4)}{(x-4)(x+2)}$

$=\dfrac{3x^2-13x+4}{(x-4)(x+2)}=\dfrac{(3x-1)(x-4)}{(x-4)(x+2)}=\dfrac{3x-1}{x+2}$

37. $f(x)=\dfrac{2x}{x-2}$

$g(x)=\dfrac{x}{x-3}$

(a) $f(4)+g(4)=(4)+(4)=8$

(b) $h(x)=f(x)+g(x)=\dfrac{2x}{x-2}+\dfrac{x}{x-3}$

$=\dfrac{2x(x-3)+x(x-2)}{(x-2)(x-3)}$

$=\dfrac{3x^2-8x}{(x-2)(x-3)}=\dfrac{x(3x-8)}{(x-2)(x-3)}$

(c) $(4,h(4))=(4,8)$

39. $\dfrac{\frac{x^3}{15}}{\frac{x^5}{10}}=\dfrac{x^3}{15}\div\dfrac{x^5}{10}=\dfrac{x^3}{15}\cdot\dfrac{10}{x^5}=\dfrac{2}{3x^2}$

41. $\dfrac{1+\frac{a}{b}}{1-\frac{a}{b}}=\dfrac{b\left(1+\frac{a}{b}\right)}{b\left(1-\frac{a}{b}\right)}=\dfrac{b+a}{b-a}$

43. $\dfrac{\frac{1}{r^2}-\frac{1}{s^2}}{\frac{1}{r}-\frac{1}{s}}=\dfrac{\left(\frac{1}{r}+\frac{1}{s}\right)\left(\frac{1}{r}-\frac{1}{s}\right)}{\frac{1}{r}-\frac{1}{s}}=\dfrac{1}{r}+\dfrac{1}{s}=\dfrac{r+s}{rs}$

45. $\dfrac{1-\dfrac{2}{x-1}}{x+\dfrac{3}{x-4}}=\dfrac{(x-1)(x-4)-2(x-4)}{x(x-1)(x-4)+3(x-1)}$

$=\dfrac{x^2-7x+12}{(x-1)(x^2-4x+3)}$

$=\dfrac{(x-3)(x-4)}{(x-1)(x-3)(x-1)}=\dfrac{x-4}{(x-1)^2}$

47. $\dfrac{1}{1-\dfrac{1}{1-\dfrac{1}{y-1}}}=\dfrac{1}{1-\dfrac{1}{\dfrac{y-1}{y-1}-\dfrac{1}{y-1}}}=\dfrac{1}{1-\dfrac{1}{\dfrac{y-2}{y-1}}}$

$=\dfrac{1}{1-\dfrac{y-1}{y-2}}=\dfrac{1}{\dfrac{y-2}{y-2}-\dfrac{y-1}{y-2}}$

$=\dfrac{1}{\dfrac{-1}{y-2}}=\dfrac{y-2}{-1}=-y+2$

49. $\dfrac{1-\dfrac{1}{x-1}}{x-\dfrac{8}{x+2}}=\dfrac{\dfrac{x-2}{x-1}}{\dfrac{x(x+2)-8}{x+2}}$

$=\dfrac{x-2}{x-1}\bullet\dfrac{x+2}{x^2+2x-8}$

$=\dfrac{x+2}{(x-1)(x+4)}$

51. $f(x)=\dfrac{x-3}{x+4}$; Domain $\{x|x\neq -4\}$

53. $f(x)=\dfrac{8x}{x-2}$; Domain $\{x|x\neq 2\}$

55. $f(x)=\dfrac{2x-6}{x+1}$; Domain $\{x|x\neq -1\}$

57. $f(x)=\dfrac{x-3}{x+4}$

$f(0)=\dfrac{0-3}{0+4}=-\dfrac{3}{4}$; y-intercept $\left(0,-\dfrac{3}{4}\right)$

$0=\dfrac{x-3}{x+4}$ has solution $x=3$, so the x-intercept is (3, 0).

59. $f(x)=\dfrac{8x}{x-2}$

$f(0)=\dfrac{8(0)}{0-2}=\dfrac{0}{-2}=0$; y-intercept (0, 0)

$0=\dfrac{8x}{x-2}$ has solution $x=0$, so the x-intercept is (0, 0).

61. $f(x)=\dfrac{2x-6}{x+1}$

$f(0)=\dfrac{2(0)-6}{0+1}=\dfrac{-6}{1}=-6$

The y-intercept is (0, –6).

$0=\dfrac{2x-6}{x+1}=2x-6$

$6=2x$

$3=x$

The x-intercept is (3, 0).

63. $f(x)=\dfrac{x-3}{x+4}$ has vertical asymptote $x=-4$ since the function is undefined at that value. The horizontal asymptote is $y=1$ since the y-values get closer and closer to 1 when larger and larger x-values are tried.

65. $f(x)=\dfrac{8x}{x-2}$ has vertical asymptote $x=2$ since the function is undefined at that value. The horizontal asymptote is $y=8$ since the y-values get closer and closer to 8 when larger and larger x-values are tried.

67. $f(x)=\dfrac{2x-6}{x+1}$ has vertical asymptote $x=-1$ since the function is undefined at that value. The horizontal asymptote is $y=2$ since the y-values get closer and closer to 2 when larger and larger x-values are tried.

69. $f(x)=\dfrac{x-3}{x+4}$

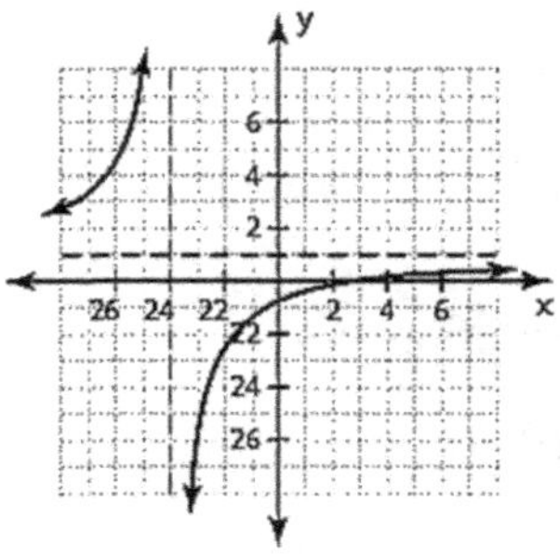

71. $f(x)=\dfrac{2x-6}{x+1}$

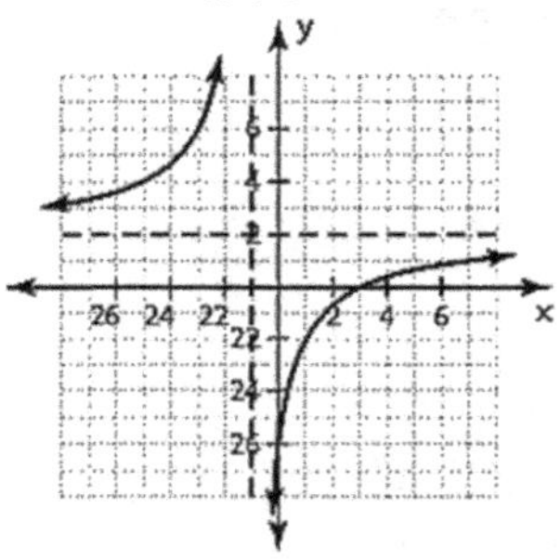

73. $\text{LCD}=6x$

$$\frac{1}{2x}+\frac{1}{3x}=\frac{1}{6}$$
$$6x\left(\frac{1}{2x}+\frac{1}{3x}\right)=6x\bullet\frac{1}{6}$$
$$6x\bullet\frac{1}{2x}+6x\bullet\frac{1}{3x}=x$$
$$3+2=x$$
$$x=5;\{5\}$$

75. $\text{LCD}=x-2$

$$\frac{x}{x-2}+1=\frac{x+4}{x-2}$$
$$x+(x-2)=x+4$$
$$x=6;\{6\}$$

77. $\text{LCD}=(3x+1)(x+2)$

$$\frac{2}{3x+1}=\frac{1}{x+2}$$
$$2(x+2)=3x+1$$
$$3=x;\{3\}$$

79. $\text{LCD}=(x-1)(3x-7)$

$$\frac{4}{x-1}-\frac{5}{3x-7}=\frac{3}{x-1}$$
$$4(3x-7)-5(x-1)=3(3x-7)$$
$$7x-23=9x-21$$
$$x=-1;\{-1\}$$

81. $\text{LCD}=(x+3)(x-3)$

$$\frac{2}{x-3}-\frac{11}{x^2-9}=\frac{3}{x+3}$$
$$2(x+3)-11=3(x-3)$$
$$2x-5=3x-9$$
$$x=4;\{4\}$$

83. $\text{LCD}=(x-2)(x-4)$

$$\frac{2}{x-4}=\frac{x}{x-2}-\frac{x+4}{x^2-6x+8}$$
$$2(x-2)=x(x-4)-(x+4)$$
$$2x-4=x^2-5x-4$$
$$x^2-7x=0$$
$$x(x-7)=0$$
$$\{0, 7\}$$

85.
$$\frac{x}{3x-1}=\frac{x+4}{5x+2}$$
$$x(5x+2)=(x+4)(3x-1)$$
$$5x^2+2x=3x^2+11x-4$$
$$2x^2-9x+4=0$$
$$(2x-1)(x-4)=0$$
$x=4$; $3x-1=11$; $x+4=8$; $5x+2=22$
The sides are 4 and 11 on the first triangle and the sides are 8 and 22 on the second triangle.

Note: $x=\frac{1}{2}$ is not an integer solution.

87. Let x be the integer. Then $x+1$ is the next integer.
$$\frac{1}{x}+\frac{1}{x+1}=\frac{11}{x(x+1)}$$
$$(x+1)+(x)=11$$
$$2x=10$$
$$x=5$$
$$x+1=6$$
The integers are 5 and 6.

89. Let t be the time it takes, working together, to wire the house.
$$\frac{t}{20}+\frac{t}{30}=1$$
$$60\left(\frac{t}{20}+\frac{t}{30}\right)=60\bullet 1$$
$$3t+2t=60$$
$$5t=60$$
$$t=12$$
It would take 12 hr, working together, to wire the house.

Chapter Test 9

1. $$\frac{m^2-3m}{m^2-9}\div\frac{4m}{m^2-m-12}$$
$$=\frac{m(m-3)}{(m+3)(m-3)}\bullet\frac{(m-4)(m+3)}{4m}$$
$$=\frac{m-4}{4}$$

3. $$\frac{3-\frac{x}{y}}{9-\frac{x^2}{y^2}}=\frac{3-\frac{x}{y}}{\left(3+\frac{x}{y}\right)\left(3-\frac{x}{y}\right)}=\frac{1}{3+\frac{x}{y}}$$
$$=\frac{y\bullet 1}{y\left(3+\frac{x}{y}\right)}=\frac{y}{3y+x}$$

5. $$\frac{3}{x^2-3x-4}+\frac{5}{x^2-16}$$
$$=\frac{3}{(x-4)(x+1)}+\frac{5}{(x-4)(x+4)}$$
$$=\frac{3(x+4)+5(x+1)}{(x+1)(x+4)(x-4)}$$
$$=\frac{8x+17}{(x+1)(x+4)(x-4)}$$

7. $$\frac{9x^2-9x-4}{6x^2-11x+3}\bullet\frac{15-10x}{3x-4}$$
$$=\frac{(3x-4)(3x+1)\bullet 5(3-2x)}{(2x-3)(3x-1)\bullet(3x-4)}=-\frac{5(3x+1)}{3x-1}$$

9. $$\frac{x^2+3xy}{2x^3-x^2y}\div\frac{x^2+6xy+9y^2}{4x^2-y^2}$$
$$=\frac{x(x+3y)}{x^2(2x-y)}\bullet\frac{(2x+y)(2x-y)}{(x+3y)^2}$$
$$=\frac{2x+y}{x(x+3y)}$$

11. $\dfrac{5}{x-2}-\dfrac{1}{x}=\dfrac{5x-(x-2)}{x(x-2)}=\dfrac{4x+2}{x(x-2)}$
$=\dfrac{2(2x+1)}{x(x-2)}$

13. $\dfrac{3w^2+w-2}{3w^2-8w+4}=\dfrac{(3w-2)(w+1)}{(3w-2)(w-2)}=\dfrac{w+1}{w-2}$

15. $\dfrac{-21x^5y^3}{28xy^5}=-\dfrac{3x^4}{4y^2}$

17. $f(x)=\dfrac{3x}{x+1}$

The domain is $\{x|x\neq -1\}$. The vertical asymptote is $x=-1$. The horizontal asymptote is $y=3$.

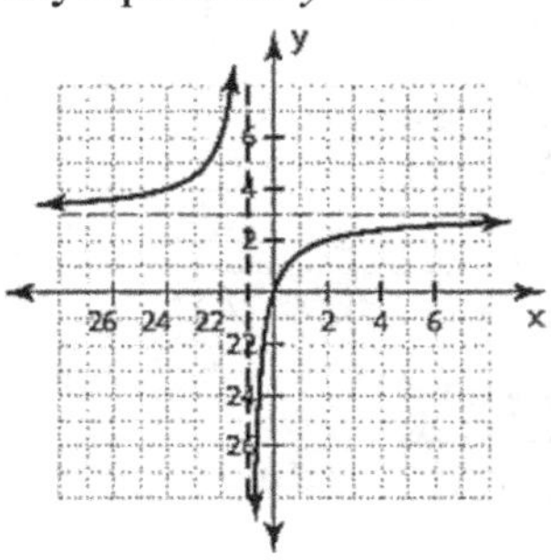

19. $f(x)=\dfrac{x^2-5x+4}{x-4}=\dfrac{(x-4)(x-1)}{x-4}=x-1;$
$x\neq 4$

Cumulative Review: Chapters 0–9

1. $5x-3(2x+6)=4-(3x-2)$
$5x-6x-18=4-3x+2$
$-x-18=-3x+6$
$2x=24$
$x=12;\{12\}$

3. $6x+7y=42$
$7y=-6x+42$
$y=-\dfrac{6}{7}x+6$
$m=-\dfrac{6}{7}$

The equation can be written $y=-\dfrac{6}{7}x-3$ or $6x+7y=-21$.

5. $f(x)=3x-2[x-(3x-1)]+6x(x-2)$
$=3x-2(-2x+1)+6x^2-12x$
$=6x^2-5x-2$

7. $f(x)=\dfrac{x^2+x}{x}=\dfrac{x(x+1)}{x}$
$D=\{x|x\neq 0\}$

9. $6x^3+7x^2-3x=x(6x^2+7x-3)$
$=x(2x+3)(3x-1)$

11. $\dfrac{5}{x-1}-\dfrac{2x+6}{x^2+2x-3}=\dfrac{5(x+3)-(2x+6)}{(x-1)(x+3)}$
$=\dfrac{3x+9}{(x-1)(x+3)}=\dfrac{3}{x-1}$

13. $\dfrac{1-\dfrac{3}{x+3}}{\dfrac{1}{x^2-9}}=\left(\dfrac{x+3}{x+3}-\dfrac{3}{x+3}\right)\div\dfrac{1}{x^2-9}$
$=\dfrac{x}{x+3}\bullet\dfrac{(x+3)(x-3)}{1}=x(x-3)$
$=x^2-3x$

15. $$\begin{aligned} 7x+(x-10)&=-12(x-5)\\ 8x-10&=-12x+60\\ 8x&=-12x+70\\ 20x&=70\\ x&=\frac{70}{20}=\frac{7}{2} \end{aligned}$$
The answer is $\left\{\frac{7}{2}\right\}$.

17. $$\begin{aligned} -4(7x+6)&=8(5x+12)\\ -28x-24&=40x+96\\ -28x&=40x+120\\ -68x&=120\\ x&=\frac{120}{-68}=-\frac{30}{17} \end{aligned}$$
The answer is $\left\{-\frac{30}{17}\right\}$.

19. $$\begin{aligned} \frac{5}{x}&=\frac{2}{x+3}\\ 5(x+3)&=2x\\ 5x+15&=2x\\ 15&=-3x\\ -5&=x \end{aligned}$$
The answer is $\{-5\}$.

21. $$\begin{aligned} x^2+15^2&=22^2\\ x^2+225&=484\\ x^2&=259\\ x&=\sqrt{259}\approx 16.1 \text{ cm} \end{aligned}$$

23. $$\left(\frac{a^{-2}b}{a^3b^{-2}}\right)^2=\frac{a^{-4}b^2}{a^6b^{-4}}=a^{-4-6}b^{2-(-4)}=a^{-10}b^6$$
$$=\frac{b^6}{a^{10}}$$

25. Let W be the width of the rectangle. Then $2W-2$ is its length.
$$\begin{aligned} W(2W-2)&=180\\ 2W^2-2W-180&=0\\ W^2-W-90&=0\\ (W-10)(W+9)&=0\\ W&=10\\ 2W-2&=18 \end{aligned}$$
The rectangle is 10 cm by 18 cm.

Chapter 10
Exponential and Logarithmic Functions

Exercises 10.1

< Objectives 1 and 2 >

1. $(f+g)(2)=f(2)+g(2)=3+4=7$

3. $(k-g)(7)=k(7)-g(7)=-3-(-1)=-2$

5. $(f+k)(2)=f(2)+k(2)$

$k(2)$ does not exist. Therefore, $(f+k)(2)$ does not exist.

7. $D=\{0,2\}$

9. $D=\{0,2\}$

11. $f(x)=-4x+5$

$g(x)=7x-4$

(a) $(f+g)(x)=f(x)+g(x)$
$=(-4x+5)+(7x-4)$
$=3x+1$

(b) $(f-g)(x)=f(x)-g(x)$
$=(-4x+5)-(7x-4)$
$=-4x+5-7x+4$
$=-11x+9$

(c) $(f+g)(3)=3(3)+1=10$

(d) $(f-g)(2)=-11(2)+9=-13$

13. $f(x)=8x-2$

$g(x)=-5x+6$

(a) $(f+g)(x)=(8x-2)+(-5x+6)$
$=3x+4$

(b) $(f-g)(x)=(8x-2)-(-5x+6)$
$=8x-2+5x-6=13x-8$

(c) $(f+g)(3)=3(3)+4=13$

(d) $(f-g)(2)=13(2)-8=18$

15. $f(x)=x^2+x-1$

$g(x)=-3x^2-2x+5$

(a) $(f+g)(x)$
$=(x^2+x-1)+(-3x^2-2x+5)$
$=-2x^2-x+4$

(b) $(f-g)(x)$
$=(x^2+x-1)-(-3x^2-2x+5)$
$=x^2+x-1+3x^2+2x-5$
$=4x^2+3x-6$

(c) $(f+g)(3)=-2(3)^2-(3)+4=-17$

(d) $(f-g)(2)=4(2)^2+3(2)-6=16$

17. $f(x)=-x^3-5x+8$
$g(x)=2x^2+3x-4$

(a) $(f+g)(x)$
$=(-x^3-5x+8)+(2x^2+3x-4)$
$=-x^3+2x^2-2x+4$

(b) $(f-g)(x)$
$=(-x^3-5x+8)-(2x^2+3x-4)$
$=-x^3-5x+8-2x^2-3x+4$
$=-x^3-2x^2-8x+12$

(c) $(f+g)(3)=-(3)^3+2(3)^2-2(3)+4$
$=-11$

(d) $(f-g)(2)$
$=-(2)^3-2(2)^2-8(2)+12=-20$

19. (a) $(f+g)(x)=f(x)+g(x)$
$=(-9x+11)+(15x-7)$
$=6x+4$

(b) $D=\mathbb{R}$

20. (a) $(f+g)(x)=f(x)+g(x)$
$=(-11x+3)+(8x-5)$
$=-3x-2$

(b) $D=\mathbb{R}$

21. (a) $(f+g)(x)=(3x+2)+\left(\frac{1}{x-2}\right)$
$=3x+2+\frac{1}{x-2}$

(b) $D=\{x|x\neq 2\}$

23. (a) $(f+g)(x)=x^2+x-5+\frac{2}{3x+1}$

(b) $3x+1\neq 0$
$x\neq -\frac{1}{3}$
$D=\left\{x\middle|x\neq -\frac{1}{3}\right\}$

< Objectives 3 and 4 >

25. $(f\bullet g)(2)=f(2)\bullet g(2)=(5)(-3)=-15$

27. $(f\bullet k)(0)=f(0)\bullet k(0)=(-4)(0)=0$

29. $(f\div h)(-3)=f(-3)\div h(-3)=(15)\div(-3)$
$=-5$

31. $(g\div k)(0)=g(0)\div k(0)=(12)\div(0)$ is undefined.

33. $D=\{0,2\}$

35. $f(x)=2x-1$
$g(x)=x-3$

(a) $(f\bullet g)(x)=f(x)\bullet g(x)$
$=(2x-1)(x-3)$
$=2x^2-7x+3$

(b) $(f\div g)(x)=f(x)\div g(x)=\frac{2x-1}{x-3}$

(c) $D=\{x|x\neq 3\}$

37. $f(x)=3x+2$
$g(x)=2x-1$

(a) $(f\bullet g)(x)=(3x+2)(2x-1)$
$=6x^2+x-2$

(b) $(f\div g)(x)=\frac{3x+2}{2x-1}$

(c) $2x-1\neq 0$
$x\neq\frac{1}{2}$
$D=\left\{x\middle|x\neq\frac{1}{2}\right\}$

39. $f(x)=2-x$

$g(x)=5+2x$

(a) $(f\bullet g)(x)=(2-x)(5+2x)$

$=-2x^2-x+10$

(b) $(f\div g)(x)=\dfrac{2-x}{5+2x}$

(c) $5+2x\neq 0$

$x\neq -\dfrac{5}{2}$

$D=\left\{x\middle|x\neq -\dfrac{5}{2}\right\}$

41. $P(x)=R(x)-C(x)$

$=(25x)-(x^2+4x+50)$

$=-x^2+21x-50$

The profit is given by $P(x)=-x^2+21x-50$.

43. $V(t)=V_0+g\bullet t=10-9.8t$

The velocity (in m/s) is given by $V(t)=10-9.8t$.

45. $R(x)=x\bullet p(x)=x(119-6x)=119x-6x^2$

46. $R(x)=x\bullet p(x)=x(1{,}190-36x)$

$=1{,}190x-36x^2$

47. Selling price of the ice cream maker = \$89.85

Wholesale purchase price of the ice cream maker = \$58.47

Weekly fixed cost associated with the ice cream makers = \$225

(a) Let the store sell x units of ice cream maker in a week. Therefore, the weekly revenue of the store $=\$89.85\times x$.

Store's weekly revenue function:

$R(x)=89.95x$

Store's weekly cost = purchase cost of x ice cream makers + weekly fixed cost

$=\$58.47\times x+\$225$

Store's weekly cost function:

$C(x)=58.47x+225$

(b) We evaluate the revenue and cost for $x=15$.

$R(15)=89.95(15)=1{,}349.25$

$C(15)=58.47(15)+225=877.05+225$

$=1{,}102.05$

If they sell 15 ice cream makers in a week, they earn \$1,349.25 while incurring \$1,102.05 in costs.

(c) Profit = revenue – cost

$=1{,}349.25-1{,}102.05=247.20$

The profit earned on the sale of 15 ice cream makers = \$247.20.

49. Selling price of pairs of rattan escrima sticks = \$20.99
Wholesale purchase price of pairs of rattan escrima sticks = \$8.40
Weekly fixed cost associated with the pairs of rattan escrima sticks = \$295

(a) Let the store sell x pairs of rattan escrima sticks in a week. Therefore, the weekly revenue of the store $= \$20.99 \times x$.
Store's weekly revenue function:
$R(x) = 20.99x$
Store's weekly cost = purchase cost of x pairs of rattan escrima sticks + weekly fixed cost
$= \$8.40 \times x + \$295$
Store's weekly cost function:
$C(x) = 8.4x + 295$

(b) We evaluate the revenue and cost for $x = 30$.
$R(30) = 20.99(30) = 629.70$
$C(30) = 8.4(30) + 295 = 252 + 295 = 547$
If they sell 30 pairs of rattan escrima sticks in a week, they earn \$629.70 while incurring \$547 in costs.

(c) Profit = revenue − cost
$= 629.70 - 547 = 82.7$
The profit earned on the sale of 30 pairs of rattan escrima sticks = \$82.70.

(d) Profit = revenue − cost
$P(x) = R(x) - C(x) = 20.99x - (8.4x + 295)$
$= 12.59x - 295$
The store's profit function is:
$P(x) = 12.59x - 295$.

(e) We evaluate profit for $x = 30$.
$P(30) = 12.59(30) - 295 = 377.7 - 295 = 82.70$

The profit is \$82.70.

51. The domain of $f + g$ is sometimes the same as the domain of f.

53. If f and g are linear functions then the function $f \bullet g$ will always be a second-degree (quadratic) function.

Exercises 10.2

< Objective 1 >

1. $(f \circ g)(4) = f(g(4)) = f(3) = -3$

3. $(h \circ g)(1) = h(g(1)) = h(-2) = 5$

5. $(g \circ h)(3) = g(h(3)) = g(6) = 2$

7. $(h \circ k)(0) = h(k(0)) = h(4)$
$h(4)$ does not exist.. Therefore, $(h \circ k)(0)$ does not exist.

9. $(f \circ h)(3) = f(h(3)) = f(6)$
$f(6)$ does not exist. Therefore, $(f \circ h)(3)$ does not exist.

11. $(k \circ k)(2) = k(k(2)) = k(3) = -4$

< Objective 2 >

13. $f(x) = x - 3$
$g(x) = 2x + 1$

(a) $(f \circ g)(0) = f(g(0)) = f(1) = -2$

(b) $(f \circ g)(-2) = f(g(-2)) = f(-3) = -6$

(c) $(f \circ g)(3) = f(g(3)) = f(7) = 4$

(d) $(f \circ g)(x) = f(g(x)) = f(2x + 1)$
$= (2x + 1) - 3 = 2x - 2$

15. $f(x) = 3x + 1$
$g(x) = 4x - 3$

(a) $(f \circ g)(0) = f(g(0)) = f(-3) = -8$

(b) $(f \circ g)(-2) = f(g(-2)) = f(-11)$
$= -32$

(c) $(g \circ f)(3) = g(f(3)) = g(10) = 37$

(d) $(g \circ f)(x) = g(f(x)) = g(3x + 1)$
$= 4(3x + 1) - 3 = 12x + 1$

17. $f(x)=x^2$

$g(x)=x+3$

(a) $(f\circ g)(0)=f(g(0))=f(3)=9$

(b) $(f\circ g)(-2)=f(g(-2))=f(1)=1$

(c) $(g\circ f)(3)=g(f(3))=g(9)=12$

(d) $(g\circ f)(x)=g(f(x))=g(x^2)$
$=x^2+3$

19. $f(x)=2x^2-1$

$g(x)=-2x$

(a) $(g\circ f)(0)=g(f(0))=g(2(0)^2-1)$
$=g(-1)=2$

(b) $(g\circ f)(-2)=g(f(-2))$
$=g(2(-2)^2-1)=g(7)$
$=-14$

(c) $(f\circ g)(3)=f(g(3))=f(-6)=71$

(d) $(f\circ g)(x)=f(g(x))=f(-2x)$
$=2(-2x)^2-1=8x^2-1$

< Objective 3 >

21. $h(x)=(g\circ f)(x)$

23. $h(x)=(g\circ f)(x)$

25. $h(x)=(g\circ f)(x)$

27. $h(x)=(g\circ f)(x)$

29. $h(x)=(f\circ g)(x)$

< Objective 4 >

31. (a) $N=n(v)=0.8v$

(b) $C=c(N)=40N+1{,}000$

(c) $C=(c\circ n)(v)=40(0.8v)+1{,}000$
$=32v+1{,}000$

33. Let Bichvan have x dollars.
Amount of British pounds she gets when she arrives London $= 0.6221x$
Amount of Icelandic kronas she gets when she arrives Reykjavik
$=197.3749(0.6221x)=122.786x$.
Therefore, the exchange rate of 1 dollar is 122.786 Icelandic kronas. If she has 12,000 she would receive
$\frac{12{,}000}{122.786}$ dollars $=97.73$ dollars.

35. False

37. To compute $(f\circ g)(5)$, we must always find $g(5)$ first.

Exercises 10.3

< Objective 1 >

1. $f(x)=3x+5$

$y=3x+5$

$x=3y+5$

$y=\frac{x-5}{3}$

$f^{-1}(x)=\frac{x-5}{3}$

3. $f(x)=\frac{x-1}{2}$

$y=\frac{x-1}{2}$

$x=\frac{y-1}{2}$

$2x=y-1$

$y=2x+1$

$f^{-1}(x)=2x+1$

5. $f(x)=2x-3$

$y=2x-3$

$x=2y-3$

$y=\frac{x+3}{2}$

$f^{-1}(x)=\frac{x+3}{2}=3x-1$

7. $f(x)=\dfrac{x+4}{3}$

$y=\dfrac{x+4}{3}$

$x=\dfrac{y+4}{3}$

$3x=y+4$

$3x-4=y$

$f^{-1}(x)=3x-4$

9. $f(x)=\dfrac{x}{3}+5$

$y=\dfrac{x}{3}+5$

$x=\dfrac{y}{3}+5$

$3(x-5)=y$

$f^{-1}(x)=3(x-5)=3x-15$

11. $f(x)=\dfrac{6-5x}{3}$

$y=\dfrac{6-5x}{3}$

$x=\dfrac{6-5y}{3}$

$3x=6-5y$

$5y=-3x+6$

$y=\dfrac{-3x+6}{5}$

$f^{-1}(x)=\dfrac{6-3x}{5}$

$=-\dfrac{3}{5}(x-2)$

< Objective 2 >

13. $f(x)=\dfrac{11x}{5}+2$

$y=\dfrac{11x}{5}+2$

$x=\dfrac{11y}{5}+2$

$x-2=\dfrac{11y}{5}$

$y=\dfrac{5}{11}(x-2)$

$f^{-1}(x)=\dfrac{5x-10}{11}$

15. $f^{-1}=\{(-5,-2),(3,4),(4,3),(6,2)\}$ is a function.

17. $f^{-1}=\{(-4,7),(3,-4),(5,3),(7,-2)\}$ is a function.

19. $f^{-1}=\{(3,2),(4,3),(5,4)\}$ is a function.

21. $f^{-1}=\{(5,1),(5,2),(5,3)\}$ is not a function.

23. $f^{-1}=\{(3,2),(5,3),(7,4)\}$ is a function.

25. $f(x)=3x-6$

$f^{-1}(x)=\dfrac{x+6}{3}=\dfrac{x}{3}+2$

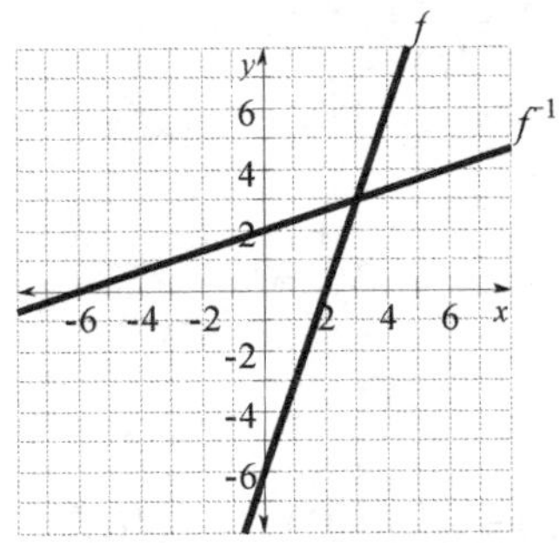

27. $f(x)=-2x+6$

$f^{-1}(x)=\dfrac{x-6}{-2}=-\dfrac{x}{2}+3$

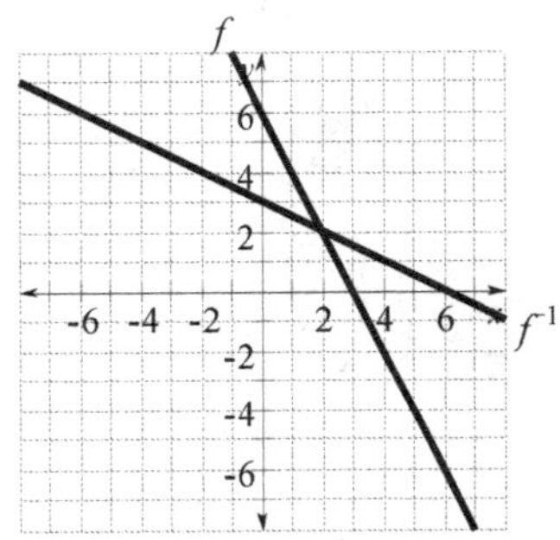

< Objective 3 >

29. f is not one-to-one; its inverse relation is not a function.

31. f is one-to-one; its inverse relation is a function.

33. The function is not one-to-one; its inverse relation is not a function.

35. The function is one-to-one; its inverse relation is a function.

37. $f(6)=3(6)-6=12$

39. $f(f^{-1}(6))=6$

41. $f\left(f^{-1}(x)\right)=x$

43. $g(3)=\dfrac{(3)+1}{2}=2$

45. $g\left(g^{-1}(3)\right)=3$

47. $g\left(g^{-1}(x)\right)=x$

49. $h(4)=16$

51. $h\left(h^{-1}(4)\right)=4$

53. $h\left(h^{-1}(x)\right)=x$

55. If $f(5)=7$, then $f^{-1}(7)=5$.

57. $f(x)=mx+b$

$$y=mx+b$$
$$x=my+b$$
$$x-b=my$$
$$y=\frac{x-b}{m}$$

$$f^{-1}(x)=\frac{x-b}{m}=\frac{1}{m}x-\frac{b}{m}$$

59. The slope of f^{-1} is $\dfrac{5}{2}$.

61. $f^{-1}(x)=\dfrac{x-b}{m}$
For *f*, $m=-2$, then $f=-2x+b$.
As *y*-intercept of *f* is (0, 5) then $b=5$.
Now, $f^{-1}(x)=\dfrac{x-5}{-2}$.
To find *y*-intercept put $x=0$.
y-intercept of $f^{-1}=\left(0,\dfrac{5}{2}\right)$

63. True

65. False; the graphs of a function and its inverse are symmetric to each other over the line $y=x$.

67. The inverse of a function will sometimes be a function.

69. Above and Beyond

Exercises 10.4

< Objective 1 >

1. **(c)** $y=2^x$

3. **(b)** $y=2x-1$

5. **(h)** $x=y^2$

7. **(f)** $y=5^x$

9. $f(0)=4^{(0)}=1$

11. $f(2)=4^{(2)}=16$

13. $g(-1)=4^{(-1)+1}=4^0=1$

15. $g(2)=4^{(2)+1}=4^3=64$

17. $h(-1)=4^{(-1)}+1=\dfrac{1}{4}+1=\dfrac{5}{4}$

19. $h(2)=4^{(2)}+1=17$

21. $f(1)=\left(\dfrac{1}{4}\right)^{(1)}=\dfrac{1}{4}$

23. $f(0)=\left(\dfrac{1}{4}\right)^{(0)}=1$

25. $y = 4^x$

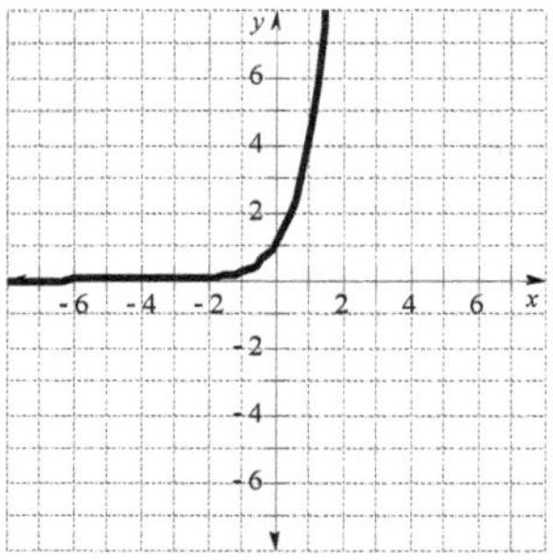

27. $y = \left(\frac{2}{3}\right)^x$

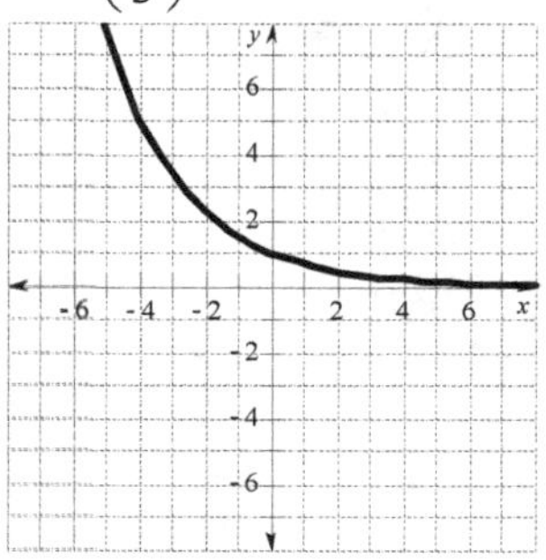

29. $y = 3 \cdot 2^x$

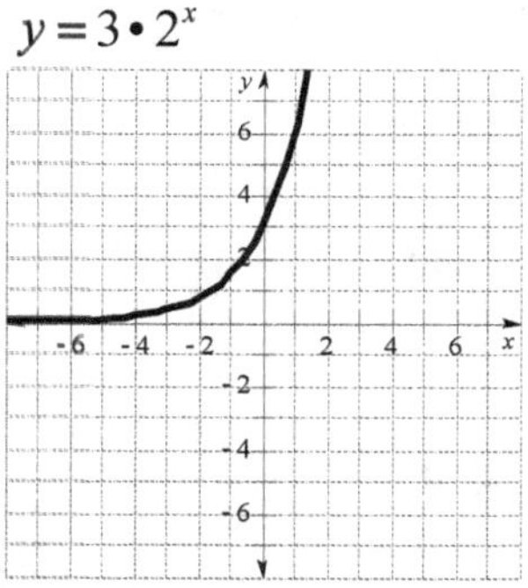

31. $y = 3^x$

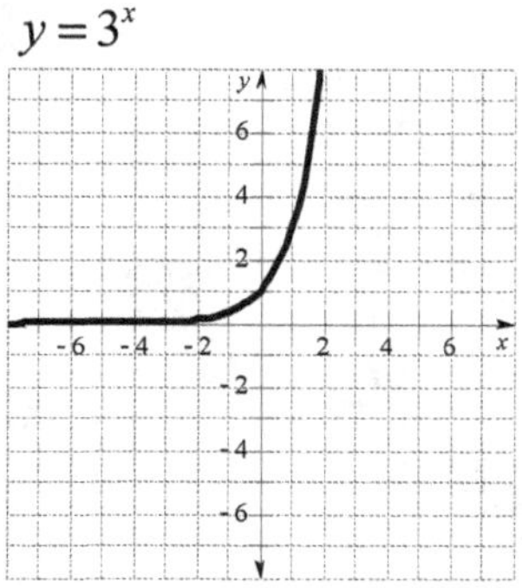

33. $y = 2^{2x}$

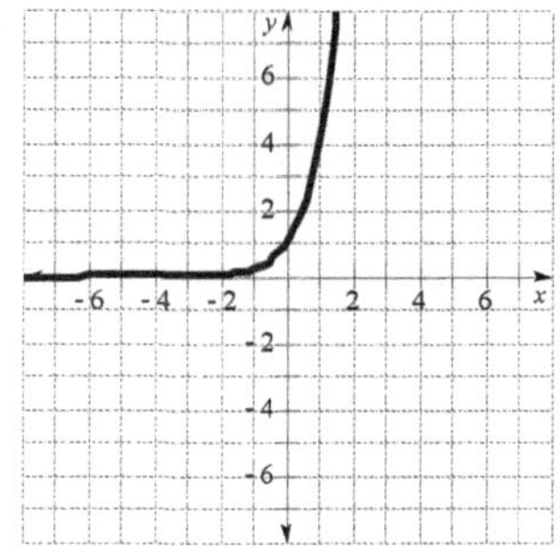

35. $y = e^{-x}$

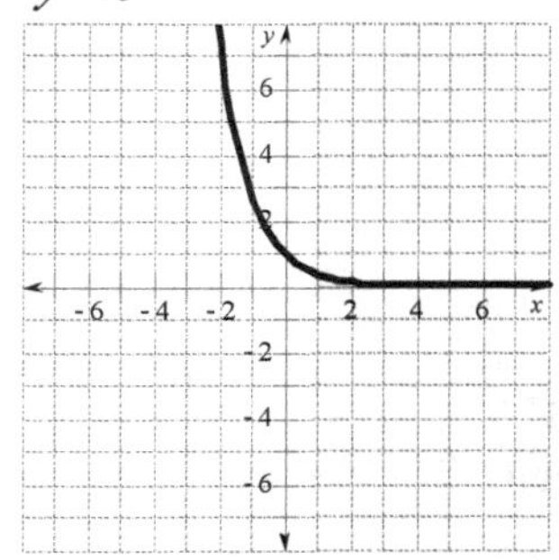

37. $y = 3^x + 2 = 2$

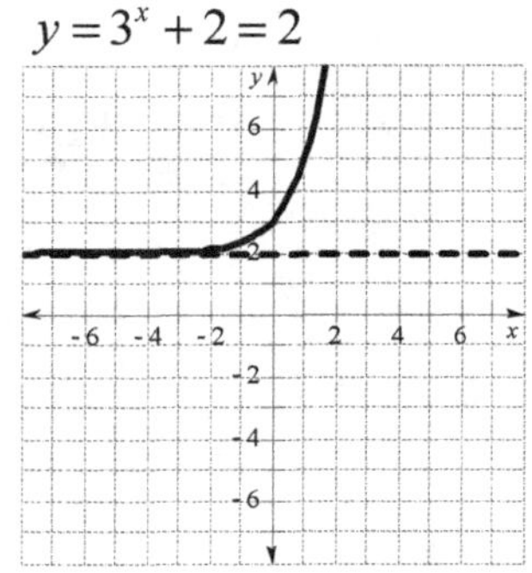

< Objective 3 >

39. $2^x = 128 = 2^7$

$x = 7; \{7\}$

41. $10^x = 10{,}000 = 10^4$

$x = 4; \{4\}$

43. $3^x = \frac{1}{9} = \frac{1}{3^2} = 3^{-2}$

$x = -2; \{-2\}$

45. $4^{2x} = 64 = 4^3$

$2x = 3$

$x = \frac{3}{2}; \left\{\frac{3}{2}\right\}$

47. $3^{x-1} = 81 = 3^4$

$x - 1 = 4$

$x = 5; \{5\}$

49. $3^{x-1} = \frac{1}{27} = 3^{-3}$

$x - 1 = -3$

$x = -2; \{-2\}$

< Objective 2 >

51. $N(2) = 100 \bullet 2^{(2)} = 400$

There are 400 bacteria present in the culture after 2 h.

53. $N(5) = 100 \bullet 2^{(5)} = 3{,}200$

There are 3,200 bacteria present in the culture after 5 h.

55. $R(1{,}690) = 64 \bullet 2^{-\frac{1690}{1690}} = 64 \bullet 2^{-1} = 64 \bullet \frac{1}{2}$

$= 32$

There are 32 g of radium left after 1,690 yr.

57. $R(5{,}070) = 64 \bullet 2^{-\frac{5070}{1690}} = 64 \bullet 2^{-3} = 64 \bullet \frac{1}{8} = 8$

There are 8 g of radium left after 5,070 yr.

59. $A(2) = 1{,}000(1.08)^2 \approx 1{,}166.40$

After 2 yr, the account balance is $1,166.40.

61. $A(9) = 1{,}000(1.08)^9 \approx 1{,}999.00$

After 9 yr, the account balance is $1,999.

63. **(a)** $N(10) = 80\left(1 - e^{-0.06(10)}\right) \approx 36$

A person can type 36 words per minute after 10 weeks of training.

(b) $N(20) = 80\left(1 - e^{-0.06(20)}\right) \approx 56$

A person can type 56 words per minute after 20 weeks of training.

(c) $N(30) = 80\left(1 - e^{-0.06(30)}\right) \approx 67$

A person can type 67 words per minute after 30 weeks of training.

65. Let x be the year since 2003

We begin by clearing data list.

[STAT] **4 : ClrList** [2nd] [L1] [,] [2nd] [L2]

[ENTER]

Then we enter data from list 1 and 2

[STAT] **1 : Edit** and type the numbers.

Make a scatterplot:

[2nd] [STAT PLOT] [ENTER]

We find best fitting exponential function by

[STAT] **CALC0 : ExpReg** [ENTER]

To four decimal places, we have

$y = 10.595(1.1532)^x$

67. $y = (10.0645)(1.0491)^x$

69. The function $g(x) = b^x$ is sometimes a decay function.

71. The graph of $f(x) = b^x$ never intersects the x-axis.

73. **(a)** $B(t) = (1.132)^t = (1.132)^{80} \approx 20{,}310$

After 80 days of growth, the biomass in a 1-acre field is 20,310 lb.

(b) $B(t) = (1.132)^t = (1.132)^{90} \approx 70{,}174$

$70{,}174 - 20{,}310 = 49{,}864$

In a 1-acre field, the biomass will increase 49,864 lb between day 80 and day 90.

75. $V_c(t) = V_{ps}\left(1 - e^{-\frac{t}{\tau}}\right) = 14\left(1 - e^{-\frac{150}{103}}\right) \approx 11$

The stored capacitive voltage is 11 V.

77. Above and Beyond

79. $\left(1 + \frac{1}{(100)}\right)^{(100)} \approx 2.7048$

81. $\left(1 + \frac{1}{(10{,}000)}\right)^{(10{,}000)} \approx 2.7181$

83. $\left(1 + \frac{1}{(1{,}000{,}000)}\right)^{(1{,}000{,}000)} \approx 2.71828$

85. $y = 2^x$

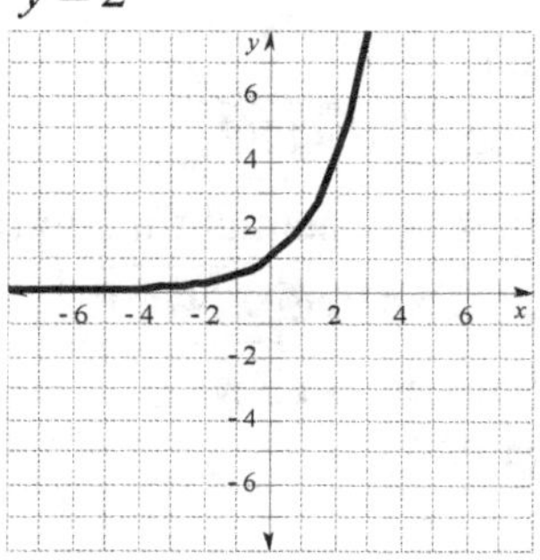

87. **(a)** $h(n) = 0.003 \cdot 2^n$

(b) $0.003 \cdot 2^n > 8$

Enter the function into a graphing calculator and determine when h is greater than 8.

$h(11) \approx 6.1$

$h(12) \approx 12.3$

After the 12th tear, the height of the stack will exceed 8 in.

(c) $h(15) \approx 98.3$

After the 15th tear, the height is 98.3 in., or nearly 8.2 ft.

Exercises 10.5

< Objective 1 >

1. $y = \log_4 x$

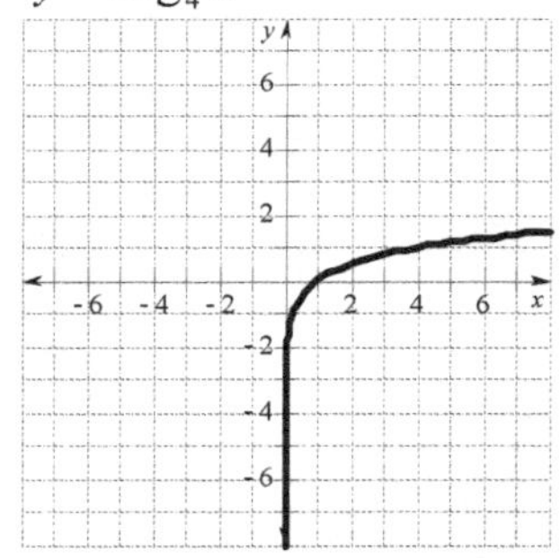

3. $y = \log_2(x-1)$

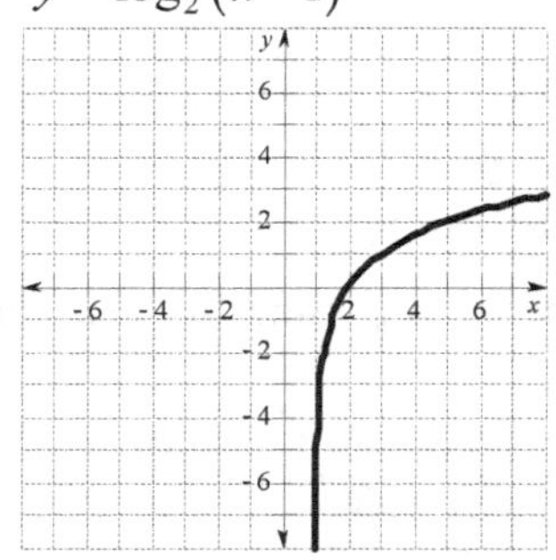

5. $y = \log_8 x$

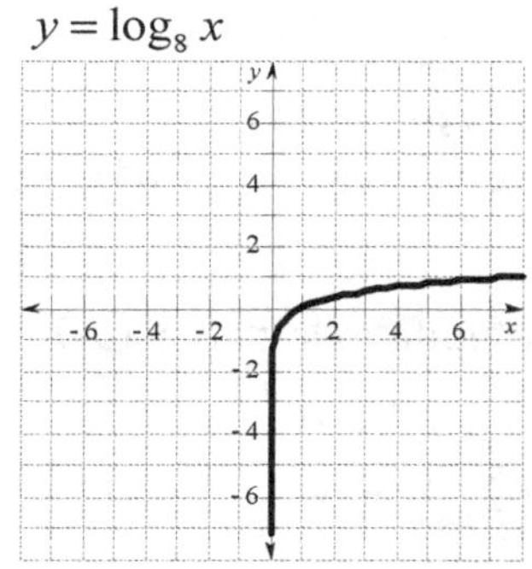

< Objective 2 >

7. $2^5 = 32$

$\log_2 32 = 5$

9. $10^2 = 100$

$\log_{10} 100 = 2$

11. $3^0 = 1$

$\log_3 1 = 0$

13. $4^{-2} = \frac{1}{16}$

$\log_4 \frac{1}{16} = -2$

15. $10^{-2} = \frac{1}{100}$

$\log_{10} \frac{1}{100} = -2$

17. $16^{\frac{1}{2}} = 4$

$\log_{16} 4 = \frac{1}{2}$

19. $64^{-\frac{1}{3}} = \frac{1}{4}$

$\log_{64} \frac{1}{4} = -\frac{1}{3}$

21. $27^{\frac{2}{3}} = 9$

$\log_{27} 9 = \frac{2}{3}$

23. $27^{-\frac{2}{3}} = \frac{1}{9}$

$\log_{27} \frac{1}{9} = -\frac{2}{3}$

25. $\log_2 16 = 4$

$2^4 = 16$

27. $\log_5 1 = 0$

$5^0 = 1$

29. $\log_{10} 10 = 1$

$10^1 = 10$

31. $\log_5 125 = 3$

$5^3 = 125$

33. $\log_3 \frac{1}{27} = -3$

$3^{-3} = \frac{1}{27}$

35. $\log_{10} 0.001 = -3$

$10^{-3} = 0.001$

37. $\log_{16} 4 = \frac{1}{2}$

$16^{\frac{1}{2}} = 4$

39. $\log_8 4 = \frac{2}{3}$

$8^{\frac{2}{3}} = 4$

41. $\log_{25} \frac{1}{5} = -\frac{1}{2}$

$25^{-\frac{1}{2}} = \frac{1}{5}$

< Objective 3 >

43. $\log_2 64 = \log_2 2^6 = 6$

45. $\log_4 64 = \log_4 4^3 = 3$

47. $\log_3 \frac{1}{81} = \log_3 3^{-4} = -4$

49. $\log_{10} \frac{1}{100} = \log_{10} 10^{-2} = -2$

51. $\log_{25} 5 = \log_{25} 25^{\frac{1}{2}} = \frac{1}{2}$

< Objective 4 >

53. $y = \log_5 25$

$5^y = 25 = 5^2$

$y = 2; \{2\}$

55. $\log_b 256 = 4$

$b^4 = 256 = 4^4$

$b = 4; \{4\}$

57. $\log_{10} x = 2$

$x = 10^2 = 100; \{100\}$

59. $y = \log_5 5$

$5^y = 5 = 5^1$

$y = 1; \{1\}$

61. $\log_{\frac{3}{2}} x = 3$

$x = \left(\frac{3}{2}\right)^3 = \frac{27}{8}; \left\{\frac{27}{8}\right\}$

63. $\log_b \frac{1}{25} = -2$

$b^{-2} = \frac{1}{25} = 5^{-2}$

$b = 5; \{5\}$

65. $\log_{10} x = -3$

$x = 10^{-3} = \frac{1}{1,000}; \left\{\frac{1}{1,000}\right\}$

67. $y = \log_8 \frac{1}{64}$

$8^y = \frac{1}{64} = 8^{-2}$

$y = -2; \{-2\}$

69. $\log_{27} x = \frac{1}{3}$

$x = 27^{\frac{1}{3}} = 3; \{3\}$

71. $\log_b 5 = \frac{1}{2}$

$b^{\frac{1}{2}} = 5 = 25^{\frac{1}{2}}$

$b = 25; \{25\}$

73. $y = \log_{27} \frac{1}{9}$

$27^y = \frac{1}{9} = 27^{-\frac{2}{3}}$

$y = -\frac{2}{3}; \left\{-\frac{2}{3}\right\}$

< Objective 5 >

75. $L = 10\log_{10} \frac{(10^{-11})}{10^{-16}} = 10\log_{10} 10^5 = 10 \cdot 5$

$= 50$

The commercial's rating is 50 dB.

77. $L = 10\log_{10} \frac{(10^{-9})}{10^{-16}} = 10\log_{10} 10^7 = 10 \cdot 7$

$= 70$

The printer's rating is 70 dB.

79. $I = I_0 \cdot 10^{\frac{L}{10}} = 10^{-16} \cdot 10^{\frac{80}{10}} = 10^{-16} \cdot 10^8 = 10^{-8}$

The area's intensity is 10^{-8} W/cm^2.

81. $I = I_0 \cdot 10^{\frac{L}{10}}$

$\frac{10^{-8}}{10^{-9}} = 10$

The ratio is 10.

83. $I = I_0 \cdot 10^{\frac{L}{10}}$

$I(70) = 10^{-16} \cdot 10^{\frac{70}{10}} = 10^{-16} \cdot 10^7 = 10^{-9}$

$I(40) = 10^{-16} \cdot 10^{\frac{40}{10}} = 10^{-16} \cdot 10^4 = 10^{-12}$

$\frac{10^{-9}}{10^{-12}} = 10^3 = 1,000$

The ratio is 1,000.

85. $M = \log_{10} \frac{(10^6 \cdot a_o)}{a_0} = \log_{10} 10^6 = 6$

The magnitude is 6.

87. $5 = \log_{10}\dfrac{a}{a_0}$

$\dfrac{a}{a_0} = 10^5$

$a = 10^5 a_0$

89. False; $m = \log_k n$ means the same as $k^m = n$.

91. The graph of $f(x) = \log_b x$ always has a vertical asymptote.

93. $\left[H^+\right] = (0.15)(0.65) = 0.0975$

$\text{pH} = -\log(0.0975) \approx 1.011$

The pH of the solution is 1.011.

95. $\text{dB} = 10\log\left(\dfrac{I}{I_0}\right) = 10\log\left(\dfrac{3.2\times10^{-9}}{10^{-12}}\right)$

$= 10\log(3.2\times10^3) \approx 35$

The decibel level in the factory is 35 dB.

97. Above and Beyond

99. $x = -\dfrac{1}{0.000029}\log_e\dfrac{1}{2} \approx 23{,}902$

The half-life of Plutonium-239 is 23,902 yr.

101. $x = -\dfrac{1}{0.000009}\log_e\dfrac{1}{2} \approx 77{,}016$

The half-life of Thorium-230 is 77,016 yr.

103. Multiply each of the above by 10.
Pu-239: 239,020 yr
Sr-90: 280 yr
Th-230: 770,160 yr
Cs-135: 19,804,210 yr

105. **(a)** $\log_3(9\cdot81) = \log_3 729 = \log_3 3^6 = 6$

(b) $\log_3 9 = \log_3 3^2 = 2$

$\log_3 81 = \log_3 3^4 = 4$

(c) $\log_3(9\cdot81) = \log_3 9 + \log_3 81$

Exercises 10.6

< Objective 1 >

1. $\log_b 5x = \log_b 5 + \log_b x$

3. $\log_6\dfrac{x}{7} = \log_6 x - \log_6 7$

5. $\log_3 a^2 = 2\log_3 a$

7. $\log_5\sqrt{x} = \log_5 x^{\frac{1}{2}} = \dfrac{1}{2}\log_5 x$

9. $\log_b x^2y^4 = \log_b x^2 + \log_b y^4$

$= 2\log_b x + 4\log_b y$

11. $\log_4 y^2\sqrt{x} = \log_4 y^2 + \log_4 x^{\frac{1}{2}}$

$= 2\log_4 y + \dfrac{1}{2}\log_4 x$

13. $\log_b\dfrac{x^2y}{z} = \log_b x^2y - \log_b z$

$= \log_b x^2 + \log_b y - \log_b z$

$= 2\log_b x + \log_b y - \log_b z$

15. $\log\dfrac{xy^2}{\sqrt{z}} = \log xy^2 - \log\sqrt{z}$

$= \log x + \log y^2 - \log z^{\frac{1}{2}}$

$= \log x + 2\log y - \dfrac{1}{2}\log z$

17. $\log_5\sqrt[3]{\dfrac{xy}{z^2}} = \log_5\left(\dfrac{xy}{z^2}\right)^{\frac{1}{3}} = \dfrac{1}{3}\log_5\dfrac{xy}{z^2}$

$= \dfrac{1}{3}\left(\log_5 xy - \log_5 z^2\right)$

$= \dfrac{1}{3}\left(\log_5 x + \log_5 y - 2\log_5 z\right)$

19. $\log_b x + \log_b y = \log_b xy$

21. $3\log_5 x - 2\log_5 y = \log_5 x^3 - \log_5 y^2$

$= \log_5 \dfrac{x^3}{y^2}$

23. $\log_b x + \dfrac{1}{2}\log_b y = \log_b x + \log_b y^{\frac{1}{2}}$

$= \log_b xy^{\frac{1}{2}} = \log_b x\sqrt{y}$

25. $\log_b x - 2\log_b y - \log_b z$

$= \log_b x - \left(\log_b y^2 + \log_b z\right)$

$= \log_b x - \log_b y^2 z = \log_b \dfrac{x}{y^2 z}$

27. $\dfrac{1}{2}\log_6 y - 3\log_6 z = \log_6 y^{\frac{1}{2}} - \log_6 z^3$

$= \log_6 \dfrac{\sqrt{y}}{z^3}$

29. $\dfrac{1}{3}(2\log_b x + \log_b y - \log_b z) = \dfrac{1}{3}\log_b \dfrac{x^2 y}{z}$

$= \log_b \left(\dfrac{x^2 y}{z}\right)^{\frac{1}{3}}$

$= \log_b \sqrt[3]{\dfrac{x^2 y}{z}}$

< Objective 2 >

31. $\log 24 = \log\left(2^3 \bullet 3\right) = 3\log 2 + \log 3$

$= 3 \bullet 0.301 + 0.477 = 1.380$

33. $\log 8 = \log 2^3 = 3\log 2 = 3 \bullet 0.301 = 0.903$

35. $\log\sqrt{2} = \log 2^{\frac{1}{2}} = \dfrac{1}{2}\log 2 = \dfrac{1}{2} \bullet 0.301 = 0.151$

37. $\log\dfrac{1}{4} = \log 2^{-2} = -2\log 2 = -2 \bullet 0.301$

$= -0.602$

39. $10^{\log 8.2} = 8.2$

41. $\ln e^{5.8} = 5.8$

43. $3^2 = 9$; $3^3 = 27$

$\log_3 25$ is between 2 and 3.

45. $2^6 = 64$; $2^7 = 128$

$\log_2 70$ is between 6 and 7.

47. $10^2 = 100$; $10^3 = 1{,}000$

log680 is between 2 and 3.

49. $\log 5 + \log 2 = \log(5 \bullet 2) = \log 10 = 1$

51. $\log_3 45 - \log_3 5 = \log_3 \dfrac{45}{5} = \log_3 9 = 2$

53. True

55. False; $\log_m m = 1$

57. $\log 7.3 \approx 0.863$

59. $\log 680 \approx 2.833$

61. $\log 0.72 \approx -0.143$

63. $\ln 2 \approx 0.693$

65. $\ln 10 \approx 2.303$

< Objective 4 >

67. $10^{0.749} \approx 5.610$

69. $10^{3.749} \approx 5{,}610.480$

71. $e^{1.238} \approx 3.449$

73. $e^{-0.786} \approx 0.456$

< Objective 3 >

75. $\text{pH} = -\log\left(3.8 \times 10^{-8}\right) \approx 7.42$

Blood is basic.

77. $\mathrm{pH} = 4.7$

$\left[\mathrm{H}^+\right] = 10^{-4.7} \approx 2\times10^{-5}$

79. $S(3) = 85 - 8\ln\left[(3)+1\right] \approx 74$

After 3 weeks, the average score on retests is 74.

< Objective 5 >

81. $\log_3 25 = \dfrac{\log 25}{\log 3} \approx 2.930$

83. $\log_2 10 = \dfrac{\log 10}{\log 2} = 3.322$

85. $\log_{12} 8 = \dfrac{\log 8}{\log 12} = 0.837$

87. $y = -4.2465 + 6.2220\ln x$

89. $A(50{,}000) = e^{-0.000029\bullet(50{,}000)+\ln 24} \approx 5.6$

5.6 kg of Pu-239 remain.

91.
$$15 = e^{-0.024755\bullet(56)+\ln A_0}$$
$$\ln 15 = -0.024755 \bullet 56 + \ln A_0$$
$$\ln A_0 = \ln 15 + 0.024755 \bullet 56$$
$$A_0 = e^{\ln 15 + 0.024755\bullet 56} \approx 60$$

Initially, there were 60 kg of Sr-90.

93. Above and Beyond

95. **(a)** $\log_3 8 = \dfrac{\log 8}{\log 3} \approx 1.893$

(b) $\log_3 8 = \dfrac{\ln 8}{\ln 3} \approx 1.893$

(c) The answers are the same.

Exercises 10.7

< Objective 1 >

1.
$$\log_5 x = 3$$
$$x = 5^3 = 125; \{125\}$$

3.
$$\log(x+1) = 2$$
$$x+1 = 10^2$$
$$x = 99; \{99\}$$

5.
$$\log_2 x + \log_2 8 = 6$$
$$\log_2 8x = 6$$
$$8x = 2^6 = 64$$
$$x = 8; \{8\}$$

7.
$$\log_3 x - \log_3 6 = 3$$
$$\log_3 \frac{x}{6} = 3$$
$$\frac{x}{6} = 3^3 = 27$$
$$x = 6 \bullet 27 = 162; \{162\}$$

9.
$$\log_2 x + \log_2(x+2) = 3$$
$$\log_2\left[x(x+2)\right] = 3$$
$$x(x+2) = 8$$
$$x^2 + 2x - 8 = 0$$
$$(x+4)(x-2) = 0$$

{2}

Note: –4 is an extraneous solution.

11.
$$\log_7(x+1) + \log_7(x-5) = 1$$
$$\log_7\left[(x+1)(x-5)\right] = 1$$
$$x^2 - 4x - 5 = 7$$
$$x^2 - 4x - 12 = 0$$
$$(x-6)(x+2) = 0$$

{6}

Note: –2 is an extraneous solution.

13. $\log x - \log(x-2) = 1$

$$\log \frac{x}{x-2} = 1$$
$$\frac{x}{x-2} = 10$$
$$x = 10(x-2) = \frac{20}{9}; \left\{\frac{20}{9}\right\}$$

15. $\log_3(x+1) - \log_3(x-2) = 2$

$$\frac{x+1}{x-2} = 9$$
$$x+1 = 9(x-2)$$
$$x = \frac{19}{8}; \left\{\frac{19}{8}\right\}$$

17. $\log(x+5) - \log(x-2) = \log 5$

$$\log \frac{x+5}{x-2} = \log 5$$
$$\frac{x+5}{x-2} = 5$$
$$x+5 = 5(x-2)$$
$$x = \frac{15}{4}; \left\{\frac{15}{4}\right\}$$

19. $\log_2(x^2-1) - \log_2(x-2) = 3$

$$\log_2 \frac{x^2-1}{x-2} = 3$$
$$\frac{x^2-1}{x-2} = 8$$
$$x^2 - 1 = 8(x-2)$$
$$x^2 - 8x + 15 = 0$$
$$(x-3)(x-5) = 0$$

{3, 5}
Note: Both solutions are valid (neither is extraneous).

< Objective 2 >

21. $6^x = 1296 = 6^4$

$x = 4; \{4\}$

23. $2^{x+1} = \frac{1}{8} = 2^{-3}$

$x + 1 = -3$

$x = -4; \{-4\}$

25. $8^x = 2 = 8^{\frac{1}{3}}$

$x = \frac{1}{3}; \left\{\frac{1}{3}\right\}$

27. $3^x = 7$

$$x = \log_3 7 = \frac{\log 7}{\log 3} \approx 1.771; \{1.771\}$$

29. $4^{x+1} = 12$

$x + 1 = \log_4 12$

$$x = -1 + \frac{\log 12}{\log 4} \approx 0.792; \{0.792\}$$

31. $7^{3x} = 50$

$3x = \log_7 50$

$$x = \frac{\log 50}{3\log 7} \approx 0.670; \{0.670\}$$

33. $5^{3x-1} = 15$

$3x - 1 = \log_5 15$

$$3x = 1 + \frac{\log 15}{\log 5}$$
$$x = \frac{1}{3} + \frac{\log 15}{3\log 5} \approx 0.894; \{0.894\}$$

35.
$$4^x = 3^{x+1}$$
$$\log 4^x = \log 3^{x+1}$$
$$x\log 4 = (x+1)\log 3$$
$$x\log 4 - x\log 3 = \log 3$$
$$x(\log 4 - \log 3) = \log 3$$
$$x = \frac{\log 3}{\log 4 - \log 3} \approx 3.819$$

{3.819}

37.
$$2^{x+1}=3^{x-1}$$
$$(x+1)\log 2=(x-1)\log 3$$
$$x\log 2+\log 2=x\log 3-\log 3$$
$$x\log 2-x\log 3=-\log 3-\log 2$$
$$x(\log 2-\log 3)=-(\log 3+\log 2)$$
$$x=\frac{-(\log 3+\log 2)}{\log 2-\log 3}$$
$$=\frac{\log 3+\log 2}{\log 3-\log 2}\approx 4.419$$
{4.419}

< Objective 3 >

39. $P=5{,}000$; $A=10{,}000$; $r=0.09$; $n=1$
$$10{,}000=5{,}000(1+0.09)^t$$
$$2=1.09^t$$
$$\log 2=\log 1.09^t=t\log 1.09$$
$$t=\frac{\log 2}{\log 1.09}\approx 8.04$$
The account balance will double in 8.04 yr.

41. $P=5{,}000$; $A=10{,}000$; $r=0.09$; $n=4$
$$10{,}000=5{,}000\left(1+\frac{0.09}{4}\right)^{4t}$$
$$2=1.0225^{4t}$$
$$\log 2=\log 1.0225^{4t}=4t\log 1.0225$$
$$t=\frac{\log 2}{4\log 1.0225}\approx 7.79$$
The account balance will double in 7.79 yr.

43. $N(t)=N_0\bullet 2^{\frac{t}{2}}$
$$2^{\frac{t}{2}}=\frac{N}{N_0}$$
$$\frac{t}{2}\log 2=\log\frac{N}{N_0}=\log N-\log N_0$$
$$t=\frac{2(\log N-\log N_0)}{\log 2}$$
$$N_0=12{,}000$$
$$N=20{,}000$$
$$t=\frac{2(\log 20{,}000-\log 12{,}000)}{\log 2}\approx 1.47$$
It will take 1.47 h.

45.
$$N(t)=N_0\bullet 2^{\frac{t}{2}}$$
$$2^{\frac{t}{2}}=\frac{N}{N_0}$$
$$\frac{t}{2}\log 2=\log\frac{N}{N_0}=\log N-\log N_0$$
$$t=\frac{2(\log N-\log N_0)}{\log 2}$$
$$\log N-\log N_0=\log\frac{N}{N_0}$$
If $N=3N_0$, then $\log\frac{N}{N_0}=\log\frac{3N_0}{N_0}=\log 3$.
$$t=\frac{2\log 3}{\log 2}\approx 3.17$$
The bacteria will triple in 3.17 h.

47. $$A = A_0\left(\frac{1}{2}\right)^{-\frac{t}{28}} = A_0 \cdot 2^{-\frac{t}{28}}$$

$$2^{-\frac{t}{28}} = \frac{A}{A_0}$$

$$-\frac{t}{28}\log 2 = \log\left(\frac{A}{A_0}\right) = \log A - \log A_0$$

$$t = \frac{-28(\log A - \log A_0)}{\log 2} = \frac{28(\log A_0 - \log A)}{\log 2}$$

$A_0 = 100$; $A = 60$

$$t = \frac{28(\log 100 - \log 60)}{\log 2} \approx 20.6$$

It takes 20.6 yr.

49. $$A = A_0\left(\frac{1}{2}\right)^{-\frac{t}{28}} = A_0 \cdot 2^{-\frac{t}{28}}$$

$$2^{-\frac{t}{28}} = \frac{A}{A_0}$$

$$-\frac{t}{28}\log 2 = \log\left(\frac{A}{A_0}\right) = \log A - \log A_0$$

$$t = \frac{-28(\log A - \log A_0)}{\log 2} = \frac{28(\log A_0 - \log A)}{\log 2}$$

$$\log A_0 - \log A = \log\frac{A_0}{A}$$

If $A = 0.75A_0$, then

$$\log\frac{A_0}{A} = \log\frac{A_0}{0.75A_0} = \log\frac{1}{0.75} = \log\frac{4}{3}.$$

$$t = \frac{28\log\frac{4}{3}}{\log 2} \approx 11.6$$

It takes 11.6 yr.

51. $$P = 25{,}000e^{0.045t}$$

$$e^{0.045t} = \frac{P}{25{,}000}$$

$$0.045t = \ln\left(\frac{P}{25{,}000}\right)$$

$$t = \frac{\ln\left(\frac{P}{25{,}000}\right)}{0.045} = \frac{\ln\left(\frac{35{,}000}{25{,}000}\right)}{0.045} \approx 7.5$$

In 7.5 yr, the population will reach 35,000. The population has doubled when $P = 50{,}000$.

53. $$N = N_0e^{0.03t}$$

$$e^{0.03t} = \frac{N}{N_0}$$

$$0.03t = \ln\frac{N}{N_0}$$

$$t = \frac{\ln\frac{N}{N_0}}{0.03}$$

$$N = 2N_0 \to \frac{N}{N_0} = 2$$

$$t = \frac{\ln 2}{0.03} \approx 23.1$$

The culture will double in 23.1 h.

55. $$P = 30e^{-0.00004h}$$

$$e^{-0.00004h} = \frac{P}{30}$$

$$-0.00004h = \ln\left(\frac{P}{30}\right)$$

$$h = \frac{\ln\left(\frac{P}{30}\right)}{-0.00004} = \frac{\ln\frac{25}{30}}{-0.00004} \approx 4{,}558$$

The altitude is 4,558 ft.

57. $A = A_0 e^{-0.000124t}$

$e^{-0.000124t} = \frac{A}{A_0}$

$-0.000124t = \ln \frac{A}{A_0}$

$t = \frac{\ln \frac{A}{A_0}}{-0.000124}$

If $A = 0.7A_0$, then $\frac{A}{A_0} = 0.7$.

$t = \frac{\ln 0.7}{-0.000124} \approx 2{,}876$

70% remains after 2,876 yr.

59. $\text{pH} = -\log\left(\left[\text{H}^+\right]\right)$

$2.4 = -\log\left(\left[\text{H}^+\right]\right)$

$-2.4 = \log\left(\left[\text{H}^+\right]\right)$

$\left[\text{H}^+\right] = 10^{-2.4} \approx 3.98 \times 10^{-3}$

The hydrogen ion concentration of the rainfall was 3.98×10^{-3} M.

61. **(a)** $L(t) = a(0.57435)^t = 30(0.57435)^{\frac{1}{2}}$

≈ 23

After 6 months, there should be 23 "live" links.

$L(t) = a(0.57435)^t = 30(0.57435)^5 \approx 2$

After 5 years, there should be 2 "live" links.

(b) $L(t) = a(0.57435)^t$

$12 = a(0.57435)^2$

$a = \frac{12}{(0.57435)^2} \approx 36$

After 2 years, there should be $36 - 12 = 24$ invalid links.

63. Above and Beyond

65. The graph is the same as the graph of $y = x$. The two functions undo each other.

67. The graph is the same as the graph of $y = x$. The two functions undo each other.

69. $3^x = 2^{x+1}$

$\ln 3^x = \ln 2^{x+1}$

$x \ln 3 = (x+1)\ln 2 = x\ln 2 + \ln 2$

$x \ln 3 - x \ln 2 = \ln 2$

$x(\ln 3 - \ln 2) = \ln 2$

$x = \frac{\ln 2}{\ln 3 - \ln 2} \approx 1.710; \{1.710\}$

The answer is the same as that given in Example 6 (in the text).

Summary Exercises

1. $(f+g)(-2) = f(-2) + g(-2) = (3) + (-1)$

$= 2$

3. $(f-g)(-1) = f(-1) - g(-1)$

$g(-1)$ does not exist. Therefore, $(f-g)(-1)$ does not exist.

5. $(f \div g)(7) = f(7) \div g(7) = (-6) \div (0)$ does not exist.

7. $D = \{-2, 3, 7\}$

9. $(f+g)(x)$

$= f(x) + g(x)$

$= (4x^2 + 5x - 3) + (-2x^2 + x - 5)$

$= 2x^2 + 6x - 8$

11. $(f+g)(x)$

$= (2x^4 + 4x^2 + 5) + (x^3 - 5x^2 + 6x)$

$= 2x^4 + x^3 - x^2 + 6x + 5$

13. $(f-g)(x)$

$= f(x) - g(x)$

$= (7x^2 - 2x + 3) - (2x^2 - 5x - 7)$

$= 7x^2 - 2x + 3 - 2x^2 + 5x + 7$

$= 5x^2 + 3x + 10$

15. $(f-g)(x)=(8x^2+5x)-(4x^2-3x)$
$=8x^2+5x-4x^2+3x=4x^2+8x$

17. $(f\bullet g)(x)=f(x)\bullet g(x)=(2x)(3x-5)$
$=6x^2-10x$

19. $(f\bullet g)(x)=(3x)(x^2)=3x^3$

21. $(f\div g)(x)=f(x)\div g(x)=(2x)\div(x-3)$
$=\dfrac{2x}{x-3}$
$D=\{x|x\neq 3\}$

23. $(f\div g)(x)=(3x)\div(x^2)=\dfrac{3x}{x^2}=\dfrac{3}{x}$
$D=\{x|x\neq 0\}$

25. $(f\circ g)(-2)=f(g(-2))=f(-1)=8$

27. $(g\circ f)(-2)=g(f(-2))=g(3)=2$

29. $(g\circ g)(-4)=g(g(-4))=g(5)=-3$

31. $f(x)=x-3$
$g(x)=3x+1$
(a) $(f\circ g)(0)=f(g(0))=f(3(0)+1)$
$=f(1)=(1)-3=-2$
(b) $(f\circ g)(-2)=f(g(-2))$
$=f(3(-2)+1)$
$=f(-5)=(-5)-3=-8$
(c) $(f\circ g)(3)=f(g(3))=f(10)=7$
(d) $(f\circ g)(x)=f(g(x))=f(3x+1)$
$=(3x+1)-3=3x-2$

33. $f(x)=x^2$
$g(x)=x-5$
(a) $(f\circ g)(0)=f(g(0))=f(-5)=(-5)^2$
$=25$
(b) $(f\circ g)(-2)=f(g(-2))=f(-7)$
$=(-7)^2=49$
(c) $(g\circ f)(3)=g(f(3))=g(9)=4$
(d) $(g\circ f)(x)=g(f(x))=g(x^2)$
$=x^2-5$

35. $h(x)=(f\circ g)(x)$

37. $f(x)=-2x+3$
$y=-2x+3$
$x=-2y+3$
$y=\dfrac{3-x}{2}$
$f^{-1}(x)=\dfrac{3-x}{2}$

39. $f(x)=\dfrac{x-3}{4}$
$y=\dfrac{x-3}{4}$
$x=\dfrac{y-3}{4}$
$4x=y-3$
$y=4x+3$
$f^{-1}(x)=4x+3$

41. $f^{-1}=\{(1,-3),(2,-1),(3,2),(4,3)\}$ is a function.

43. $f^{-1}=\{(5,1),(7,2),(9,3)\}$ is a function.

45. $f^{-1}=\{(3,4),(4,2),(4,6)\}$ is not a function.

47. $f(x)=5x+3$

$f^{-1}(x)=\dfrac{x-3}{5}$

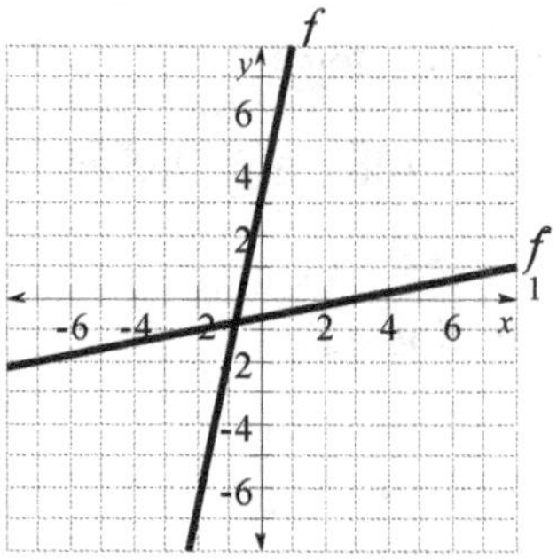

49. $f(x)=\dfrac{x-4}{5}$

$f^{-1}(x)=5x+4$

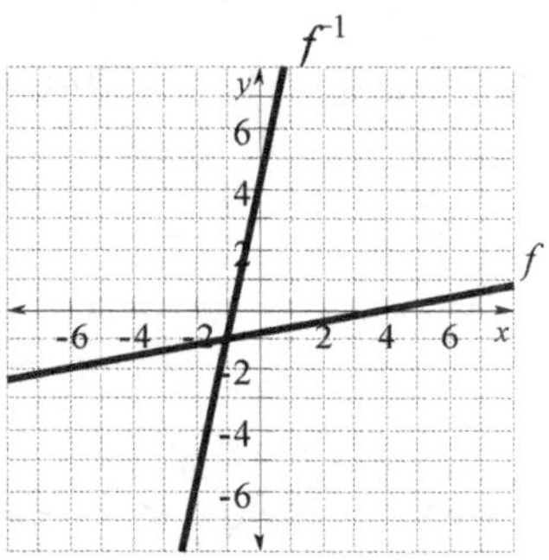

51. $f=\{(-1,2),(1,3),(2,5),(4,7)\}$ is one-to-one, therefore f^{-1} is a function.

53. f is one-to-one, therefore f^{-1} is a function.

55. f is not one-to-one, therefore f^{-1} is not a function.

57. $f(-2)=4(-2)+12=4$

59. $f\left(f^{-1}(8)\right)=8$

61. $f\left(f^{-1}(x)\right)=x$

63. $y=3^x$

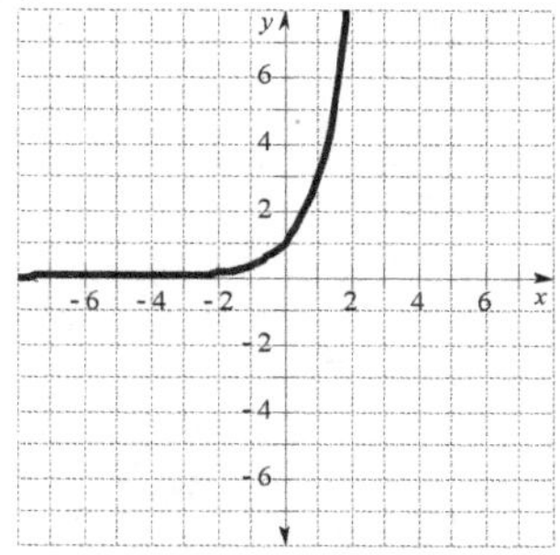

65. $5^x=125=5^3$

$x=3;\{3\}$

67. $3^{x-1}=\dfrac{1}{9}=3^{-2}$

$x-1=-2$

$x=-1;\{-1\}$

69. $N(12)=1{,}000\bullet 2^{12/2}=1{,}000\cdot 64=64{,}000$

There are 64,000 bacteria after 12 h.

71. $y=\log_3 x=\dfrac{\log x}{\log 3}$

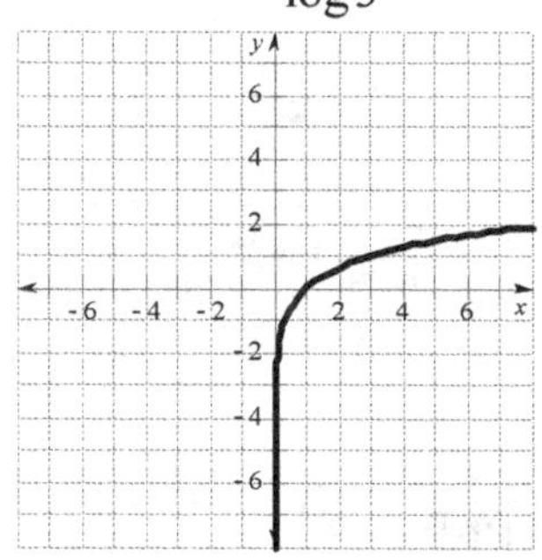

73. $2^5=32$

$\log_2 32=5$

75. $5^0=1$

$\log_5 1=0$

77. $25^{\frac{1}{2}}=5$

$\log_{25} 5=\dfrac{1}{2}$

79. $\log_4 64 = 3$

$4^3 = 64$

81. $\log_{81} 9 = \frac{1}{2}$

$81^{1/2} = 9$

83. $\log 0.001 = -3$

$10^{-3} = 0.001$

85. $y = \log_5 125$

$5^y = 125 = 5^3$

$y = 3; \{3\}$

87. $\log_7 x = 2$

$x = 7^2 = 49; \{49\}$

89. $\log_b 3 = \frac{1}{2}$

$b^{1/2} = 3 = 9^{1/2}$

$b = 9; \{9\}$

81. $y = \log_8 2$

$8^y = 2 = 8^{1/3}$

$y = \frac{1}{3}; \left\{\frac{1}{3}\right\}$

93. $L\left(10^{-6}\right) = 10\log\frac{10^{-6}}{10^{-16}} = 10\log 10^{10} = 10 \bullet 10$

$= 100$

The saw has a rating of 100 dB.

95. $I(60) = 10^{-16} \bullet 10^{60/10} = 10^{-16}10^{6} = 10^{-10}$

$I(50) = 10^{-16} \bullet 10^{50/10} = 10^{-16}10^{5} = 10^{-11}$

$\frac{10^{-10}}{10^{-11}} = 10$

The ratio is 10.

97. $M\left(10^{8.4}a_0\right) = \log\frac{\left(10^{8.4}a_0\right)}{a_0} = \log 10^{8.4} = 8.4$

The magnitude was 8.4.

99. $\log_b x^2 y = \log_b x^2 + \log_b y$

$= 2\log_b x + \log_b y$

101. $\log_5 \frac{x^2 y}{z^3} = \log_5 x^2 y + \log_5 z^3$

$= 2\log_5 x + \log_5 y - 3\log_5 z$

103. $\log\frac{xy}{\sqrt{z}} = \log xy - \log\sqrt{z}$

$= \log x + \log y - \log z^{\frac{1}{2}}$

$= \log x + \log y - \frac{1}{2}\log z$

105. $\log x + 2\log y = \log x + \log y^2 = \log xy^2$

107. $\log_b x + \log_b y - \log_b z = \log_b xy - \log_b z$

$= \log_b \frac{xy}{z}$

109. $\log x - \frac{1}{2}\log y = \log x - \log y^{1/2}$

$= \log x - \log\sqrt{y} = \log\frac{x}{\sqrt{y}}$

111. $\log 18 = \log\left(2 \bullet 3^2\right) = \log 2 + \log 3^2$

$= \log 2 + 2\log 3 = 0.301 + 2 \bullet 0.477$

$= 1.255$

113. $\log\frac{1}{8} = \log 2^{-3} = -3\log 2 = -3 \bullet 0.301$

$= -0.903$

115. $\text{pH} = -\log\left(5\times 10^{-6}\right) \approx 5.3$

Coffee is acidic.

117. $\left[\text{H}^+\right] = 10^{-3.5} \approx 3.2\times 10^{-4}$

119. $S(5) = 81 - 6\ln\left[(5)+1\right] \approx 71$

After 5 weeks, the average retest score will be 71.

121. $S(15) = 81 - 6\ln[(15)+1] \approx 64$

After 15 weeks, the average retest score will be 64.

123. $\log_4 20 = \dfrac{\log 20}{\log 4} \approx 2.161$

125.
$$\begin{aligned} \log_3 x + \log_3 5 &= 3 \\ \log_3 5x &= 3 \\ 5x &= 3^3 = 27 \\ x &= \frac{27}{5}; \left\{\frac{27}{5}\right\} \end{aligned}$$

127.
$$\begin{aligned} \log_3 x + \log_3 (x+6) &= 3 \\ \log_3 [x(x+6)] &= 3 \\ x(x+6) &= 3^3 \\ x^2 + 6x - 27 &= 0 \\ (x+9)(x-3) &= 0 \end{aligned}$$

$\{3\}$

Note: If $x = -9$, then $\log_3 x$ does not exist. Therefore, –9 is an extraneous solution.

129.
$$\begin{aligned} \log x - \log(x-1) &= 1 \\ \log \frac{x}{x-1} &= 1 \\ \frac{x}{x-1} &= 10^1 \\ x &= 10(x-1) = 10x - 10 \\ &= \frac{10}{9}; \left\{\frac{10}{9}\right\} \end{aligned}$$

131. $3^x = 243 = 3^5$

$x = 5; \{5\}$

133. $5^x = 10$

$x = \log_5 10 = \dfrac{\log 10}{\log 5} = \dfrac{1}{\log 5} \approx 1.431$

$\{1.431\}$

135.
$$\begin{aligned} 6^x &= 2^{2x+1} \\ \log 6^x &= \log 2^{2x+1} \\ x \log 6 &= (2x+1)\log 2 \\ x\log 6 - 2x \log 2 &= \log 2 \\ x(\log 6 - 2\log 2) &= \log 2 \\ x &= \frac{\log 2}{\log 6 - 2\log 2} \approx 1.710 \end{aligned}$$

$\{1.710\}$

137. $P = 1{,}000$; $n = 4$; $A = 2{,}000$

$$\begin{aligned} A(t) &= P\left(1 + \frac{0.12}{n}\right)^{nt} \\ \left(1 + \frac{0.12}{n}\right)^{nt} &= \frac{A}{P} \\ nt \log\left(1 + \frac{0.12}{n}\right) &= \log\left(\frac{A}{P}\right) \\ t &= \frac{\log\left(\frac{A}{P}\right)}{n \log\left(1 + \frac{0.12}{n}\right)} \\ &= \frac{\log\left(\frac{2{,}000}{1{,}000}\right)}{4\log\left(1 + \frac{0.12}{4}\right)} \approx 5.86 \end{aligned}$$

The account will double in 5.86 yr.

139. $A_0 = 100$; $A = 40$

$$\begin{aligned} A(t) &= A_0 2^{-\frac{t}{50}} \\ 2^{-t/50} &= \frac{A}{A_0} \\ -\frac{t}{50}\log 2 &= \log\left(\frac{A}{A_0}\right) \\ t &= -\frac{50\log\left(\frac{A}{A_0}\right)}{\log 2} = -\frac{50\log\left(\frac{40}{100}\right)}{\log 2} \approx 66 \end{aligned}$$

40 mg remains after 66 yr.

141. $P = 50{,}000e^{0.08t}$

$$e^{0.08t} = \frac{P}{50{,}000}$$

$$0.08t = \ln\left(\frac{P}{50{,}000}\right)$$

$$t = \frac{\ln\left(\frac{P}{50{,}000}\right)}{0.08} = \frac{\ln\left(\frac{70{,}000}{50{,}000}\right)}{0.08} \approx 4.2$$

In 4.2 yr, the population will reach 70,000.

143. $P = 30e^{-0.021h}$

$$e^{-0.021h} = \frac{P}{30}$$

$$-0.021h = \ln\left(\frac{P}{30}\right)$$

$$h = \frac{\ln\left(\frac{P}{30}\right)}{-0.021} = \frac{\ln\left(\frac{27.7}{30}\right)}{-0.021} \approx 3.8$$

The altitude atop Mt. McKinley is 3.8 mi.

Chapter Test 10

1. $10^4 = 10{,}000$

$\log 10{,}000 = 4$

3. $f\left(f^{-1}(x)\right) = 5\left(\frac{x+7}{5}\right) - 7 = (x+7) - 7 = x$

5. $\log_5 \sqrt{\frac{xy^2}{x}} = \log_5\left(\frac{xy^2}{z}\right)^{1/2} = \frac{1}{2}\log_5 \frac{xy^2}{z}$

$$= \frac{1}{2}\left(\log_5 x + \log_5 y^2 - \log_5 z\right)$$

$$= \frac{1}{2}\left(\log_5 x + 2\log_5 y - \log_5 z\right)$$

7. $(g \div f)(x) = g(x) \div f(x)$

$$= (3x-2) \div \left(x^2 - 1\right) = \frac{3x-2}{x^2-1}$$

$$x^2 - 1 \neq 0$$

$$(x+1)(x-1) \neq 0$$

$$x \neq \pm 1$$

$$D = \{x \mid x \neq \pm 1\}$$

9. $(g \circ f)(x) = g(f(x)) = g\left(x^2 - 1\right)$

$$= 3\left(x^2 - 1\right) - 2 = 3x^2 - 5$$

11. $3^{2x-1} = 81 = 3^4$

$2x - 1 = 4$

$x = \frac{5}{2}; \left\{\frac{5}{2}\right\}$

13. $\log 0.01 = -2$

$10^{-2} = 0.01$

15. f is one-to-one, therefore f^{-1} is a function.

17. $3^{x+1} = 4$

$$x + 1 = \log_3 4 = \frac{\log 4}{\log 3}$$

$$x = -1 + \frac{\log 4}{\log 3} \approx 0.262; \{0.262\}$$

19. $f(x) = -3x^3 + 5x^2 - 2x - 7$

$g(x) = -2x^2 + 7x - 2$

(a) $(f+g)(x)$

$$= \left(-3x^3 + 5x^2 - 2x - 7\right) + \left(-2x^2 + 7x - 2\right)$$

$$= -3x^3 + 3x^2 + 5x - 9$$

(b) $(f-g)(x)$

$$= \left(-3x^3 + 5x^2 - 2x - 7\right) - \left(-2x^2 + 7x - 2\right)$$

$$= -3x^3 + 5x^2 - 2x - 7 + 2x^2 - 7x + 2$$

$$= -3x^3 + 7x^2 - 9x - 5$$

21. $\log_6(x+1)+\log_6(x-4)=2$

$$\log_6[(x+1)(x-4)]=2$$
$$(x+1)(x-4)=6^2$$
$$x^2-3x-40=0$$
$$(x-8)(x+5)=0$$

{8}
Note: –5 is an extraneous solution.

23. $f(x)=\frac{x-3}{5}$

$f^{-1}(x)=5x+3$

25. $f^{-1}=\{(-2,5),(1,-3),(2,4)\}$

27. $y=\left(\frac{2}{3}\right)^x$

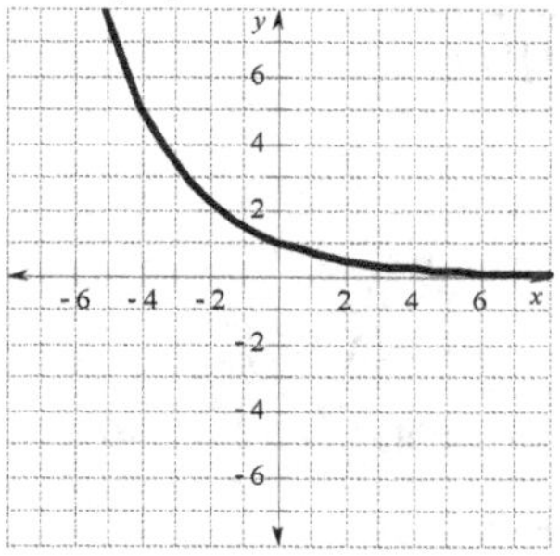

29. $\log_b\frac{1}{16}=-2$

$$b^{-2}=\frac{1}{16}=4^{-2}$$
$$b=4;\{4\}$$

Cumulative Review: Chapters 0–10

1. $2x-3(x+2)=4(5-x)+7$

$$2x-3x-6=20-4x+7$$
$$3x=33$$
$$x=11;\{11\}$$

3. $\log x-\log(x-1)=1$

$$\log\frac{x}{x-1}=1$$
$$\frac{x}{x-1}=10^1$$
$$x=10(x-1)$$
$$x=\frac{10}{9};\left\{\frac{10}{9}\right\}$$

5. $-8(2-x)\geq y$

$$y\leq 8x-16$$

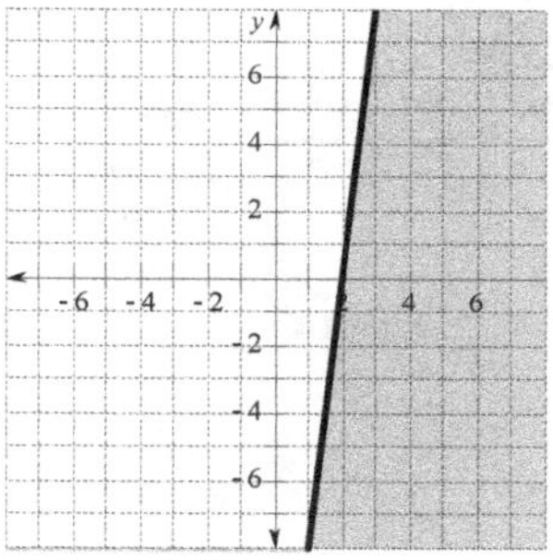

7. $3x-2(x-5)\geq 20$

$$3x-2x+10\geq 20$$
$$x\geq 10$$

$\{x|x\geq 10\}$

9. $(3x+1)(2x-5)=6x^2-13x-5$

11. $25x^3-16xy^2=x(25x^2-16y^2)$

$$=x(5x+4y)(5x-4y)$$

13. $\frac{x^2-x-6}{x^2+2x-15}\div\frac{x-2}{x+5}=\frac{(x-3)(x+2)}{(x+5)(x-3)}\bullet\frac{x+5}{x-2}$

$$=\frac{x+2}{x-2}$$

15. $(3\sqrt{2}+2)(3\sqrt{2}+2)=9\bullet 2+2\bullet 2\bullet 3\sqrt{2}+2^2$

$$=22+12\sqrt{2}$$

17. Let x, $x+2$, and $x+4$ be three consecutive odd integers.

$$\begin{aligned} x+(x+2)+(x+4)&=237\\ 3x+6&=237\\ 3x&=231\\ x&=77 \end{aligned}$$

The numbers are 77, 79, and 81.

19. $2x^2-6x-5=0$

$a=2$; $b=-6$; $c=-5$

$$\begin{aligned} x&=\frac{-(-6)\pm\sqrt{(-6)^2-4(2)(-5)}}{2(2)}\\ &=\frac{6\pm\sqrt{36+40}}{4}=\frac{6\pm2\sqrt{19}}{4}=\frac{3\pm\sqrt{19}}{2} \end{aligned}$$

$$\left\{\frac{3\pm\sqrt{19}}{2}\right\}$$

21. $f(x)=3x-1$

$g(x)=x^2-x-6$

$$\begin{aligned} (f\cdot g)(x)&=f(x)\cdot g(x)\\ &=(3x-1)(x^2-x-6)\\ &=3x^3-3x^2-18x-x^2+x+6\\ &=3x^3-4x^2-17x+6 \end{aligned}$$

$D=\mathbb{R}$

23. $f(x)=2x-1$

$g(x)=-3x+2$

(a) $$\begin{aligned}(f\circ g)(x)&=f(g(x))=f(-3x+2)\\ &=2(-3x+2)-1=-6x+3\end{aligned}$$

(b) $$\begin{aligned}(f\circ g)(5)&=f(g(5))=f(-3(5)+2)\\ &=f(-15+2)=f(-13)\\ &=2(-13)-1=-26-1\\ &=-27\end{aligned}$$

25. $14a^2b^2-21a^2b+35ab^2$

$=7ab(2ab-3a+5b)$

27. $25c^2-64d^2=(5c-8d)(5c+8d)$

29. $$\begin{aligned}16a^4+2ab^3&=2a(8a^3+b^3)\\ &=2a(2a+b)(4a^2-2ab+b^2)\end{aligned}$$

31. $10x^2-39x+14=(5x-2)(2x-7)$

33. $$\begin{aligned}\frac{x^3y}{4xy^2}\div\frac{xy}{12xy^2}&=\frac{x^3y}{4xy^2}\bullet\frac{12xy^2}{xy}=\frac{12x^4y^3}{4x^2y^3}\\ &=3x^2\end{aligned}$$

35. $$\begin{aligned}&\frac{4}{m^2-9}-\frac{3}{m^2-4m+3}\\ &=\frac{4}{(m+3)(m-3)}-\frac{3}{(m-3)(m-1)}\\ &=\frac{4(m-1)-3(m+3)}{(m+3)(m-3)(m-1)}\\ &=\frac{m-13}{(m+3)(m-3)(m-1)}\end{aligned}$$

37. Let x, $2x+4$, and $2x+6$ be the lengths of the sides of the triangle.

$$\begin{aligned} x^2+(2x+4)^2&=(2x+6)^2\\ x^2+4x^2+16x+16&=4x^2+24x+36\\ x^2-8x-20&=0\\ (x-10)(x+2)&=0 \end{aligned}$$

The lengths are 10 in., 24 in., and 26 in.
Note: If $x=-2$, then the triangle does not exist. Therefore, –2 is an extraneous solution.

39. $$\begin{aligned} \sqrt{2x-1}+x&=8\\ \sqrt{2x-1}&=8-x\\ 2x-1&=x^2-16x+64\\ x^2-18x+65&=0\\ (x-5)(x-13)&=0 \end{aligned}$$

Check: $\sqrt{2(5)-1}+(5)=8$

$\sqrt{2(13)-1}+(13)=18\neq8$

13 is an extraneous solution.
The solution set is $\{5\}$.

Appendices

Exercises A.1

< Objective 1 >

1. $2x < 8$

$f(x) = 2x$

$g(x) = 8$

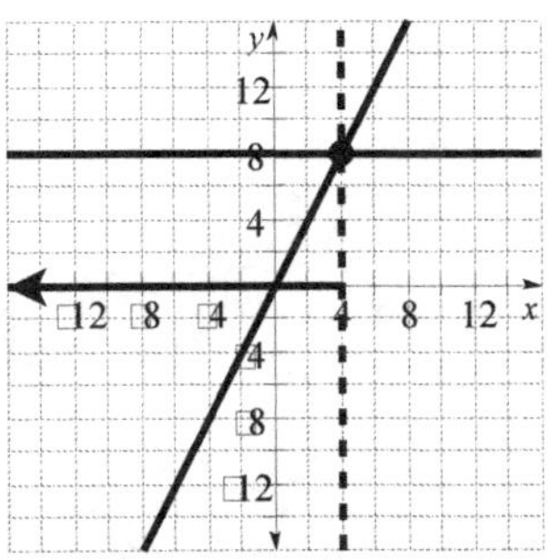

The point of intersection is (4, 8).
The solution set is $\{x | x < 4\}$.

3. $\frac{x+3}{2} < -1$

$f(x) = \frac{x+3}{2} = \frac{x}{2} + \frac{3}{2}$

$g(x) = -1$

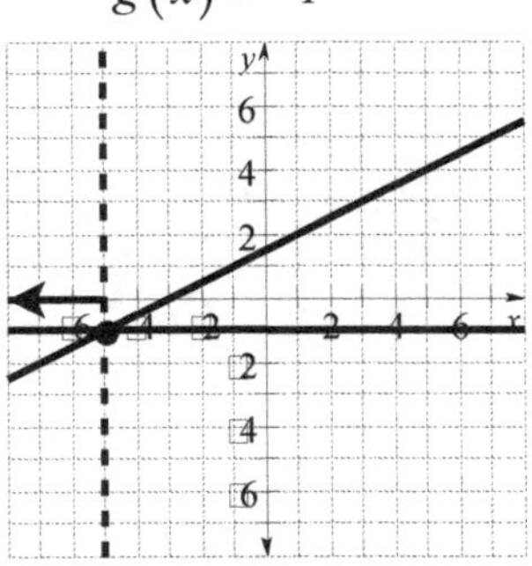

The point of intersection is (−5, 1).
The solution set is $\{x | x < -5\}$.

5. $7x - 7 < -2x + 2$

$f(x) = 7x - 7$

$g(x) = -2x + 2$

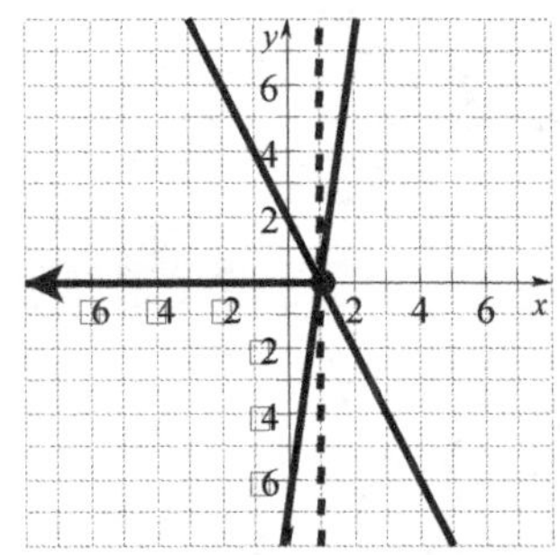

The point of intersection is (1, 0).
The solution set is $\{x | x < 1\}$.

7. $6(1 + x) \geq 2(3x - 5)$

$f(x) = 6(1 + x) = 6x + 6$

$g(x) = 2(3x - 5) = 6x - 10$

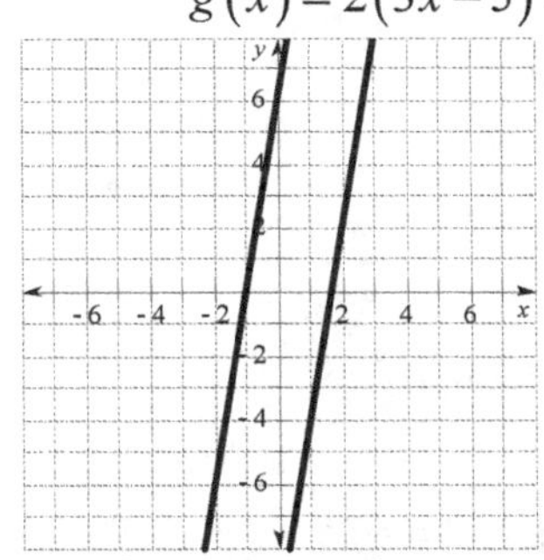

The lines do not intersect.
f is always above *g*.
The solution set is $\mathbb{R}$.

9. $7x > \dfrac{9x-5}{2}$

$f(x) = 7x$

$g(x) = \dfrac{9x-5}{2} = \dfrac{9}{2}x - \dfrac{5}{2}$

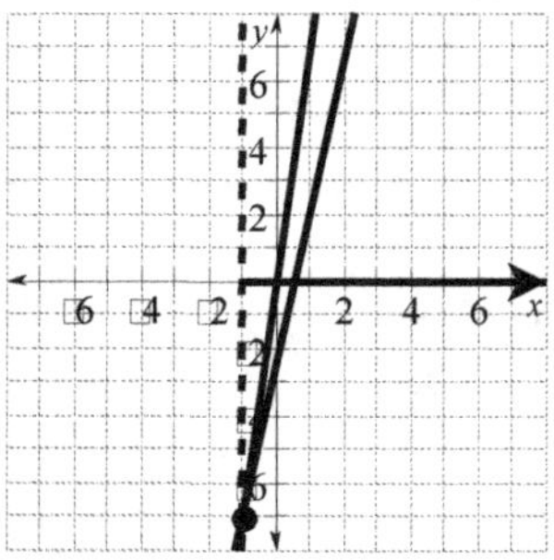

The point of intersection is (−1, 7).
The solution set is $\{x \mid x > -1\}$.

Exercises A.2

< Objective 1 >

1. $|15| = 15$

3. $|8-3| = |5| = 5$

5. $|-12-19| = |-31| = 31$

7. $|-13| + |12| = 13 + 12 = 25$

9. $-|-13| - |-12| = -13 - 12 = -25$

< Objective 2 >

11. $|x| = 5$

$x = 5$ or $x = -5$

$\{\pm 5\}$

13. $|2x-1| = 6$

$2x-1 = 6$ or $2x-1 = -6$

$2x = 7$ $\quad$ $2x = -5$

$x = \dfrac{7}{2}$ $\quad$ $x = -\dfrac{5}{2}$

$\left\{-\dfrac{5}{2}, \dfrac{7}{2}\right\}$

15. $|5x-2| = -3$

No solutions; $\varnothing$

17. $8 - |x-4| = 5$

$-|x-4| = -3$

$|x-4| = 3$

$x-4 = 3$ or $x-4 = -3$

$x = 7$ $\quad$ $x = 1$

$\{1, 7\}$

19. $|5x-2| = |2x-4|$

$5x-2 = 2x-4$ or $5x-2 = -(2x-4)$

$3x = -2$ $\quad$ $5x-2 = -2x+4$

$x = -\dfrac{2}{3}$ $\quad$ $7x = 6$

$x = \dfrac{6}{7}$

$\left\{-\dfrac{2}{3}, \dfrac{6}{7}\right\}$

Exercises A.3

< Objective 1 >

1. $f(x) = |x-3|$

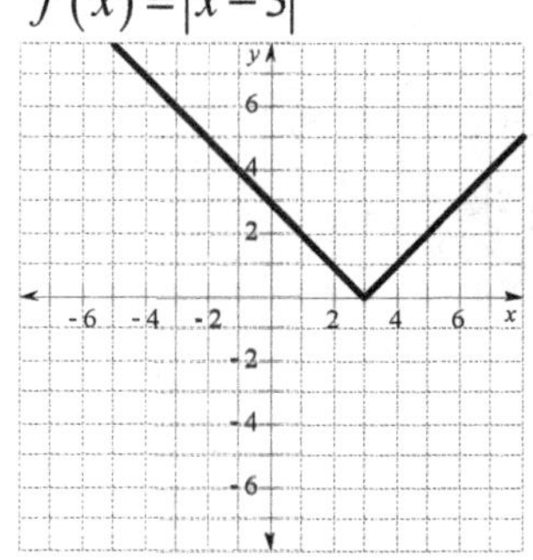

3. $f(x) = |x+3|$

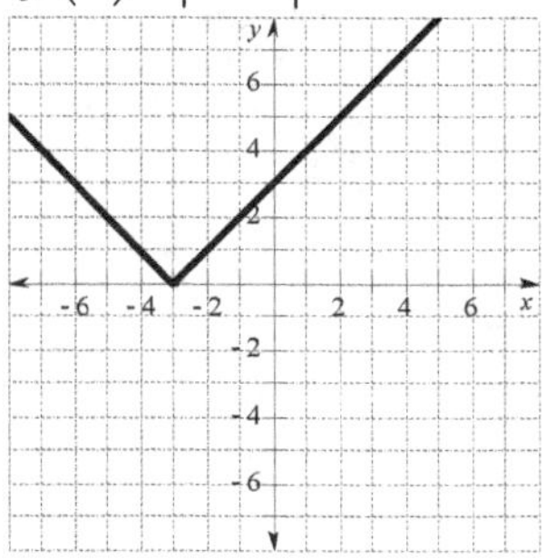

< Objective 2 >

5. $|x| = 3$

$f(x) = |x|$

$g(x) = 3$

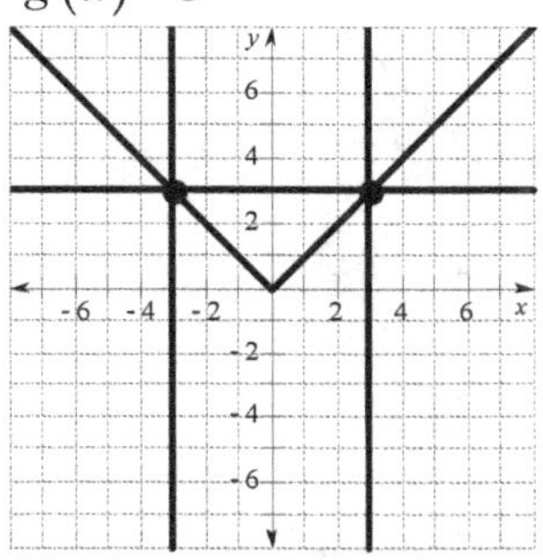

The intersection points are $(-3, 3)$ and $(3, 3)$. The solution set is $\{\pm 3\}$.

7. $|x-2| = 5$

$f(x) = |x-2|$

$g(x) = 5$

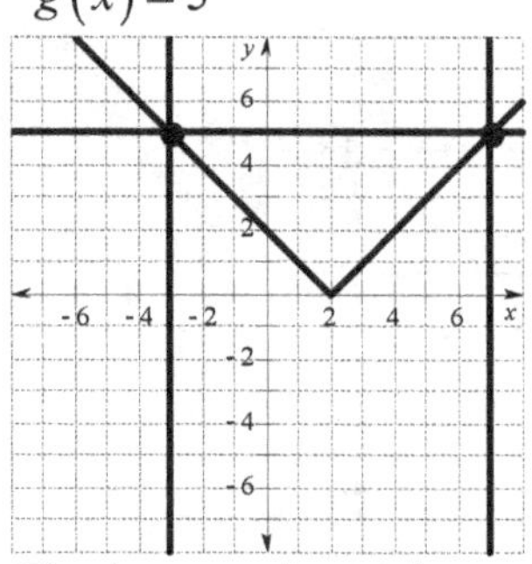

The intersection points are $(-3, 5)$ and $(7, 5)$ and. The solution set is $\{-3, 7\}$.

9. $|x-3| = 5 - \frac{1}{3}x$

$f(x) = |x-3|$

$g(x) = 5 - \frac{1}{3}x$

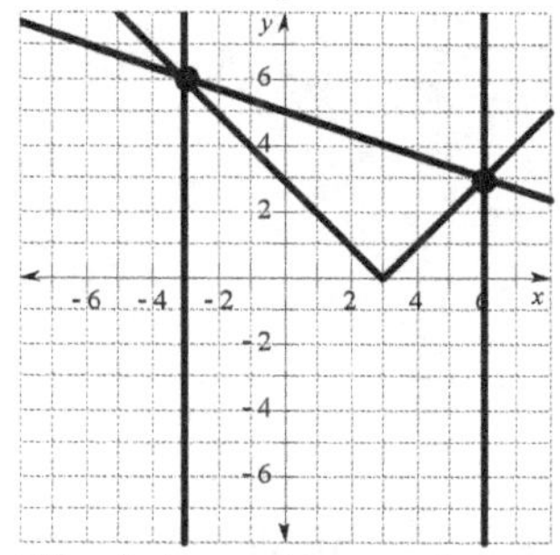

The intersection points are $(-3, 6)$ and $(6, 3)$. The solution set is $\{-3, 6\}$.

11. The absolute value function, shifted right 2 units, is given by $f(x) = |x-2|$.

Exercises A.4

< Objective 1 >

1. $4 \le x+1 \le 7$

$3 \le x \le 6$

$\{x | 3 \le x \le 6\}$

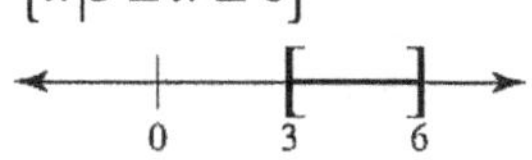

3. $1 \le 2x-3 \le 6$

$4 \le 2x \le 9$

$2 \le x \le \frac{9}{2}$

$\left\{x \middle| 2 \le x \le \frac{9}{2}\right\}$

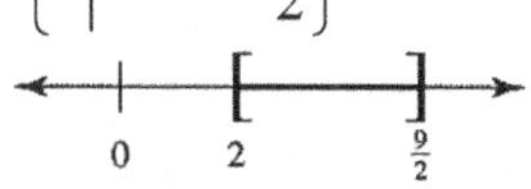

5. $x-1 < -3$ or $x-1 > 3$

$x < -2$ $\quad$ $x > 4$

$\{x | x < -2 \text{ or } x > 4\}$

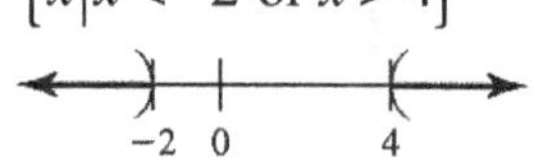

< Objective 2 >

7. $|x|<5$

$-5<x<5$

$\{x|-5<x<5\}$

9. $|x+6|\le 4$

$-4\le x+6\le 4$

$-10\le x\le -2$

$\{x|-10\le x\le -2\}$

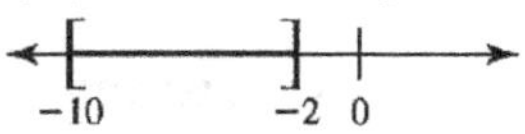

11. $|3x+4|\ge 5$

$3x+4\le -5$ or $3x+4\ge 5$

$3x\le -9$ $\quad$ $3x\ge 1$

$x\le -3$ $\quad$ $x\ge \frac{1}{3}$

$\left\{x\middle|x\le -3 \text{ or } x\ge \frac{1}{3}\right\}$

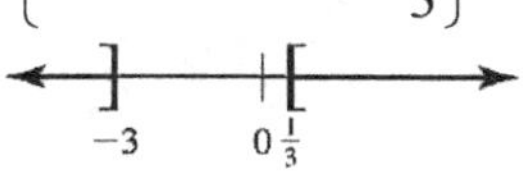

Exercises A.5

< Objective 2 >

1. $|x|<5$

$\{x|-5<x<5\}$

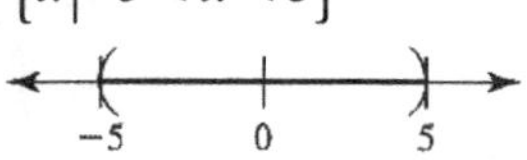

3. $|x+6|\le 4$

$-4\le x+6\le 4$

$-10\le x\le -2$

$\{x|-10\le x\le -2\}$

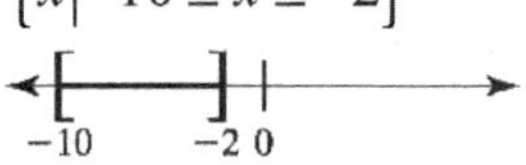

5. $|3x+4|\ge 5$

$3x+4\ge 5$ or $3x+4\le -5$

$x\ge \frac{1}{3}$ $\quad$ $x\le -3$

$\left\{x\middle|x\le -3 \text{ or } x\ge \frac{1}{3}\right\}$

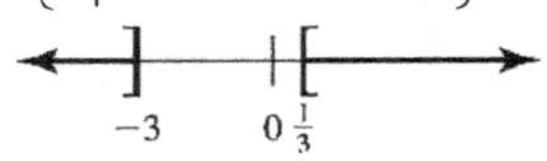

< Objective 2 >

7. $|x|<4$

$f(x)=|x|$

$g(x)=4$

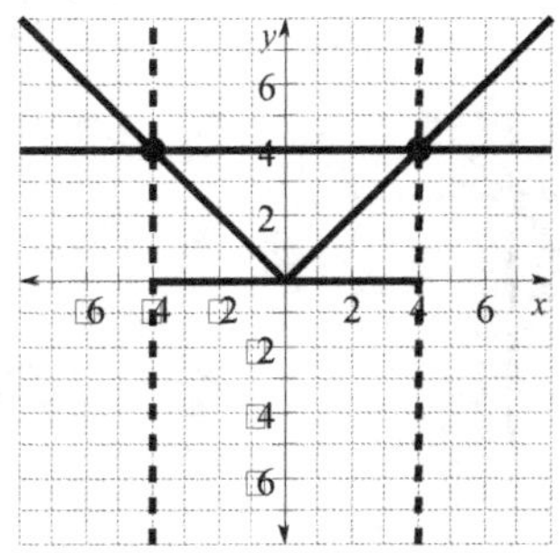

$\{x|-4<x<4\}$

9. $|x-3|<4$

$f(x)=|x-3|$

$g(x)=4$

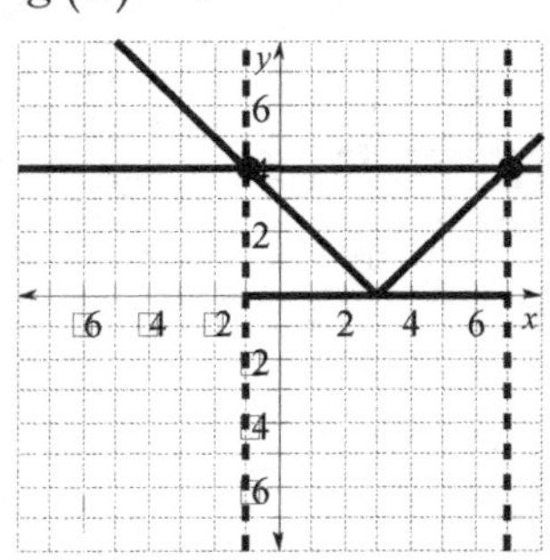

$\{x|-1<x<7\}$

11. $|x+1| \le 5$

$f(x) = |x+1|$

$g(x) = 5$

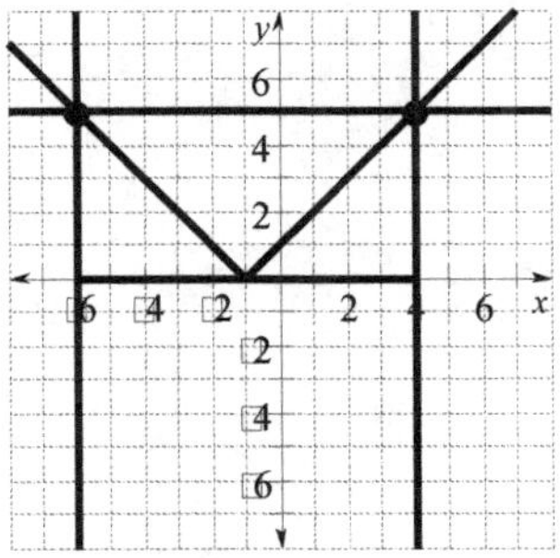

$\{x \mid -6 \le x \le 4\}$

NOTES

NOTES

NOTES